Logic, Methodology and Philosophy of Science
Proceedings of the Thirteenth International Congress
Volume 2

Logic, Methodology and Philosophy of Science

Proceedings of the Thirteenth International Congress

Volume 2

edited by

Shushan Cai,

Guoping Zeng,

and

Ning Wang

© Individual author and College Publications 2011. All rights reserved.

ISBN 978-1-84890-045-5

College Publications
Scientific Director: Dov Gabbay
Managing Director: Jane Spurr
Department of Computer Science
King's College London, Strand, London WC2R 2LS, UK

http://www.collegepublications.co.uk

Original cover design by orchid creative   www.orchidcreative.co.uk
Printed by Lightning Source, Milton Keynes, UK

All rights reserved. No part of this publication may be reproduced, stored in a retrieval system or transmitted in any form, or by any means, electronic, mechanical, photocopying, recording or otherwise without prior permission, in writing, from the publisher.

Contents

Preface   i

A   Logic   1

1   Gao Aihua: The Relationship between the *Hetuvidya* and the *Wen Xin Diao Lon*   3

2   Wang Wenfang, et al.: A Perspective Semantics for Subjunctive Conditionals   7

3   Qin Weiyuan, et al.: Review of Wittgenstein's "Logic Picture Theory"   12

4   Li Xiaowu: Logics for Knowing an Individual   17

5   Wu Hongzhi, et al: Argument Schemes   23

6   Chienkuo Mi: Predication and the Slingshot   29

7   Hou, Richard Wei Tzu: A Liar's Thorn in the Minimalist's Side   38

8   Zhang Hansheng: A Brief Introduction to *Para-consistent logic and AI*   42

9   Ren Xiaoming, et al: The Rise and Meaning of the Paraconsistent Model of the Philosophy of Science from the Perspective of Logic   50

10   Zhou Xunwei: Second-level Hypothetical Inference and its Applications   54

11   Zhai Jincheng: The Importance and Research Methodology of Chinese Logic   58

12   Long Xiaoping: On the Nature of Logical Truth based on Quine's Critique   62

13   Ning Lina: On the Philosophical Foundations of the Evolution of Logical Methods   67

14   Liu Shanshan, et al: Kripke's Research on the Theory of Proper Names – About Essentialism   72

15   Zou Chongli et. al.: A Theory of Causative Constructions in Chinese Based on Type-Logical Grammar   76

B General Philosophy of Science   80

16 Zhang Zhongyi: Researches on Deduction in Chinese Logic   81

17 Yan Kunru, et al: Cognitive Context and Scientific Explanation 85

18 Wei Yidong: Presupposition, Evidence and Model-based Reasoning in Science   89

19 Cheng Rui: Contextual Realism and Relativism   94

20 Chen Xiaoping: Bayesian Test and Kuhn's Paradigm   98

21 Lv Shirong, et al: Does the "Grue" Paradox Have a Solution? How to Solve It?   103

22 Wang Wei: A Defense of Ceteris Paribus Laws   108

23 Hu Guang: On the Evolutional Structure of Science Theory   114

24 Yuann Jeu-Jenq: The Extension of Vienna Circle Protocol Sentences Debates: A Comparative Study of W. Quine and P. Feyerabend   119

25 Zheng Huizi: Sociology of Science and its Limitation to the Rational Justification for Science   125

26 Li Jingjing: A Scientific and Philosophical Study of Fengshui Theory and Practice   129

27 Shang Zhicong: New Epistemological Ideas Formed by Sino-Western Academic Interaction in Late Ming Times   135

28 Hong Xiaonan: Research on Front Issues in the Cultural Philosophy of Science   140

29 Hu Mingyan: Walk out from Imer Lakatos' Dilemma   146

30 He Huaqin: A Project against Traditional Scientific Realism and Anti-realism – An Appraisal of Rouse's Practical Realism of Science   152

31 Yang, Jing: The Significance of Leibniz's View of Science in the Academy and for Public Happiness   158

32 Wang, Xiaojing: Research on Modal Terms in the Mohist Canon 162

CONTENTS v

C Philosophy Issues of Particular Sciences 167

33 Tzu Keng Fu: Does the Spirit of Universal Logic Coincide with Logical Pluralism? 169

34 Zhang Xiaoxiang: Thoughts on Sinicizing the Hetuvidya 174

35 He Xiangdong, et al: On Sinologic and Sinomathematics in Ancient China 177

36 Liu Yetao, et al: Kripke on Necessity and the Essence of Individual Objects 187

37 Xia Sumin: A Study of the "Deontic Paradox" 192

38 Bai Tongdong: The Cushing Thesis and Underdetermination 197

39 Wu Xinzhong: The History of Quantum Theory and Quantum Curvature Interpretation 205

40 Li Hongfang: The Schröedinger's Cat Paradox, Entanglement and Decoherence 210

41 Zhao Guoqiu, et al: The Descriptive Object of the Schrödinger Equation and Wave Particle Unification in Quantum Mechanics Curvature Interpretation 215

42 Zhao Guoqiu: An Afterthought on the Interpretation of Quantum Decoherence 221

43 Chen Gang: Hierarchy, Form, and Reality 227

44 Wan Xiaolong: From Determination, Indetermination, Extrinsic Under-determination, to Intrinsic Under-determination : A Suggestion about the Curvature Interpretation of Quantum Mechanics 234

D Science and Society 240

45 Wu Shuxian: Logic and Higher-Order Cognition 241

46 Zhou Yuncheng: The Problem with the Deflationary Account of Self-consciousness 247

47 Qiu Huili: A New Image of Science 253

48 Qin Yulin, et al: ACT-R, a Computational Cognitive Architecture Model and its Methodological Reflection 258

49 Ma Yongjun: A New Scheme for Embodied Cognition: from Chaos to Order   263

50 Jerry J. Yang: Toward a Naturalistic Account of Self-Knowledge 268

51 Yang Xiaolu: The Ontogenesis of Language: Evidence from the Early Grammatical Development of Chinese   274

52 Zhou Jianshe: The Philosophy of Language in the Pre-Qin Era 279

53 Wang Lihui: Context and Relevance   284

54 Wang Zhidong: Pragmatics and the Frame Problem   289

55 Wang Xiaohong: A Philosophical Exploration into the Cognitive Construction of Scientific Discovery   297

56 Wang Simin: Can a Machine Lie?   303

57 Li Bocong, et al: On Engineering Rationality   309

58 Qu Maisheng: A New Interdisciplinary Study of Economics and Logic   313

59 Song Chunyan: A Model for the Construction of Institutional Facts by Speech Acts   319

60 Xu Zhu: Laws, Causality and the Intentional Explanation of Action   324

61 Yan Wei, et al.: A Probe into the Question of Ethical Restrictions on Military Technologies   328

62 Niu Junmei: The Ethical Connotations of I. Bernard Cohen's Thoughts on Scientific Revolution   332

63 Zhu Fengqing: Ethical Issues Posed by Nanotechnology in the Workplace   336

64 Qi Yanxia, et al.: On Ethical Problems in Realistic Engineering Activity   341

65 Xu Zhili: On Ethical-Value Tension between Science / Technology and Politics   345

66 Wang Qian, et al.: On Guarantee-Systems of Science and Technology Ethical Practice in the Present Age   351

67 Wang Hailong: Critical Thinking and Education   358

CONTENTS

68 Wen Jianying: Rediscovery of the Connotation of Interaction Between Science, Technology and Society   362

69 Sun Hongxia, et al.: The 'Atomic Bomb Project': A Study Comparing China with the USA from a Nuclear Physics Perspective 366

A
Logic

The Relationship between the *Hetuvidya* and the *Wen Xin Diao Lon*[1]

Gao Aihua
College of Continuing Education
aihuaggao@yahoo.com.cn

ABSTRACT. Liu Xie's *Wen Xin Diao Long* is a masterpiece of literary theory and criticism from ancient China. Scholars have studied many aspects of it, but there has been scarcely any research into the ways in which the logical thought of the Indian *Hetuvidya* shaped the content and principles used in the *Wen Xin Diao Long*. This paper discusses three aspects of this relationship: the time when the *Hetuvidya* was first introduced into ancient China and how Liu Xie could have come in contact with it; the effect of the mode of argumentation developed in it on the *Wen Xin Diao Long*; and the impact of the *Hetuvidya's* theory of fallacy on the *Wen Xin Diao Long*.

There are three ancient traditions which prescribe rules of logic for thinking and argumentation. Each systematically establishes a scientific foundation for the logical use of thought patterns and prescribes a set of rules for correct thinking as early as the period between the fifth century B.C. and the fourth century A.D. These three ancient traditions originate in ancient India, ancient Greece and ancient China. These three traditions collectively comprise the roots of world logic. Ancient Greek logic is based on the logical works of Aristotle. The ancient Indian tradition of logic stems from the *Hetuvidya*. The Chinese tradition of logic is derived in part from Chinese translations of the *Hetuvidya* and the translation of the Buddhist sutras from Sanskrit into the Chinese vernacular. In Chinese, *hetuvidaya* is translated as *yin ming*, in which *yin* is a semantic translation of *hetu* which means 'grounds of argument' and *ming* translates *vidaya* which means 'learning'. Thus *yin ming (hetuvidya)* refers to knowledge of reasoning. The *Hetuvidya* itself was introduced into China during the Wei-Jin Southern and Northern Dynasty period. At that time, scholars were translating the Buddhist sutras from Sanskrit into Chinese. The introduction of the *Hetuvidya* into ancient China enhanced the level of reasoning and argumentation and also pushed forward the development of literary theory and criticism. The *Wen Xin Diao Long* itself was composed in the early sixth century. Scholars have studied many aspects of the *Wen Xin Diao Long*, but there has been little study of how the *Hetuvidya* influenced its composition and the content of its thought. This paper studies three aspects of this influence and the relationship between the *Hetuvidya* and the *Wen Xin Diao Long*.

First, the time at which the *Hetuvidya* was introduced into ancient China is a controversial issue in academic circles. There are two opposing arguments. According to the first viewpoint, the *Hetuvidya* was introduced into ancient China

---

[1] The author is with Yanshan University, Qinhuangdao, Hebei, China.

at the beginning of the Tang Dynasty at around the same time that the famous Chinese monk Xuanzang translated *Sankarasvamin's Nyayadvarata-rakasastra* in 647 AD and *Dignaga's Nyaya-Mukha* in 649 A.D. Both works are important landmarks in the history of Buddhist thought. It is known that Xuanzang used the *Hetuvidya* as a text in teaching his disciples. It became an intensely discussed and studied text during these years. His disciples in turn wrote many articles and commentaries on the *Hetuvidya*. According to these scholars, the *Hetuvidya* exerted its strongest influence on ancient Chinese thought in the years after Xuanzang retruned from India in 645 A.D. According to the opposing viewpoint, the *Hetuvidya* was introduced to ancient China much earlier, during the Wei-Jin and Southern and Northern Dynasty period. These scholars argue that many sutras introduced into China at this time and translated into Chinese discuss the *Hetuvidya* in significant detail. The scholars cite, for example, the Indian monk Tan Wuchan's *Da Ban Nie Pan Jing* in 421, which in volume 17 notably applies the method of argumentation described in the *Hetuvidy*a. After the *Da Ban Nie Pan Jing*, three other important works were introduced into ancient China around the same time. They were the *Upayakauslyahrdaysastra*, the *Vigraha Vyavartani Karika* and the *Tarkasastra*. All three show the influence of the *Hetuvidya*. Thus, these scholars argue that the *Hetuvidya* was introduced into China between 400 and 450 A.D., some two hundred years before Xuanzhuang. I will not pursue the relative merits of these arguments. Most scholars now concede that some of the most important thought content in the *Hetuvidya* had been introduced into ancient China during the Wei-Jin and Southern and Northern dynasties even if the *Navya-Hetuvidya* had not yet been introduced and translated in full. The content and influence may have seeped in with the translation and study of other ancient Buddhist texts. Some Buddhist scholars noticed it and used it in their work.

Having established the *Hetuvidya's* presence in China in some form in the early fifth century, it is then a more straightforward matter to connect it to Liu Xie and the *Wen Xin Diao Long*. The historical evidence for Liu Xie in the historical record is sparse. He is generally assumed to have been born around 465, some thirty-five years before the composition between 499 and 502 of the *Wen Xin Diao Long*. He was forced to live a monk's life for ten years because of poverty. Fortunately, he was fond of reading from early in his youth. Seng You was a well-known scholar and expert on Buddhism at that time. Since it had been introduced with the translation of such sutras as the *Da Ban Nie Pan Jing* in 421 and the *Upayakausalyahrdaysastra* in 427, the *Hetuvidya* could have been available for study by Seng You and Liu Xie. It is possible then that the composition and content of the *Wen Xin Diao Long* were influenced by the *Hetuvidya* even at this early date.

My second point on this relationship is this. The *Hetuvidya's* mode of argument has had an effect on the mode of argument of the *Wen Xin Diao Long*. The unmistakable traces of this influence can be traced clearly in the narrative style. The first chapter, the "Yuan Dao" of the *Wen Xin Diao Long*, proposes that "literature stems from Dao". This translates *siddhānta* from the *Hetuvidya*. The next four *Wen Xin Diao Long* chapters "Zheng Sheng", "Zong Jing", "Zheng Wei" and "Bian Sao", follow the general grounds of the mode of argument which corresponds to hetu. The next twenty chapters, from "Ming Shi" to "Shu Ji", provide examples which correspond to the concept of udā da in the *Hetuvidya*.

Liu Xie combines the second and third parts in the following twenty-four chapters from "Shen Si" to "Cheng Qi" to support the general argument. "Xu Zhi", the last chapter, concludes the book. Many scholars argue that the structure of the argument in the *Wen Xin Diao Long* follows the *Hetuvidya* sequence *siddhānta, hetu, udāda, upanaya and nigamana*. In addition, the method of argument exemplified in the *Upayakauslyahrdaysastra* is also applied in the *Wen Xin Diao Long*. As the *Upayakauslyahrdaysastra* prescribes, udaharana uses practical examples as comparable objects to prove a thesis. It categorizes examples as either anvaji udaharana or *vyati-eki* udaharana. Anvaji udaharana refers to an affirmative or homogeneous example while *vyati-eki* udaharana refers to a negative or heterogeneous example. The *Wen Xin Diao Long* uses quite a number of udaharana to support its mode of argumentation. It uses such images as the sun, the moon, the stars and mountains, rivers, animals and plants such unparalleled ways that they bear obvious traces from the *Hetuvidya* and related religious literature. In the "Yuan Dao" chapter, for example, Liu Xie employs the universe, animals (dragon, phoenix) and plants as *anvaji udaharana* examples. He also uses abundant *vyati-eki udaharana* examples as well. Moreover, the *Wen Xin Diao Long* also applies dialectical logic in a way which is consistent with the *Hetuvidya* definition of dialectical logic as a subject of human thoughts. Liu Xie used dialectical logic in the composition of the *Wen Xin Diao Long*. It is this influence, I would argue, that provides the foundation for his fair and just literary appraisals. This type of dialectical logic, which we nowadays term as "inference", belongs under the rubric *anumana* in the *Hetuvidya*.

My third point has to do with the impact of the *Hetuvidya's* theory of fallacies in argumentation on the *Wen Xin Diao Long*. The theory of fallacy occupies an important position in the *Hetuvidya*. Liu Xie's understanding of informal fallacies in the *Wen Xin Diao Long* has been shaped by the theory of fallacies as contained in the *Hetuvidya*. This theory of fallacy is also treated in the *Upayakauslyahrdaysastra*. The concept of the theory of fallacies in logical argumentation was just being established for the first time in China. It received commentary in the *jati* hetuvidya of the *Upayakauslyahrdaysastra* and in the Nyayasutra. Traces of the elements of the theory of fallacies show up in scattered form in Liu Xie's *Wen Xin Diao Long*. His use of it is not systematic and it is not treated in strictly categorical terms. It is not difficult, however, to pick out the echoes and traces from the *Hetuvidya*. Such fallacies as hetvabhasa and vakya-dosa which are discussed in the *Wen Xin Diao Long* are mentioned in the *Upayakauslyahrdaysastra*. Hetvabhasa is a fallacy caused by based on incomplete cognition or wrong inference and comparison. In the "Zheng Wei" chapter of the *Wen Xin Diao Long*, Liu Xie regards *Wei Shu* as fake because it does not conform with *Jing Shu* and obviously violates a number of established rules. Since the content is not authentic, the grounds of argument, he concludes, are not valid. Liu Xie also analyses the records in *Shi Zhuan* and cites the fallacies in quotation marks. He argues that the historical events can not be verified because long years have passed and the events which they purport to describe have not been scientifically recorded. Fallacies could creep in as these records get circulated and recopied. The truth of the account may drift away from the original facts. This undermines the validity of the argument. Liu Xie also discusses the fallacy of composition. This kind of fallacy arises when we infer that something is true of the whole from the fact that it is true of some part of the whole or even of every part separately. Only after comprehensive judgment can we obtain profound insights. Aside from that, logical errors of improper recognition arise because some writers make subjective and biased judgments. Liu Xie explains how this happens as he comments on literary works especially in the "Zhi Yin" chapter. He discusses common errors in literary criticism, such as placing too much authority and value in ancient works and deprecating contemporary ones or privileging ones own works over the works of others or believing in the fake and doubting the genuine. These three fallacies come about when scholars adopt the wrong attitudes toward the subjects of their study. Liu Xie cites them as obstacles to correct and appropriate literary criticism. People sometimes tend to accept or reject literary works according to their own tastes. As a result, they make fallacious or impertinent judgments about the works under consideration. These cases are examples of the fallacy hetvabhasa. In addition, Liu Xie claims in the "Zhi Xia" chapter that writers often commit logical errors because they use or adapt words improperly or use the wrong word. This leads to ambiguity, obscurity, or incorrect meaning. This type of fallacy is called *vakya-dosa* in the *Upayakauslyahrdaysastra*. It refers to the over-use or under-use of certain types of words words or the improper use of words. Liu Xie realized the importance of abiding by the rules of argument. It is one of the advances which he incorporates in his thought.

Finally, allow me to caution that we should adopt an objective attitude regarding the *Hetuvidya's* influence on the *Wen Xin Diao Long*. It would be unscientific and imprudent to completely deny the relationship. Maybe Liu Xie had not mastered completely the entire logical system of the *Hetuvidya*. In his time, he had inadequate tools to work with. Translation of the *Hetuvidya* was still at an elementary stage. There were still no adequate commentaries on the nuances of its content. The subtleties of its logical argumentation had yet to be fully worked out in Chinese translation. Its practical application was limited, therefore, by the limited level of understanding which had been achieved by 300-350 A.D. Yet even if we admit these limitations, the evidence is unmistakable that Liu Xie at least had the opportunity to learn the fundamentals of the method of argumentation contained in the *Hetuvidya* directly and through rudimentary commentaries. The *Wen Xin Diao Long* bears the ghostly traces of that influence.

Bibliography

[1] China Wen Xin Diao Long Academic Association. *Wen Xin Diao Long Study*. Baoding: Hebei University Press, 2002.

[2] Huang, Z. Principles of buddhism and the study of the "Hetuvidya". *World Religion Study*, 1, 2001.

[3] Jiang, Z. Shanghai: Shanghai Chinese Classics Publishing House, 1985.

[4] Shen, H. On several questions in the "Hetuvidya". *World Religion Study*, 3, 2004.

[5] Sheng, H. *Essentials of the Ancient "Hetuvidya"*. Taiwan: Zhizhe Press, 1978.

A Perspective Semantics for Subjunctive Conditionals[1]

Wang Wenfang
Yang Ming University
wenfwang@hotmail.com

Wang Linton
Yang Ming University
lingtonwang@phil.ccu.edu.tw

ABSTRACT. We roughly agree with Lewis[2] that a subjunctive conditional "$\phi\square \to \psi$" is true if "ψ" is true at certain ϕ-worlds. But what are these?-worlds like? Well, first of all, they have to be worlds at which "ϕ" holds - that is why they are called "ϕ-worlds" - and secondly, they must, in overall, resemble our world as much as ϕ permits it to. The notion of comparative overall similarity used here is a context-dependent one; the criterion used in deciding which world is more similar, in overall, to a second world than a third one is to the second differs from one context to another. The diversity of criteria in determining the overall similarity in different contexts is reflected in the diversity of models, but the common formal properties of all these overall similarity relations are captured within a model. Thus, an L-model for Lewis is a triple $< I, \$, [] >$, where I is a set of possible worlds, and $\$$ is an assignment to each $i \in I$ of a set $\$_i$ of sets of possible worlds such that $\{i\} \in \$_i$, $\$_i$ is nested, and is closed under both unions and nonempty intersection.

We roughly agree with Lewis (1973) that a subjunctive conditional "$\psi\square \to \psi$" is true iff "ψ" is true at certain ϕ-worlds. But what are these?-worlds like? Well, first of all, they have to be worlds at which "ϕ" holds - that is why they are called "ϕ-worlds" - and secondly, they must, in overall, resemble our world as much as "ϕ" permits it to. The notion of comparative overall similarity used here is a context-dependent one; the criterion used in deciding which world is more similar, in overall, to a second world than a third one is to the second differs from one context to another. The diversity of criteria in determining the overall similarity in different contexts is reflected in the diversity of models, but the common formal properties of all these overall similarity relations are captured within a model. Thus, an L-model for Lewis is a triple $< I, \$, [] >$, where I is a set of possible worlds, and $\$$ is an assignment to each $i \in I$ of a set $\$_i$ of sets of possible worlds such that $\{i\} \in \$_i$, $\$_i$ is nested, and is closed under both unions and nonempty intersection. For Lewis, a subjunctive conditional "$\phi\square \to \psi$" is true at a world i in an L-model $M = < I, \$, [] >$ iff either

(i) no ϕ-world belongs to any sphere S in $\$_i$ (i.e., any member of $\$_i$), or
(ii) some sphere S in $\$_i$ does contain at least one ϕ-world, and every such ϕ-world is also a ψ-world.

But there is another kind of dependency which lurks in almost every subjunctive conditional. Again, we roughly agree with Nute that, when engaging

---
[1] The authors are with Chung-Cheng University, 160 University Rd, Mingshung Chia-yi Taiwan.

in hypothetical deliberations, we sometimes want to preserve some, but not all, "of the dispositional properties possessed by objects, substances, and kinds and arrangements of these mentioned in the antecedent of [the] conditional,"[5, p. 352], and that what is supposed to be preserved may differ from one utterance to another. To see some force for such a claim, let us consider the following two sentences:

1. If this penny were asbestos, then it would not conduct electricity.
2. If this penny were asbestos, then some asbestos would conduct electricity.

As Nute rightly points out, 1 is true if we preserve the dispositional properties permanently possessed by asbestos but not those possessed by the penny in the process of hypothetical deliberation, i.e., the process of considering various hypothetical situations in the evaluation of any subjunctive conditional, and 2 is true if we preserve the dispositional properties of this penny but not those possessed by asbestos in that process.

Nute's explanation has the advantage of being able to explain how pairs of sentences like 1 and 2 can both be true, though from different perspectives, when uttered, while other accounts may or may not do so. However, unlike Nute but more like Gabbay[1], we believe that the properties being preserved in the process of hypothetical deliberation need not always be dispositional properties and need not always be properties of entities involved in the antecedent either. Consequently, while Nute thinks that what is supposedly to be preserved in the process of hypothetical deliberation is a function of the antecedent alone, we think rather that it is a function of both the antecedent and the consequent of the subjunctive conditional.

To see why things are what we believe them to be, consider the following two sentences:

3. If there were a can of beer in my refrigerator, there would have been something wrong with my eyes (since I would not see it).
4. If there were a can of beer in my refrigerator, I would be able to see it.

As Nute would certainly agree, 3 is true if we preserve the dispositional properties permanently possessed by cans of beer but not those permanently possessed by my eyes in the process of hypothetical deliberation, and 4 is true if we preserve the dispositional properties of both in that process. But then it is obvious that we need to preserve properties of some object not mentioned in the antecedent in order to make 4 true. This example also shows that what is supposedly to be preserved in the process of hypothetical deliberation is a function of both the antecedent and the consequent of a subjunctive conditional. As an example showing that the properties being preserved need not be "dispositional", consider the case where two objects A and B are both very small according to the normal standard and the following two utterances:

5. If A were ten times larger than B, A would not be very small.
6. If A were ten times larger than B, A would still be very small (and B would be even smaller).

Since both A and B are actually very small, 5 is true if we preserve the supposedly non-dispositional properties of B's size in the process of hypothetical

A Perspective Semantics for Subjunctive Conditionals   9

deliberation, and 6 is true if we preserve the supposedly non-dispositional properties of A's size in that process. In view of examples like 3-6, we conclude that, when evaluating a subjunctive conditional for its truth value, we sometimes need to "narrow down" the scope of our deliberation and select for our considerations only those worlds at which some, but not all, of the properties possessed by some selected entities mentioned in the conditional are preserved, and the selection is a function both of the antecedent and of the consequent (and others perhaps) of that conditional.

It would be a mistake to conflate the aforementioned two kinds of dependency. The first kind of dependency tells us that, when evaluating the truth value of a subjunctive conditional at a world i, whether a possible world should be selected for consideration depends on whether it is more similar to i in overall than other possible worlds to i, whereas the second kind of dependency tells us that it depends on whether the entities and their properties being focused or emphasized by the conditional are preserved in that possible world. A speaker who utters 6 may agree completely with one who utter 5 on the overall similarity relations among all possible worlds, but, due to differences in perspectives and emphases, they may still disagree on the truth values of utterances of 5 and 6. Such a disagreement is possible only because the second kind of dependency may sometimes lead one to ignore some possible worlds that are as similar to i as some other possible worlds to i. For these reasons, we shall call these two kinds of dependency separately as "similarity-dependency" and "perspective-dependency".

Putting both insights from Lewis and Nute (or Gabbay) together, we propose the following revision of Lewis's semantics as a more adequate semantics for subjunctive conditionals. A W-model for a language L suitable for expressing subjunctive conditionals is a quadruple $< I, \$, s, [] >$, where I and $\$$ are as before. s is a selection function from pairs of sentences and a possible world i to sets of possible worlds such that, if there is a sphere S of $\$_i$ such that $S' \cap [\phi] \neq \emptyset$, then (i) $s(\phi, \psi, i) \subseteq [\phi]$ and (ii) $s(\phi, \psi, i) \cap S' \neq \emptyset$ for every S' of $\$_i$ such that $S' \cap [\phi] \neq \emptyset$; and $s(\phi, \psi, i) = \emptyset$ if otherwise. A subjunctive conditional "$\phi \square \to \psi$" is true at a world i w.r.t. s in a W-model $M = < I, \$, s, [] >$ iff there is an S of $\$_i$ such that $S \cap s(\phi, \psi, i) \subseteq [\psi]$. The notions validity-in-a-W-model and validity are defined in the usual way.

One salient feature of our semantics, which is not to be had by any other well-known semantics for subjunctive conditionals, is that it allows for the possibility that a pair of sentences of the forms "$\phi \square \to \psi$" and "$\phi \square \to \neg \psi$" can *be both true at the same world w.r.t. the same s in the same model*. Sentence 1-6 are but some examples of such schematic forms, and other examples abound in the literatures. All such sentences are true in a model in which some sphere S of $\$_i$ intersects with $[\phi]$, but the intersection is divided into two groups: a group of ϕ-and-ψ-worlds and a group of ϕ-but-ψ-worlds.

A second feature of our semantics is that it is hyperintensional in the sense that two logically equivalent sentences may not be interchangeable within a subjunctive conditional without affecting its truth value. This is so only because our selection function s takes sentences (and a possible world), rather than propositions, as its arguments, and hence $s(\phi, \chi, i)$ and $s(\psi, \chi, i)$ may select different sets of worlds even if "ϕ" and "ψ" are logically equivalent. To be sure, we can decide to make it a function of propositions if we like to, but we think that there are good reasons for regarding contexts of subjective conditionals as hyperintensional contexts. Consider the following two pairs of sentences (7)-(10):

7. If Superman were I, he would not be able to fly.
8. If I were Superman, I would be able to fly.
9. If New York City were in Georgia, New York City would be in the South.
10. If Georgia included New York City, Georgia would not be entirely in the South (as New York City would remain in the North).

When one utters sentences 7-10, his or her utterances are likely to be taken as true in normal circumstances. However, the antecedents of 7 and 8, as well as the antecedents of 9 and 10, are logically equivalent, but if we switch the antecedents of 7 and 8, or 9 and 10, we are likely to get utterances that will be regarded as false in normal situations. Examples such as these show that a hyperintensional logic for subjunctive conditionals is to be preferred than intensional ones that are more commonly seen.

A third feature of our semantics is that it should be "essentially" first-order in the sense that a "more adequate" semantics for subjunctive conditionals should have a more delicate structure than the one we gave above which will allow us to be able to decide *which entities and which properties* about them are preserved in a model. However, since the selected entities and their properties will unavoidably and jointly form a set of propositions or, equivalently, a long conjunction of all these propositions, we can still represent them by a set of possible worlds in our model and capture this formal feature by our selection function s at the propositional level.

Now it is easy to see that the logic determined by the set of all W-models is a proper sub-logic of Lewis's **VC**. Call a W-model an "L-counterpart-model" if $s(\phi, \psi, i) = (\cup \$_i) \cap [\phi]$ when there is an S of $\$_i$ such that $S \cap [\phi] \neq \emptyset$, and $s(\phi, \psi, i) = \emptyset$ if otherwise, then the logic determined by the set of all L-models is exactly the same as that determined by the set of all L-counterpart-models. However, the set of all L-counterpart-models is only a proper subset of the set of all W-models. It follows that whatever is valid in our semantics is also valid in Lewis's original semantics. Hence, the logic determined by the set of all W-models is a sub-logic of Lewis's **VC**. On the other hand, every sentence of the form "$\phi\Box \to \psi$" that is valid in Lewis's semantic is still valid in the our semantics (we will omit the proof here). However, this does not mean that Lewis's semantics and ours determine the same logic or, equivalently, the same set of theorems. For it is not excluded that a much more complex sentence not of the form "$\phi\Box \to \psi$" may be valid in Lewis's semantics but invalid in ours. One example of such a formula is "$(\Diamond p \supset \neg[(p\Box \to q)\&(p\Box \to \neg q)])$", where "$\Diamond p$" is true at a world L in an L-model or a W-model M iff "p" is true at some world of $\cup \$_i$. "$(\Diamond p \supset \neg[(p\Box \to q)\&(p\Box \to \neg q)])$" is valid in Lewis's semantics because if $[p]$ intersect with some S of $\$_i$, then it is impossible that there is an S of $\$_i$ such that $S \cap [p] \subseteq [q]$ and another S' of $\$_i$ such that $S \cap [p] \subseteq [\neg q]$. However, such a sentence is possibly false at some world in some W-model. We therefore conclude that the logic determined by the set of all W-models is only a proper sub-logic of Lewis's **VC**. The exact sound and complete axiomatization of those sentences that are valid in our semantics still awaits further research.

The logic determined by W-models may be too weak or too strong when further considerations are taken into account. If that were the case, we might want to impose some more stringent requirements or to release some existing ones regarding what to count as a W-model. Accordingly, the logic resulting from such a revision will be strengthened or weakened depending on how we choose the revision. If, for example, we choose to drop the requirement that a sphere assignment must be centered out of our definition, then the schemata "$(\phi \& \psi) \supset (\phi\square \to \psi)$" and "$(\phi\square \to \psi) \supset (\phi\square \to \psi)$", which are valid in Lewis semantics as well as ours, will become invalid in the new semantics. Similar things can be said about the other direction for revising our semantics.

Bibliography

[1] Gabbay, D. M. A general theory of the conditional in terms of a ternary operation. *Theoria*, 38:97–104, 1972.

[2] Lewis, D. *Counterfactuals*. Blcakwell Publishers, 1973.

[3] Lewis, D. *On the Plurality of Worlds*. Basil Blackwell Inc, 1986.

[4] Nute, D. *Topic in Conditional Logic*. D. Redel Publishing Company, 1980.

[5] Nute, D. Causes, laws, and law statements. *Synthese*, 48:347–369, 1981.

Review of Wittgenstein's "Logic Picture Theory"[1]

Qin Weiyuan

Mianyang Normal University

qinweiyuan@163.com

zjm9611@163.com[0.2cm] Sun Yan

Mianyang Normal University

sunyan943@yahoo.com.cn

Zi Jianmin

Mianyang Normal University

ABSTRACT. "Logic Picture Theory" is the main proposition in Ludwig Wittgenstein's *Tractatus Logico-Philosophicus*. Since he proposed it, the academic community has tried various explanations for it; however, scholars have not completely grasped what Wittgenstein really means. Wittgenstein once thought his theory was misunderstood by even those who claimed they were his believers. Therefore, it is very important to review his proposition once more in order to get closer to the meaning of his thought, even when we admit in advance that it may not even be possible to grasp completely the meaning of his thought.

1 The Proposal of "Logic Picture Theory"

The proposal of the concept of the "Logic Picture Theory" is the major theme of the famous philosopher Ludwig Wittgenstein's *Tractatus Logico-Philosophicus*. It embodies his early stage philosophical thought. It is said that one cannot successfully understand Wittgenstein's philosophical thought in the early stage without comprehending the "Logic Picture Theory". Russell remarks, "In order to understand Mr. Wittgenstein's book, it is necessary to realize what is the problem with which he is concerned [2]". It is obvious that it is necessary to mention the following story when talking about the "Logic Picture Theory".

It was autumn in 1914 on the eastern front. Wittgenstein was reading in a magazine about a lawsuit in Paris concerning an automobile accident. At the trial a miniature model of the accident was presented before the court. The model served as a proposition; that is, it depicted a possible state of affairs. It has this function owing to a correspondence between the parts of the model (the miniature-houses, -cars, -people) and things (houses, cars, people) in reality. It occurred to Wittgenstein that one might reverse the analogy and say that a proposition serves as a model or picture, by virtue of a similar correspondence between its parts and the world. The way in which the parts of the proposition

---

[1] The authors are with Mianyang Normal University, Mianyang, Sichuan, 621000.

are combined – the structure of the proposition – depicts a possible combination of elements in reality, a possible state of affairs. [3]

It is said that Wittgenstein himself once related the incident to Norman Malcolm and G. H. von Wright. Previously, he had realized that the connection between language and reality is just like the connection between real images and retinal images. However, what brought forth the "Logic Picture Theory" is the illumination of H. Hertz's scientific view. According to Hertz, the objects in the external world are firstly some specious pictures in our minds which we will validate afterwards when they get clear with the help of logical rules, relevant scientific rules and experiences we have commanded. Wittgenstein expresses the similar thought that "The picture is a model of reality" (2.12)[1]. Moreover, he directly refers to Hertz's *Mechanics Principle* in 4.04. Therefore, the direct theoretical source of Wittgenstein's "Logic Picture Theory" comes from Hertz.

2 The formation of the theme of the "Logic Picture Theory"

According to Wittgenstein, the "automobile" in the automobile accident and the "automobile" in a proposition (sentence) are elements of his so-called propositional marks although the latter "automobile" is unreal compared with the former. However, as one element of propositional marks, it can express a corresponding event in the real world when it is in a particular form and is related with other elements. In this case, it has the same function with a corresponding picture, music, sculpture, photography, etc. although the most important and frequently-used means is language. Wittgenstein thinks that all these media, which express our thought, just like thought itself have representing relation with facts or state of affairs which the media represent. How does the representing relation form? This is what the "Logic Picture Theory" intends to answer.

Firstly, in Wittgenstein's opinion, facts or state of affairs can only be represented by facts because "The picture is a fact" (2.141). Any picture includes the following three levels: the combination of existence (direct), the combination of abstract signs/marks (indirect), and recessive psychological components penetrating in the abstract signs (more indirect). All these are facts. Also, because facts or state of affairs can be deconstructed into combinations of objects, in return, the representing relation exists only in the things that combine in a particular form.

Secondly, the things of facts or state of affairs have a corresponding relation with the elements of pictures.2.13; 2.131; 2.1514; 2.1511

Nevertheless, according to Wittgenstein, the cooperation relation between the things of facts or state of affairs and the elements of the picture and the representing relation based on this belong to the picture. (2.1513)

Thirdly, the number of the elements a picture contains must be the same as that of the things of facts or state of affairs the picture represents. They must have the same structure, namely, they should be in symmetry or in accordance with, which serves as the base of Wittgenstein's "Logic Picture Theory".

Fourthly, the elements of a picture must be likely to combine with the facts or state of affairs they correspond with, i.e., they must have the same form.

Fifthly, the definite combination of the elements of a picture must be the same with that of the things of the facts or state of affairs, namely, a picture must have the same structure with the facts or state of affairs it intends to represent, and further, must have the same kind of representation. (2.15; 2.151; 2.16; 2.17)

Based on the fore-mentioned five points and the author's opinion that other means of expressing facts or state of affairs include music, sculpture and photography, it is easy to understand that:

4.015 The possibility of all similes, of all the *Bildhaftigkeit unserer Ausdrucksweise, ruht in der Logik der Abbildung.* Then, Wittgenstein introduces the concept "logic picture" after explaining the picture concept and the logical form of reality (2.18). With 2.17 and 2.18 compared, it is easy to observe that the logical form is more essential than the form of representation because the picture of a fact or state of affairs can have different forms of representation which must have the same logical form. More exactly, the form of representation can be various but the logical form is only one which is shared by all the forms of representation. (2.181; 2.182)

Why does Wittgenstein put the word "logic" before the word "picture"? In his opinion, a "picture" doesn't have to be a drawing or a photo which is clear and easy to be seen. Instead, it should be a corresponding relation between proposition and reality. This relation is called "logic picture" or "logic model" by Wittgenstein and compared to hieroglyph or mapping in geometry.

When I remark that "Proposition is the logic picture of facts", I mean that I can insert a picture – more exactly, a drawing picture (*ein gezeichnetes bild*) –into a proposition and apply it. Thus, I can use a picture the same way I use a proposition. Is it possible? The answer is that these two concepts have something in common in some aspects and what is shared is called picture. In this case, the meaning of "picture" has been amplified. I obtain the concept from two sources. One is from a drawing picture and the other is from mathematicians.

Moreover, Wittgenstein distinguishes between general "picture" and "logic picture". Wittgenstein's "logic picture" is just the "picture" in the sense of "proposition – picture"(4.01; 4.012; 4.03; 4.1)

Obviously, it can be easily understood when basic proposition is referred to because any of these propositions is compound and in real existence, namely, a basic fact. Therefore,

4.21 The simplest proposition, the elementary proposition, asserts the existence of an atomic fact.

3.142 Only facts can express a sense, a class of names cannot.

4.032 The proposition is a picture of its state of affairs only in so far as it is logically articulated. Furthermore, given any basic fact or state of affairs, we can find basic ones in all the propositions with the same numbers of constituents. The basic propositions have to:

4.04 In the proposition there must be exactly as many things distinguishable as there are in the state of affairs, which it represents. They must both possess the same logical (mathematical) multiplicity.

On the other hand, in those basic propositions which have the same numbers of constituents with their basic facts or state of affairs, we can find their corresponding basic propositions. Moreover, we can find some basic propositions which are likely to combine with the constituents of basic facts or state of affairs. Therefore, the basic proposition, which is the logic picture of basic fact or state of affairs, can not only represent the basic fact or state of affairs but also the logic picture of basic fact or state of affairs.

It is difficult to find in compound propositions. However, we cannot deny its quality of being a picture. (4.011; 4.013; 4.014) In addition, we can understand the quality of a proposition being a picture from the perspective of how a pictograph describes facts.

4.016 In order to understand the essence of the proposition, consider hieroglyphic writing, which pictures the facts it describes. And from it came the alphabet without the essence of the representation being lost.

Up to this point, it is easy to understand Wittgenstein's "Logic Picture Theory" which does not try to tell how a proposition should be interpreted to a picture directly, but there is the content of a proposition in a picture. Because, to some extent, we can regard a proposition as a picture and vice versa, we can understand this in the opposite way. In fact, this thought pattern is necessary in order to comprehend Wittgenstein's "Logic Picture Theory".

3 Bewilderment on the "Logic Picture Theory"

3.12 The sign through which we express the thought I call the propositional sign. And the proposition is the propositional sign in its projective relation to the world.

3.323 In the language of everyday life it very often happens that the same word signifies in two different ways – and therefore belongs to two different symbols – or that two words, which signify in different ways, are apparently applied in the same way in the proposition.

3.324 Thus there easily arise the most fundamental confusions (of which the whole of philosophy is full) Because daily language "is a part of the human organism and is not less complicated than it" (4.002), in order to avoid these kinds of confusion,

3.325 We must employ a symbolism which excludes them ... which obeys the rules of *logical grammar* – of logical syntax.

Where can we find directly this kind of appropriate symbolic language? We can only find it in Wittgenstein's historical context. The symbolic language in *Principia Mathematica* by Russell and Whitehead is conceived as a general language. However, Wittgenstein thinks that this sort of language does not exclude all possible mistakes. Russell's language is one that depends on propositional connectives and unit words. Then Wittgenstein's concern is how we can construct such a language to explain all propositions as pictures. He deems that he can assume quantified proposition as maximized conjunction and disjunction. Thus, a quantified proposition of existential formula – $(\exists X)S[x]$ – is equivalent to such a disjunction as $S[a] \bigvee S[b] \bigvee S[c]$ ... (a, b, c ... are indicated to different objects of particular logic form). In the later stage of his philosophical career, Wittgenstein realized that this solution to quantified proposition is the most absurd mistake in *Tractatus Logico-Philosophicus*. Nevertheless, he couldn't cast off this problem.

What's more, according to Wittgenstein, those propositions used to describe imaginary objects in the real world and those used to describe qualities or relations between objects are regarded as meaningless. The most obvious reason is that they do not have one-to-one corresponding relation with the real world, namely, they are not structurally similar. There are plenty of negative propositions in logic. What are the corresponding facts? Could they be those facts which never happen? What sort of facts are they when they never happen? This is a typical self-contradiction. In this case, Wittgenstein's proposition is not soundly based. As the theoretical foundation of "Logic Picture Theory", it is not convincing, which is not predicted by Wittgenstein himself.

Compared with the two fore-mentioned points, the most arresting is what Wittgenstein calls "logical form". The fore-mentioned explanation helps us to realize that the logical form is one indispensable element in Wittgenstein's "Logic Picture Theory". (4.12) Can it be possible? Since the logical form cannot be described in a proposition, the "Logic Picture Theory" forces Wittgenstein to put the logical form beyond the limit of language. And thus, there is the difference between "speakable" and "unspeakable". This results in a mysterious field in the "Logical Picture Theory".

Bibliography

[1] Kolak, D. *Wittgenstein's Tractatus*. Belmont: Mayfield, 1998.

[2] Ludwig, W. *TRACTATUS LOGICO-PHILOSOPHICUS*. Routledge and Kegan Paul Ltd, 1974.

[3] Norman, M., Ludwig, W. *A Memoir*. Oxford: Oxford UP, 1958.

Logics for Knowing an Individual[1]

Li Xiaowu
Sun Yat-sen University
LXW121@yahoo.com.cn

We often utter sentences of the form "I know i", but what exactly does a sentence of the form, say, "I know Everest" mean? In what follows, I will assume that Everest is a member of an arbitrarily chosen universe of discourse (or individual domain) denoted by Ind. The current proposal is that "I know Everest" is ambiguous in many different ways. In one sense of it, it means that I know at least something about Everest, i.e., I know that it has a certain property w.r.t. Ind (for example, it is the highest mountain on Earth) or that it has a certain relation with other individuals in Ind, (for example, it is higher than Alps). In another, ideal sense of it, I know an individual i w.r.t. Ind iff I know everything about it, i.e., I know every property it has w.r.t. Ind and every relation holding between it and other individuals in Ind. Obviously, the sense it actually expresses will depend on the choice of individual domains: for example, when the individual domain Ind consists of purely geometrical entities, a more restricted sense will normally be preferred.

In Li Xiaowu [4], [5], [6], [7] and [8], I have presented several classes of epistemic logic systems characterizing the phrases "I know i", provided the corresponding classes of semantics, and proved that each system is sound and complete with respect to its corresponding semantics. In the present paper, I will continue that research.

For simplicity, I will just consider the case of a single agent and omit the proofs for the following results. Detailed proofs and precise definitions for the following undefined notations are to be found in Li Xiaowu [8].

1 A logic KI_ω for knowing an individual

Definition 1 *1. Let $PS = \{P_1, \cdots, P_n, \cdots\}$ be a countable set of relation (predicate) symbols such that $\rho : PS \to (N - \{0\})$; so ρ assigns to each $P \in PS$ a finite arity. Let $Ind = \{i_1, \cdots, i_n, \cdots\}$ be a countable set of individual constant symbols.*

2. A single-agent epistemic language L is a set of formulas φ, given by the following rules:

$$\varphi := Pa_1 \cdots a_{\rho(P)} | Ki | \neg \varphi | (\varphi \wedge \psi) | K\psi$$

where $P \in PS$, and $a_1, \cdots, a_{\rho(P)}, i \in Ind$.

---
[1] Institute of Logic and Cognition, Sun Yat-sen University, Guangdong, China, 510275. Email: LXW121@yahoo.com.cn

3. For all $i \in Ind$, let

$$PA(i) := \{Pa_1 \cdots a_{n-1}ia_{n+1} \cdots a_{\rho(P)} | P \in PS \text{ and } \\ a_1, \cdots, a_{n-1}, a_{n+1}, \cdots, a_{\rho(P)} \in Ind\}$$

$$PA := \cup\{PA(i) | i \in Ind\} = \{Pa_1 \cdots a_{\rho(p)} | P \in PS \text{ and } \\ a_1, \cdots, a_{\rho(P)} \in Ind\}$$

Remark. Intuitively, "Ki" means that the current agent knows the individual i.

Definition 2 1. *A single-agent epistemic frame for L is a tuple (W, R_K) such that $W \neq \emptyset$, and R_K is a reflexive and euclidean relation on W such that $Ind \cap W = \emptyset$.*

In this paper I always use "Frame" to stand for the class of all such frames.
2. *A single-agent epistemic model for L is a tuple (W, R_K, V) such that $(W, R_K) \in Frame$ and $V : PA \to P(W)$, where $P(W)$ is the power set of W.*

In this section I use "Model" to stand for the class of all such models.

Definition 3 *(Truth Definition) Let $(W, R_K, V) \in Model$. For every compound formula $\varphi \notin PA$, the truth set $V(\varphi)$ of φ w.r.t. M is defined inductively as follows: for all $w \in W$,*

1. $w \in V(\neg\varphi) \Leftrightarrow w \notin V(\varphi)$
2. $w \in V(\varphi \wedge \psi) \Leftrightarrow w \in V(\varphi)$ and $w \in V(\psi)$
3. $w \in V(K\varphi) \Leftrightarrow R_K(w) \subseteq V(\varphi)$ where $R_K(w) := \{u \in w | wR_K u\}$,
4. $w \in V(Ki) \Leftrightarrow \forall \theta \in PA(i), w \in V(\theta \to K\theta)$ where $\varphi \to \psi$ is defined as usual.

Remark. It is easy to see that for all $w \in W$,

$$\forall \theta \in PA(i), w \in V(\theta \to K\theta) \Leftrightarrow \forall a_1, \cdots a_{n-1}, a_{n+1}, \cdots, a_{\rho(P)} \in Ind,$$
$$w \in V(Pa_1 \cdots a_{n-1}ia_{n+1} \cdots a_{\rho(P)}) \to KPa_1 \cdots a_{n-1}ia_{n+1} \cdots a_{\rho(P)}).$$

Definition 4 1. *A single-agent epistemic system $\mathbf{ES5}_\omega$ is defined as follows:*

(Ax)	$\vdash \varphi$ if φ is an instantiation of a tautology
(MP)	$\varphi, \varphi \to \psi \vdash \psi$
(Ded)	$\Phi, \varphi \vdash \psi \Rightarrow \Phi \vdash \varphi \to \psi$
(Weak)	$\Phi \vdash \varphi \Rightarrow \Phi, \Psi \vdash \varphi$
(Cut)	$\Phi \vdash \Psi$ and $\Phi, \Psi \vdash \varphi \Rightarrow \Phi \vdash \varphi$
(SN_K)	$\Phi \vdash \varphi \Rightarrow K\Phi \vdash K\varphi$ where $K\Phi := \{K\varphi | \varphi \in \Phi\}$
(T_K)	$\vdash K\varphi \to \varphi$
(5_K)	$\vdash \neg K\varphi \to K\neg K\varphi$

2. *A single-agent epistemic system $\mathbf{KI}_\omega := \mathbf{ES5}_\omega +$ the following axiom and rule*

(KI)	$\vdash Ki \to \theta \to K\theta$ for all $\theta \in PA(i)$
(Inf_{KI})	$\{\theta \to K\theta | \theta \in PA(i)\} \vdash Ki$

Theorem 1 $\mathbf{KI}_\omega$ *is strongly sound and strongly complete w.r.t. Frame.*

Logics for Knowing an Individual

I conclude this section by the following remarks:

1. Let $BL(i)$ be the smallest set closed under the following rules:
$$PA(i) \subseteq BL(i)$$
and
$$\varphi, \psi \in BL(i) \Rightarrow \neg\varphi, (\varphi \wedge \psi) \in BL(i)$$
We can replace $PA(i)$ above by $BL(i)$. For example,
$\vdash Ki \to \theta \to K\theta$ for all $\theta \in BL(i)$,
$\{\theta \to K\theta | \theta \in BL(i)\} \vdash Ki$.

2. KI and Inf_{KI} are called Positive Knowledge Axiom and Positive Knowledge Rule. We can also replace them by the following axiom and rule:

$\vdash Ki \to \neg\theta \to K\neg\theta$ for all $\theta \in PA(i)$ (Negative Knowledge Axiom)
$\{\neg\theta \to K\neg\theta | \theta \in PA(i)\} \vdash Ki$ (Negative Knowledge Rule)

Or

$\vdash Ki \to (\theta \to K\theta) \wedge (\neg\theta \to K\neg\theta)$ for all $\theta \in PA(i)$
$\{(\theta \to K\theta) \wedge (\neg\theta \to K\neg\theta) | \theta \in PA(i)\} \vdash Ki$

3. We can also add the following formulas to a suitable system as axioms:

$Ki \to KKi$ (note that it is not an instance of T_K)
$\neg Ki \to K\neg Ki$ (note that it is not an instance of 5_K)

4. We can turn the language defined in Definition 1 into a first-order epistemic language, and then establish $\mathbf{KI}_\omega$ by the classical first-order epistemic logic. Let $IV := \{x_1, \cdots, x_n, \cdots\}$ be a countable set of individual variables such that $Ind \cap IV = \emptyset$. Axiom KI can be altered as

$$\vdash Ki \to Px_1 \cdots x_{n-1} i x_{n+1} \cdots x_{\rho(P)} \to KPx_1 \cdots x_{n-1} i x_{n+1} \cdots x_{\rho(P)}$$

And Rule Inf_{KI} be altered as:

$$\{Px_1 \cdots x_{n-1} i x_{n+1} \cdots x_{\rho(P)} \to KPx_1 \cdots x_{n-1} i x_{n+1} \cdots x_{\rho(P)} | \\ Px_1 \cdots x_{n-1} i x_{n+1} \cdots x_{\rho(P)} \in PA(i)\} \vdash Ki$$

Such a system should be more refined. In the end of Chapter 6 in Li Xiaowu [8], I present a logical system $\mathbf{KIQ}$ based on the classical first-order epistemic logic.

5. Given any $I \subseteq Ind$ such that $I \neq \emptyset$, we can replace Ki above by $K_I i$, and $PA(i)$ by

$$PA(i)_I := \{Pa_1 \cdots a_{n-1} i a_{n+1} \cdots a_{\rho(P)} | P \in PS$$
$$\text{and } a_1, \cdots, a_{n-1}, i, a_{n+1}, \cdots, a_{\rho(P)} \in I\}.$$

The semantics and system thus gained characterize the concept of *knowing an individual w.r.t. I*.

2 A logic KIE$_\omega$ for knowing an individual

Definition 5 *(Truth Definition)* Let $(W, R_K, V) \in Model$. For every formula $\varphi \notin PA, V(\varphi)$ is defined as before, except
(4e) $w \in V(Ki) \Leftrightarrow \exists \theta \in PA(i), w \in V(K\theta)$ for all $w \in W$

Remark. According to (4e), the agent knows an individual i iff it knows *some* property of i w.r.t. Ind or some relation holding between i and other individuals in Ind. Hence, when I assert "I know Everest", I may just know that Everest is the highest mountain on Earth. We usually remark sentences of the form "I know Everest" in the sense of (4e) rather than that of Definition 3(4). So (4e) is more natural than Definition 3(4).

Definition 6 *A single-agent epistemic system* $\mathbf{KIE}_\omega := \mathbf{ES5}_\omega +$ *the following axiom and rule*
(Kle) $\vdash K\theta \to Ki$ for all $\theta \in PA(i)$
(Inf$_{KIe}$) $\{\neg K\theta | \theta \in PA(i)\} \vdash \neg Ki$

Theorem 2 $\mathbf{KIE}\omega$ *is strongly sound and strongly complete w.r.t. Frame.*

I conclude this section by the following remarks:

1. We can consider negative characterization of Ki

   $w \in V(Ki) \Leftrightarrow \exists \theta \in PA(i), w \in V(K\neg\theta)$ for all $w \in W$ (w.r.t. semantics)
   $\vdash K\neg\theta \to Ki$ for all $\theta \in PA(i)$,
   $\{\neg K\neg\theta | \theta \in PA(i)\} \vdash \neg Ki$ (w.r.t. system)

   Thus, I may know Everest simply because, say, I know that Everest is not in Europe.
2. As before, we can present a first-order epistemic system like $\mathbf{KIE}_\omega$.

3 A logic KI for knowing an individual

Now I will weaken $\mathbf{KI}_\omega$ using Bounded-valuation Method.

Definition 7 *Let* $\mathbf{KI} := \mathbf{ES5} + KI$ *where* $\mathbf{ES5}$ *is defined as follows:*
(Taut)   all instantiations of tautologies,
(MP) $\varphi, \varphi \to \psi / \psi$
(K_K) $K(\varphi \to \psi) \to K\varphi \to K\psi$
(RN$_K$) $\varphi / K\varphi$
(T_K) $K\varphi \to \varphi$
(5_K) $\neg K\varphi \to K\neg K\varphi$

Definition 8 *A bounded-valuation model for* $\mathbf{KI}$ *is a tuple* (W, R_K, V) *such that* $(W, R_K) \in Frame$, *and* $V : PA \cup \{Ki | i \in Ind\} \to P(W)$ *such that*

1. $V(\theta) \subseteq W$ for all $\theta \in PA$
2. $V(Ki) \subseteq \{w \in W | \forall \theta \in PA(i), w \in V(\theta) \Rightarrow R_K(w) \subseteq V(\theta)\}$ for all $i \in Ind$

I use "Model" to stand for the class of all such models.*

Logics for Knowing an Individual 21

Definition 9 *(Truth Definition)* Let $(W, R_K, V) \in Model^*$. For every $\varphi \notin PA \cup \{Ki | i \in Ind\}$, $V(\varphi)$ is defined before.

Theorem 3 **KI** *is sound and complete w.r.t. Frame (and $Model^*$).*

I conclude this paper by the following remarks:

1. We may also consider a generalization $Ki_1 \cdots i_n$ of Ki such that $i_1, \cdots, i_n \in Ind$. By "$Ki_1 \cdots i_n$" I mean that the agent knows a set of individuals $i_1, \cdots, i_n$. A trivial method to accommodate "$Ki_1 \cdots i_n$" is to have an abbreviation axiom added to a suitable system:

   $Ki_1 \cdots i_n \leftrightarrow \wedge \{Ki_k | k = 1 \cdots n\}$ where $\varphi \leftrightarrow \psi$ is defined as usual.

   Below I consider several nontrivial methods. See the following truth set definitions:

   $w \in V(Ki_1 \cdots i_n) \Leftrightarrow \forall P \in PS$ such that $\rho(P) = n, w \in V(Pi_1 \cdots i_n \to KPi_1 \cdots i_n)$

   $w \in V(Ki_1 \cdots i_n) \Leftrightarrow \forall P \in PS$ such that $\rho(P) = n, w \in V(KPi_1 \cdots i_n)$

   I know *Romeo* and *Juliet* because I know that they were in love. Given the semantics based on the truth set definitions above, we can present corresponding systems characterizing "$Ki_1 \cdots i_n$", and then show that the systems are sound and complete with respect to their semantics.

2. We can also characterize "Ki" by a dynamic epistemic logic. Let AA be a countable set of atomic actions, and let *Action* be a set of actions such that it is given by the following rules:

   $\alpha := a | (\alpha \cup \beta) | (\alpha; \beta) | \alpha^*$ where $a \in AA$.

   According to Li Xiaowu [8], for all dynamic epistemic model (W, R, R_K, V) and $w \in W$, we can consider the following characterizations:

   $w \in V(Ki) \Leftrightarrow \exists \theta \in PA(i) \exists \alpha \in Action, w \in V(K[\alpha]\theta)$

   $w \in V(Ki) \Leftrightarrow \exists \theta \in PA(i) \exists \alpha \in Action, w \in V([\alpha]K\theta)$

   In multi-agent case, we can consider the following characterizations:

   $w \in V(K_A i) \Leftrightarrow \exists \theta \in PA(i) \exists \alpha \in Action, w \in V(K_A[\alpha]_A \theta)$

   $w \in V(K_A i) \Leftrightarrow \exists \theta \in PA(i) \exists \alpha \in Action, w \in V([\alpha]_A K_A \theta)$

   $w \in V(K_A i) \Leftrightarrow \exists \alpha \in Action \exists B \in Agent, w \in V(K_A D_B \alpha i)$ where *Agent* is a set of agents.[2]

I want to thank Mr. Wen-Fang Wang of the Department of Philosophy at Chung Cheng University for read over the present paper and saving me from a number of mistakes.

---

[2] By "$D_B \alpha i$" I mean that an agent B has done an action α on an individual i. Thus I know Everest because I know that, say, somebody has climbed Everest.

Bibliography

[1] Fagin, R., Halpern, J., Moses, Y., Vardi, M. *Reasoning about Knowledge*. The MIT Press, 1995.

[2] Harel, D., Kozen, D., Tiuryn, J. *Dynamic Logic, Foundation of Computing*. Cambrigde, Massachussetts: MIT Press, 2000.

[3] Kooi, B., de Lavalette, G. R., Verbrugge, R. Hybrid logics with infinitary proof systems. *Journal of Logic and Computation*, 16(2):161–175, 2006.

[4] Li, X. W. Report 2005 [J/OL]. *Luoji Yu Renzhi(Logic and Cognition)*, 4:17–71, 2005.

[5] Li, X. W. Some applications of dynamic epistemic logic. *Luoji Yu Renzhi(Logic and Cognition)*, 1:60–86, 2006.

[6] Li, X. W. *A logic **KI** characterizing knowing an individual*. 2007. URL http://logic.sysu.edu.cn/logic/english/ShowArticle.asp?ArticleID=19.

[7] Li, X. W. *A logic **KIe** characterizing knowing an individual*. 2007. URL http://logic.sysu.edu.cn/logic/english/ShowArticle.asp?ArticleID=19.

[8] Li, X. W. *Monographic Study in Dynamic Epistemic Logic*. 2007. URL http://logic.sysu.edu.cn/logic/english/ShowArticle.asp?ArticleID=19.

Argument Schemes[1]

Wu Hongzhi Zhang Zhimin
Yan'an University Yan'an University
yaydwhz@126.com yazzm@sohu.com

ABSTRACT. Argument scheme represents the common plausible argument pattern used in everyday arguments. It is a kind of pragmatic structure. These schemes are co-inferential rules with the form of deductive argument. The normative force of argument scheme stems from the rational requirement of avoiding pragmatic inconsistency. This force is not only related to the structure of a certain kind of scheme, but also ensured by the circumstances in which this scheme is applied. These circumstantial conditions are depicted by the matched critical questions. In dialogue the party which appropriately use the argument scheme will shift the burden of proof to the other party. However, if the circumstantial conditions for the use of the argument scheme can not be ensured, argument scheme then becomes a fallacy.

Given the cognitive goals typically set by practical agents, Gabbay and Woods point out that validity and inductive strength are typically not appropriate (or possible) standards for their attainment. This, rather than computational costs, is the deep reason that practical agents do not in the main execute systems of deductive or inductive logic as classically conceived [3]. Argument schemes have been a major focus of investigation in informal logic, new rhetoric and argumentation theory. Argument schemes not only belong to one of the four problem areas of pragmatic elements in the study of argumentation, but they are also closely related to the other three elements [10]. Modern research into argument schemes can be traced far back to Aristotle's Topics and directly derived from Toulmin's "warrant" and Perelman's "technique of argumentation". Afterwards, there were many important research inquiries into this problem. It is noteworthy that the argument scheme has drawn the attention of scholars in the cross-disciplinary field of artificial intelligence. Scholars who show interest in non-deductive reasoning or defeasible reasoning provide a more prudent analysis for the argument scheme in the AI field. The argumentation research group in the Computer College of the University of Dundee in Britain has an ARG: Dundee research program and the analysis for the argumentation scheme developed by this group used in the structure analysis for the general discussion. The project of "the argumentation scheme in natural and artificial communication" by Walton got the aid of the Social Sciences and Humanities Research Council of Canada. The famous British research fund Leverhulme Trust also set up an award for the related research of argumentation scheme.

---

[1]The authors are College of Political Science and Law and College of Foreign Languages, Yan'an University.

1 Argument form and argument scheme

The form of reasoning and argument in formal deductive logic (FDL) are all truth-function.

But, the basic reference in the argument "...thus ..." is not necessarily truth-function. Given the definition of argument form of FDL, most of the arguments in daily life would not have this truth-function or would be only analyzed as invalid form. Formal logic has every reason to exclude the non truth-function out of its vision. But other branches of logic can study this non true-value form. Does the argument of non truth-function have a form or scheme?

Early in 1996, Walton pointed out that each argumentation model will have a distinctive argumentation scheme (structure, form) [12]. To reveal the logical form of an argument is to replace all the expanded expressions except the logical terms by schematized letters. Sometimes some expressions are taken as logical terms, sometimes as non-logical terms, thus different logics are produced. By expanding logical terms, different logics came into being. We can choose so-called new logical terms, these terms are not considered as logical terms in classical logic and its expanded forms, such as authority, witness, etc., therefore, we can obtain a new logic. These logical terms are not true-value connectives, but pragmatically logical connectives, namely the evaluation result of the argument scheme formed by these connectives and variables is not only determined by the meaning of the connectives, but also by its pragmatic conditions. In new logic, rational argument is an argument which instantiates reasonable reasoning, and only the argument which instantiates this rational argument can be called rational argument. Due to the restriction of pragmatic element, a rational argument form thus includes general structure as well as the conditions which are required for rationality. As a result, the judgment of plausible reasoning in new logic is different from the judgment of the valid argument in classical logic and its expanded system. It is not the judgment of form, or meaning, but the judgment of pragmatic use. Argument scheme is popular and abstract scheme and it has inexhaustible instances. Therefore, to evaluate the use of the argument scheme, it needs to do two things: first, to distinguish different kinds of argument schemes; second, to consider special critical questions.

At the technical level the argumentation scheme which has the form of reasoning rules is not based on the meaning of logical operators, but on the rule of the principles of cognition and practical reasoning. These argument schemes classified by content can be shifted to instances of logical reasoning rules by adding a conditional premise connecting premise and conclusion, this conditional sentence can certainly be defeasible [7]. Toulmin (1958) had early observed that different elements of argument had different functions, this led to different standards for evaluation. Researchers of "AI&Law" pointed out that sentences which are not distinguished in view of formal logic may play very different roles in argument. The use of sentences not only relies on their logical forms, but also on something else, such as the nature of cognition or pragmatics. Logic which uses the abstract definition of logic validity (no matter deductive or non-monotonic) can not see this difference, thus, it should be complemented by the so-called argument-scheme method. [6]

Kienpointner includes all the reasoning forms in the concept of "scheme". The artificial intelligence scholar Verheij compared the strict form and defeasible form under the scope of general scheme. Argument scheme in the sense that the conclusion can be drawn from the given premise is a general rule for reasoning. Argument scheme is the reasonable reasoning scheme in everyday argumentation. More specifically, argument scheme expresses that given a specific premise, a specific conclusion can be drawn. But argument scheme is always defeasible (i.e., there can be exceptional circumstances in which the scheme's conclusion does not follow from its premises) and contingent (i.e., logical rules of inference seem to be neatly formalizable, necessarily valid, strict and independent of context, while pragmatic schemes are pragmatically valid or even contingent, defeasible and context-dependent.) Verheij then further points out that the reasoning pattern can be understood as a continuum: a continuum ranging from abstract reasoning patterns (like logical rules of inference, such as Modus ponens) via contextual reasoning patterns (like pragmatic argumentation schemes, such as argumentation from expert opinion) to domain rules (such as the verdict in law, it is irrelevant out of the law context). All can be given the premises-conclusion form [11].Our reasoning rule not only includes formal deductive principle, but also Toulmin's the rule of substantial inference indicated by many warranted reasons. In fact, Peirce had put forward early the idea of "the habit of thought", namely the so-called "leading principles", it is better to take these principles as inference rules [2]. Strictly speaking, due to the objective unreliability, the relevant inference rule is not included in the typical set of rule. But it is critical that inference rule can be appropriately aided (in the sense of Toulmin's backing of warrant), thus it can be fairly used in specific situations.

2 The normative force of argument scheme

With argumentation schemes, it does not seem possible to construct an algorithm to test the validity of natural language argumentation by translating it into a formal language.

Evaluating instance of argumentation schemes is best done by taking into account the context of dialogue of the given case. The soundness of argumentation can't be judged independently of semantic and pragmatic standards underlying. Consequently, the use theory of meaning can be seen as an adequate theoretical frame to understand and describe the diversity of soundness-concepts and soundness-judgments in a community [12]. Van Eemere [9, p. 98] also pointed out that under argument scheme dependent circumstances, the arguer tested the argumentation procedure in a practical way and the response to critical questions became the mature way to test the soundness of the relevant argument scheme. Being the form or structure of the argument, argument scheme has a normative force which is expressed in the following sense: when one accepts the premise in argument scheme as warranted and well organized, one (by a certain means) is forced to accept the conclusion derived from the scheme, or to provide relevant critical questions for the scheme.

Then, what is the source of the demonstration of the normative force? The explanation lies in the irrationality of accepting the premises but rejecting the conclusion of such an inference or argument in those particular circumstances. The reasoning or argument derives its cogency from the fact that to accept the premises and grant the validity of the inference using that scheme yet deny the plausibility of the conclusion, under the circumstances – without suggesting that any conditions of rebuttal exist – is pragmatically inconsistent. In all three cases, the probative force of the reasoning or argument using the scheme derives from one or another type of inconsistency involved, given that pattern of reasoning or argument and the facts of that situation, in accepting the premises, yet refusing to accept the conclusion [1]. Reconstructive deductivism (RD) or natural language deductivism (NLD) gives a higher evaluation of the normative force of argument scheme. They remain the exclusive legality of deduction, but admit that the natural language argument in real life uses the probable, plausible and acceptable premise and that there are argument schemes beyond the formal deductivism system. Yet they hold that these schemes can be restricted to deductive form in which the premise may have different scales of acceptability. Therefore, the normative force of argument scheme almost reaches the level of deductive argument [4]. In general, for each scheme, we must be able to provide a general account of why reasoning or arguing using schemes of that type is cogent. There must be some particular connection between the premise set and the conclusion in reasoning or arguments instantiating such a scheme that makes it in some way unreasonable in that kind of case and in those circumstances to deny the conclusion while granting the premises, other things being equal [1].Blair makes a distinction between descriptive scheme and normative scheme. The latter is the patterns of good reasoning. He features the normality of argument scheme in four aspects: each scheme, if the premise is true, it can provide support for the conclusion, but only if other things are equal; each scheme is a default form of good reasoning, though it may be subject to override in a variety of circumstance; each instance of normative scheme is prima facie good reasoning evidencebut whether it is good reasoning all things considered depends on the circumstance of case; the grounds that instantiate the premise-types of a given normative scheme create a presumption if favor of the proposition instantiating the conclusion. In a dialogue, the party with grounds instantiating a normative scheme thereby shifts the burden of proof to other party .But it is necessary to make a complement: the normative force of the argument scheme varies according to the changing context. The assumption derived from the argument scheme may be well proved in one context, may not be well proved in another context [5].

3 Critical questions of argument scheme

The evaluation of argument scheme is not like the evaluation of the argument form abstracted from the conversation context, it is an evaluation of syntax and semantic. For this kind of presumptive argument, it needs to weigh and make a judgment by critical questions put forward in the dialogue [13, p. 44]. The test and evaluation of the critical questions are created and elaborated systematically by Hastings (1963). Today it has become the standard approach for argument scheme evaluation.

Argument scheme is evaluated by critical questions, namely each scheme is matched with a set of critical questions. Argument scheme and the matched questions are used to evaluate the specific argument in a specific context and context related to the dialogue. The used argument is evaluated by judging the weight of the proofs of the two parties in a specific situation in which the argument is used. If all the premises used are supported by some weight of the evidence, the acceptable weight is transferred to the conclusion. The conclusion may be refuted by the appropriately raised critical questions. The core critical questions directly go to the relationship between the premise and the conclusion while other critical questions go to the attached elements which may block the transfer of the acceptability. The argument scheme is made up of the conclusion, a set of premises, exceptions of the blocking of use schemes and set of conditions–all these elements are related to the function of critical questions.

The role of critical questions includes: first, criticizing the premise in a scheme. Second, pointing out the exceptional cases in which the scheme should not be used. Third, giving the conditions for the use and satisfaction of the scheme. Fourth, pointing out the plausible arguments relevant to a conclusion of a scheme. Critical questions of argument scheme actually provide all possibilities of rational doubt for the respondent, and meanwhile provide the chief points for the arguer to exclude the rational doubts, pointing out the way for improving the quality of the argument. But, not all critical questions have the function of shifting the burden of proof [5, 7].

Informal fallacy in traditional sense can be divided into two kinds: one kind of argument scheme is always fallacious, like circular form and straw man form; the other kind of argument scheme is neutral and is fallacious if mistakenly used [8]. Both of them are argument schemes which do not satisfy the condition or critical question of argument scheme. Argument scheme its self is not fallacious. Argument scheme can be both fairly used and mistakenly. Obviously, the misuse of argument scheme namely fallacy can be defined and classified by the corresponding critical questions and sub-questions.

Bibliography

[1] Blair, J. A. Walton's argumentation schemes for presumptive reasoning:a critique and development. *Argumentation*, 15:365–379, 2001.

[2] Freeman, J. B. Relevance,warrants,backing,inductive support. *Argumentation*, 6:219–235, 1992.

[3] Gabbay, D. M., Woods, J. The practical turn in logic. In Gabbay, D. M., Guenthner, F. (Eds.), *Handbook of Philosophical logic*, vol. 13, 15–122. Springe, 2005.

[4] Groarke, L. Johnson on the metaphisics of argument. *Argumentation*, 16:277–286, 2002.

[5] Pinto, R. C. Argument schemes and the evaluation of reasoning; presumption and argument schemes. In *Argument,inference and Dialectic:collected pappers on informal logic*, 98–104, 105–112. Dordrecht:: Kluwer Academic Publishers, 2001.

[6] Prakken, H. AI & law, logic and argument schemes. *Argumentation*, 19:303–320, 2005.

[7] Prakken, H., Reed, C. R., Walton, D. N. Dialogues about the burden of proof. In *the Tenth International Conference on Artificial Intelligence and Law*, 115–124. New York: The Association for Computing Machinery (ACM), 2005.

[8] Tindale, C. W. *Fallacies and Argumen Appraisal*. Cambridge: Cambridge University Press, 2007.

[9] van Eemeren, F. H., Grootendorst, R. *Argumentation, Communication, and Fallacies A Pragma-Dialectical Perspective*. London: Erlbaum, Hillsdale, 1992.

[10] van Eemeren, F. H., Grootendorst, R., Henkemans, F. S. *Foundamentals of Argumentation Theory:A Handbook of Historical Backgrounds and Contemporary Developments*. New Jersey: Lawrence Erlbaum Associates,Inc, 1996.

[11] Verheij, B. Dialectical argumentation with argumentation schemes:an approach to legal logic. *Artificial Intelligence and Law*, 11:167–175, 2003.

[12] Walton, D. N. *Argumentation Schemes for Presumptive Reasoning*. New Jersey: Lawrence Erlbaum Associates, Inc, 1996.

[13] Walton, D. N. *Argumentation Methods for Artifiticial Intelligence in Law*. Heidelberg: Springer, 2005.

[14] Walton, D. N. *Foudamentals of Critical Argumentation*. New York: Cambridge University Press, 2006.

Predication and the Slingshot[1]

Chienkuo Mi
Soochow University & University of Iowa
cmi@scu.edu.tw

ABSTRACT. This paper aims at testifying against the correspondence theory of truth, based on the problem of predication and the slingshot argument. The correspondence theory of truth is usually characterized as the view that truth is correspondent to a fact – a view that was advocated by B. Russell, J. L. Austin, and L. Wittgenstein early in the 20th Century. But the label is usually applied much more broadly to any view explicitly embracing the idea that truth consists in a relation to reality, i.e., that truth is a relational property involving a characteristic relation (to be specified) to some portion of reality (to be specified). However, there are two main versions in virtue of how the truth-bearer corresponds to the fact. There is a direct correspondence version, in the Austinean or Wittgensteinian sense, of which the truth-bearer as a self-containing unit directly corresponds to the whole fact. There is also a structural correspondence version, in the Russellian sense, of which parts of the truth-bearer structurally corresponds to the matching parts of the fact. I will argue that the correspondence theory is confronted with a dilemma. The direct correspondence version cannot escape from the challenge of the slingshot argument, and the structural correspondence version may be able to answer the slingshot but still has to face the problem of predication. Neither version of the correspondence theory can explain satisfactorily the general concept of truth.

The purpose of this paper is to testify against the correspondence theory of truth, based on the problem of predication and the slingshot argument. I will first present two distinct versions of the correspondence theory, and then argue that neither one can escape the challenges from either one of the two horns.

The correspondence theory of truth is usually characterized as the view that truth is correspondent to a fact – a view that was advocated by B. Russell, J. L. Austin and the early Wittgenstein in the 20th century. But the label is usually applied much more broadly to any view explicitly embracing the idea that truth consists in a relation to reality, i.e., that truth is a relational property involving a characteristic relation (to be specified) to some portion of reality (to be specified). The members of this family employ various concepts for the relevant relation (correspondence, conformity, correlation, congruence, agreement, accordance, copying, picturing, representation, reference) and/or various concepts for the relevant portion of reality (facts, states of affairs, situations, events, individuals, sets, properties, tropes). However, there are two main versions which differ in the way the truth-bearer corresponds to the truth-maker. There is a direct correspondence version, in the Austinean or the Wittgensteinian sense, of which the truth-bearer as a self-containing unit directly corresponds to the whole

---

[1] The author is with Associate Professor of Philosophy, Soochow University and Visiting Associate Professor of Philosophy, University of Iowa. Email: cmi@scu.edu.tw

fact. There is also a structural correspondence version, in the Russellian sense, of which parts of the truth-bearer structurally corresponds to the matching parts of the corresponding fact. The former is sometimes called correspondence as correlation, and the latter correspondence as congruence [2].

The direct version of correspondence, or the correspondence-as-correlation (or correspondence-as-picturing) theory, can trace its root to Aristotle's remark that "To say of what is that it is not, or of what is not that it is, is false, while to say of what is that it is, or of what is not that it is not, is true." This remark implicitly invokes a direct relation between *what is said* and *what is or is not*, but it is unclear what the relation exactly is and what the relata really are. It is not until J. L. Austin that we have a clearer sense in which how the relata are directly connected. For Austin[3], a statement is said to be true when the fact to which it is correlated by the demonstrative conventions is of a type with which the sentence used in making it is correlated by the descriptive conventions. So, the sentence "Snow is white" is true when it is correlated to a certain type of situation; namely, the state of affairs that snow is white. If I use the sentence to make a statement that snow is white, the statement is true because it is correlated to a particular fact – the fact that snow is white. For Wittgenstein, proposition can be true or false only by being pictures of the reality (T. 4.06). If the elementary proposition is true, the atomic fact exists; if it is false the atomic fact does not exist (T. 4.25). So, if the proposition "Snow is white" is elementary and true, then it will picture an existing atomic fact that snow is white. Despite of the complicating issues of regarding statements/sentences and propositions as the truth bearers and taking states of affairs and (existing or non-existing) facts as truth makers, both Austin and Wittgenstein characterize the direct correspondence relation as the kind of correlation or picturing relation between the truth-bearer as a self-containing unit and the truth maker as a whole.

The structural version of correspondence, on the other hand, still holds some kind of correspondence relation between truth bearers and truth makers, but the relation held between the relata is not direct. This version claims that there is a structural isomorphism between the components of truth bearer and the constituents of truth maker. The truth bearer in question, if it is true, has its components involving a relation of congruence with the parts of the corresponding fact. We might find the original idea of the structural correspondence theory in Plato's copy (or participating) theory of Idea, but the modern version of the correspondence as congruence must be attributed to B. Russell. Russell[4] advocates a structural version of correspondence in which a belief is true when the components of the belief are congruent with objects composing a complex unity related in the same way as the matching components in the belief. For example, Michael believes that Ernest is the father of David. Michael's belief that Ernest is the father of David is true when there is a complex unity, "Ernest's being the father of David", which is composed exclusively of the objects related in the same order as those in Michaels belief: the relation (being the father of) as one

---

[2] For example, when Richard Kirkham discusses the correspondence theory of truth in his Theories of Truth: A Critical Introduction, he makes the distinction between two types of correspondence: correspondence as correlation and correspondence as congruence [7, pp.119 – 140].

[3] The brief characterization of Austin's view presented here is based on his article, "Truth" [1].

[4] Russell's multiple-relation view presented here can be found in his book, The Problems of Philosophy, Chapter 12: "Truth and Falsehood".[8]

Predication and the Slingshot

of the objects occurring now as the cement that binds together the other objects (Ernest and David in this exact order) of the belief. The point of Russell's idea lies in the structural correspondence relation that holds between truth bearers (beliefs) and truth makers (facts as the complex unities).

There are many philosophers who criticize and provide various arguments against correspondence theory. Davidson is the philosopher who usually uses the slingshot to argue against correspondence theory, and who also believes that the slingshot argument is probably the most direct and effective way of doing it. Frege's idea that the reference of a sentence is its truth value is customarily considered the origin of the slingshot, although he has never employed this idea to oppose correspondence theory. However, it is also interesting to see that Frege mentions two difficulties involved in the theory. He argues that the attempt to explain truth as correspondence will collapse in any case.[5] If the objections to a correspondence theory of truth are appealing, as both Davidson and Frege seem to be convinced, we ought to start exploring how the slingshot argument can really prove these objections, and what rebuttals and problems the argument might face.

"The slingshot argument" has been constructed based on different logical formulations and assumptions (or principles). It can also be formulated in terms of the class abstraction operator "(λx)" such that "$(\lambda x)(\cdots x \cdots)$" means "the class of all x such that $\ldots x \ldots$" (where '$\cdots x \cdots$' is thought of as any formula involving 'x'), or in terms of the definite description operator "(λx)" such that "$(\lambda x)(\cdots x \cdots)$" means "the x such that $\ldots x \ldots$". I will examine presently how Davidson's slingshot argument can be employed to argue against the two versions of the correspondence theory of truth, and then modify it into an argument form that can deal better with problems or challenges posted for Davidson's accounts of the slingshot argument.

There are two formulations of the slingshot which are involved in Davidson's works. The one is formulated in "Truth and Meaning" (the TM version), and the other in "True to the Facts" (the TF version).

1 The TM formulation

Assumption 1 *Logically equivalent singular terms have the same reference.*

Assumption 2 *A singular term does not change its reference if a contained singular term is replaced by another with the same reference.*

1. Let "R" and "S" abbreviate any two sentences alike in truth-value.
2. Suppose $\Psi(R)$, where $\Psi(\cdots)$ means that the reference of $(\cdots)$ is Ψ.
3. $\Psi[(\lambda x)(x = x.R) = (\lambda x)(x = x)]$ because "R" and "$(\lambda x)(x = x.R) = (\lambda x)(x = x)$" are logically equivalent, plus Assumption 1.
4. $\Psi[(\lambda x)(x = x.S) = (\lambda x)(x = x)]$, because "$(\lambda x)(x = x.R)$" and "$(\lambda x)(x = x.S)$" are co-referential singular terms (based on step 1), plus Assumption 2.

---

[5] Frege argues why the attempt to explain truth as correspondence will collapse by the following two reasons: first, no perfect correspondence can be made between two distinct things (e.g., sentence and fact); second, there will be an infinite regress problem for the correspondence theory because we should have to inquire whether it is true that a sentence and a fact correspond in some specified respect. [6, pp.326–327]

5. $\Psi(S)$, because "$(\lambda x)(x = x.S) = (\lambda x)(x = x)$" and "$S$" are logically equivalent, plus Assumption 1.
6. If any two sentences "R" and "S" are alike in truth-value, they will have same reference, because of 1, 2 and 5.

In this context of TM, the slingshot argument is designed to show the difficulty of identifying the meaning of a singular term with its reference, and further to prove that if the meaning of a sentence is what it refers to, then all sentences alike in truth value must be synonymous. The moral of this argument will depend on what we take the reference of a sentence to be. As Davidson has noted, "it is perhaps worth mentioning that the argument does not depend on any particular identification of the entities to which sentences are supposed to refer." [2, p. 17] If "fact" is what a sentence refers to, then all true sentences will refer to the same fact (as well as all false ones).

2 The TF formulation

Principle 1 *Logically equivalent sentences are co-referential or correspond to the same fact.*

Principle 2 *The reference (or the corresponded fact) of a sentence remains the same if an expression or singular term in the sentence is substituted by a co-referential expression or term.*

1. Let "S" and "T" be any true sentences.
2. Suppose the statement that S corresponds to the fact that S (and the statement that T corresponds to the fact that T).
3. The statement S corresponds to the fact that $(ix)(x = a.S) = (ix)(x = a)$ where a stands for any genuine proper name, because "S" and "$(ix)(x = a.S) = (ix)(x = a)$" are logically equivalent, plus Principle 1.
4. The statement S corresponds to the fact that $(ix)(x = a.T) = (ix)(x = a)$, because "$(ix)(x = a.S)$" and "$(ix)(x = a.T)$" are co-referential singular terms (based on), plus Principle 2.
5. The statement S corresponds to the fact that T, because "$(ix)(x = a.T) = (ix)(x = a)$" and "$T$" are logically equivalent, plus Principle 1.
6. Any sentence that is true corresponds to the same fact, because of , , and .

This formulation of the slingshot argument is clearly directed to trivialize the correspondence theory of truth; since aside from matters of correspondence no way of distinguishing facts has been proposed, and this argument shows further that even if a true sentence does correspond to the fact, there is exactly one fact to be corresponded to. Davidson insists that no point remains in distinguishing among various names of "The Great Fact", and we may use the single phrase "corresponds to The Great Fact" to attribute to every true sentence. Therefore any statement "S" (if true), according to the attribution above, will correspond to The Great Fact, and there will be apparently no telling this result apart from simply saying that "S" is true. The correspondence theory of truth as an attempt to be a semantic theory will be seriously damaged by this result.

Predication and the Slingshot

The reason that I coordinate Davidson's two formulations of slingshot argument here is to present precisely various formulations, structures, and assumptions (or principles) which may be used and applied in the argument. At least we can get some straightforward points in mind at the outset. First, the soundness of the slingshot argument will not be affected by whether or not we want to view sentences as singular terms. Second, the correctness of the argument will not be changed by whether we adopt the class abstraction operator "(λx)" or the definite description operator "(ιx)". Third, the credibility of the argument will not be affected by whether or not we accept that a sentence has a reference (or nominatum) or by whatever we take the reference of a sentence to be. Fourth, the logical consequence of both versions seems to be paradoxical, that is, any two sentences alike in truth value will be logically equivalent.

There are two problems with respect to the accounts of the slingshot set up by Davidson:

Problem 1 *Some might object that the expressions like*
$(\lambda x)(x = x.R) = (\lambda x)(x = x)$ *and $\underline{R}$ (in TM)*
or
$(\iota x)(x = a\square S) = (\iota x)(x = a)$ *and $\underline{S}$ (in TF)*
are not in general logically equivalent. If the logical equivalence is not obtained, the argument will not go through.

Problem 2 *Some might also question that in contexts like*
$\Psi(\cdots)$ *(in TM)*
or clauses like
the fact that ... (in TF)
*substitution **salva veritate** of logically equivalent sentences is not allowed, because the contexts may not be purely truth-functional. If PSLE cannot be applied to those opaque contexts, the argument will not go through either.*

In order to avoid these two problems, we might modify Davidson's original accounts of the slingshot into the following design of a new argument form for the slingshot argument:

1. If $\Phi(a)$ and $a = (\lambda x)(x = a\&\Phi(x))$ are both true sentences, then they correspond to the same fact.
2. Assume Fa is true, which corresponds to the fact $f1$.
3. Assume $a \neq b$ is true, which corresponds to the fact $f2$.
4. Assume Gb is true, which corresponds to the fact $f3$.
5. $a = (\lambda x)(x = a\&Fx)$ is true, which corresponds to the fact $f1$.
6. $a = (\lambda x)(x = a\&x \neq b)$ is true, which corresponds to the fact $f2$.
7. $a = (\lambda x)(x = a\&Fx)$ and $a = (\lambda x)(x = a\&x \neq b)$ correspond to the same fact. So, $f1 = f2$.
8. $b = (\lambda x)(x = b\&Gx)$ is true, which corresponds to the fact $f3$.
9. $b = (\lambda x)(x = b\&a \neq x)$ is true, which corresponds to the fact $f2$.
10. $b = (\lambda x)(x = b\&Gx)$ and $b = (\lambda x)(x = b\&a \neq x)$ correspond to the same fact. So, $f2 = f3$.
11. Since $f1 = f2$ and $f2 = f3$, we have $f1 = f3$. So, Fa and Gb correspond to the same fact.

12. Conclusion: Any two true sentences correspond to the same fact.

In this modified form of the slingshot argument, Principle 1 is no longer presupposed. However, the crucial steps involved in the argument are the steps 1, 7 and 10. One might at the very beginning object to step 1 and deny that $\Phi(a)$ and $a = (\lambda x)(x = a \& \Phi(x))$ correspond to the same fact. But as the direct version of correspondence theory must concede, if $\Phi(a)$ and $a = (\lambda x)(x = a \& \Phi(x))$ are both true, it is the fact "$\Phi(a)$" that makes them true. And if "$\Phi(a)$" is the truth maker for both $\Phi(a)$ and $a = (\lambda x)(x = a \& \Phi(x))$, then they are supposed to correspond to the same fact. If step 1 can go through, then steps 7 and 10 would not be difficult to accept. The reason that $a = (\lambda x)(x = a \& Fx)$ and $a = (\lambda x)(x = a \& x \neq b)$ correspond to the same fact and make "$f1 = f2$" is the identity between $a = (\lambda x)(x = a \& Fx)$ and $a = (\lambda x)(x = a \& x \neq b)$. Both of them, if true, express the same formula $a = a$. The same goes for step 10, because both $b = (\lambda x)(x = b \& Gx)$ and $b = (\lambda x)(x = b \& a \neq x)$ express the same formula $b = b$. So, we see no escape for the direct version of correspondence theory under the attack based on this form of argument. That is, if any true sentence corresponds to any fact at all, according to the characterization of the direct version, then all true sentences will correspond to the same fact.

Although the direct version is unable to escape the challenge of the slingshot argument, the structural version of correspondence theory can still resist the argument simply by rejecting step 1. In the structural version, $\Phi(a)$ and $a = (\lambda x)(x = a \& \Phi(x))$ do not appear to represent the same fact. This is because these two formulae contain quite different components, which in turn compose different structural facts. If step 1 is blocked, the whole argument will not go through. However, this reply can lead to some metaphysical problems. An immediate difficulty is that if we accept that it is two different facts which make the above two formulae true, then we must accept some unusual facts. If, as characterized by the structural version, the fact which makes $\Phi(a)$ true is different from the fact which makes $a = (\lambda x)(x = a \& \Phi(x))$ true, then it would appear that what makes $\Phi(a)$ true is, on the face of it, the fact "$\Phi(a)$", and what makes $a = (\lambda x)(x = a \& \Phi(x))$ true is just the fact "$a = (\lambda x)(x = a \& \Phi(x))$". But what kind of fact is "$a = (\lambda x)(x = a \& \Phi(x))$"? If we believe that "$a = (\lambda x)(x = a \& \Phi(x))$" can stand for a complex structural fact (or a complex unity as Russell would have called it), it would be an unusual fact distinct from the more ordinarily recognizable fact "$\Phi(a)$". We have to concede that this complex structural fact "$a = (\lambda x)(x = a \& \Phi(x))$" is composed not only of ordinary empirical objects, but also of logical and linguistic entities. I don't know whether correspondence theorists really want to accept these kinds of facts. But a more serious problem which we must face is, having "dissected" propositions or facts in this way, how we can restore these separate parts to their original unified form again. That is to say, any such structural version of correspondence theory must face the problem of predication – the problem of unification of a sentence or a proposition.

The problem of predication can be characterized as the problem that once plausible assignments of semantic roles have been made to the parts of sentences, the parts do not seem to compose a united whole. This problem has to do with the unification of a sentence or a proposition. It is not only a problem at the semantic level that deals with how predicates are related to names or other singular terms and contribute to the unity of a sentence, it can also be a problem at the metaphysical level that concerns how universals are related to particulars and constitute a complex structural fact.

A theory of predication should not be confused with a theory of predicates. A theory of predicates aims to provide semantics for all predicates, to explain the meaning and nature of predicates. But a theory of predication does more than that, the theory needs not only to explain the semantic role of predicates, but also to give a satisfactory account of how predicates can maintain their proper functions as parts of sentences while at the same time contribute to the unity of sentences in which they occur, the unity demanded by the fact that sentences can be true or false and can be used to express our judgments and thoughts. The problem of predication should not be confused with the problem of universals either, although the two problems are closely related. It will help us differentiate the two problems more distinctly if we make a clear distinction between the metaphysical question of how particulars are related to properties and the semantic question of how subjects and predicates are related.

What exactly are the problems? The problem of universals at the metaphysical level can find its original sin in Plato's theory of ideas or forms. It is also called the third man problem. Why third man? We all agree that an individual thing like man or flower can have many different properties and relations to other things, such as "being rational", "being red", "being a son of", or "being next to". But how can one thing have many different properties at the same time? Lets call this question the puzzle of "many over one"[6]. We all usually agree that many different individual things can have the same property, for example, many different individual men can all have the property of rationality. But how can many individual things have the same property? It is an old philosophical question and is commonly named the puzzle of "one over many". It is less controversial to admit that an individual thing is a real entity, a particular. But if we want to posit properties or relations as another kind of entity, we will get ourselves into trouble immediately. Plato should be held responsible for introducing new entities, such as ideas or forms, to explain our platitudes about the world and to answer the questions and puzzles following them. He seems to hold that the many different properties owned by a particular or the same property shared by many different particulars are ideas, and ideas are real entities. If both particulars and properties are real entities, then we need to explain how the two kinds of entities can be connected or glued together. Based on some Plato's dialogues, we may interpret his view as saying that material particulars participate in, resemble, copy, or are modeled by the ideas or forms. If particulars really participate in, resemble, copy, or are modeled by the ideas, then there must be some genuine relation held between particulars and ideas or there must be another property shared by particulars and ideas. So, the "third man" comes in. We would need to posit another idea or form for the genuine relation held between, or the property shared by, particulars and ideas. This will lead to an infinite regress. Some might continue to ask why an infinite regress is a problem at all. One can think of some situations in which an infinite regress is no problem, for example, in mathematics. But why is the infinite regress a problem for the theory of universals? The reason is simple. In order to explain our ordinary phenomena such as "Socrates is rational", philosophers of universal theory not only appeal to particulars like Socrates, but also introduce universals like being rational into

---

[6]It is called this way by Rodriguez-Pereyra, and presented in his *Resemblance Nominalism*, the Section 3: "The Many over One".

their ontology. Now when one phenomenon is analyzed and separated into two entities, there is a question of what it is that makes them cooperate to make the unity of original phenomenon. And what the regress argument is doing is saying, if you've got this congruence of entities, it's no good inventing other entities C the copula, the form, or any idea C and then just adding that in to the congruence, because then you've just got a bigger bunch of entities, and the same problem recurs again. So what the infinite regress shows is that the metaphysics of universals does not provide any useful explanation at all.

It is worth noticing that the problem of predication itself should cast no doubt on the existence of ideas, forms, universals, or tropes. The question of whether ideas, forms, universals, or tropes exist is not a central issue involved in explaining linguistic phenomena or constructing any semantic theory. The more fundamental question is whether positing properties or relations as any kind of entity can really help us explain the linguistic phenomena and the semantic role of predicates better. In order to understand the problem of predication more clearly, let's focus on the structure and nature of judgments like the judgment that Socrates is wise and that Socrates walks. We want to know what make the judgments true or false, and what the role the predicates play and contribute to the truth or falsity of the judgments. Now, we all agree that Socrates is an entity, a person, or a particular. He is wise and walks. If we decide to posit properties as entities, the properties of "being wise" and "walking" would be another kind of entity– ideas, forms, universals, or tropes that can be copied, participated, instantiated or realized by many different particulars. In our case here, Socrates is one of many particulars that copy, participate, instantiate, or realize the properties (or entities) of "being wise" and "walking". So, we are actually explaining what make the judgments that Socrates is wise and that Socrates walks true or false by saying whether or not Socrates copies, participates, instantiates, or realizes the properties of "being wise" and "walking". From what we have been describing, it is obvious that what make the judgments true, or simply the facts themselves, don't just consist of two entities – Socrates and the property of "being wise" or Socrates and the property of "walking". What makes the judgments true is that Socrates and the property of "being wise", and Socrates and the property of "walking" stand in a certain relation of copying, participating, instantiating, or realizing to each other. If we think that properties should be regarded as real entities, we don't see why we should not view relations as real entities. If we are serious about taking properties and relations as entities, then the relation of copying, participating, instantiating and realizing should all be entities too. So, the simple facts expressed by our judgment in two parts (subject and predicate) or in just two words ("Socrates" and "walk") actually involve three entities rather than two. But if the third entity is involved, it is easy to reason that there surely are more than three entities involved in both judgments that Socrates is wise and that Socrates walks. We are confronted with an infinite regress again.

It is not my concern to solve the problem of universals here. The problem of predication, viewed from the metaphysical level, involves more confusion and trouble than needs be. It is neither my focus here to answer whether there is any hope that we can solve the problem of predication. However, I still want to point out that the correspondence theory of truth, viewed as a structural version, cannot provide a satisfactory semantic theory for dealing with the problem of predication.

I have argued that the correspondence theory of truth is confronted with a dilemma. The direct correspondence version cannot escape from the challenge of the slingshot argument, and the structural correspondence version may be able to answer the slingshot but still has to face the problem of predication. Neither version of the correspondence theory satisfactorily explains the general concept of truth.

Bibliography

[1] Austin, J. L. Truth. *Proceedings of the Aristotelian Society*, 24:111–128, 1950.
[2] Davidson, D. Truth and meaning. In *[4]*, 17–36. 1967.
[3] Davidson, D. True to the facts. In *[4]*, 37–54. 1969.
[4] Davidson, D. ???? 1984.
[5] Frege, G. On sinn and beduetung. In Beaney, M. (Ed.), *The Frege Reader*, 151–171. Oxford: Blackwell, 1892. Translated by Black M.
[6] Frege, G. Thoughts. In Beaney, M. (Ed.), *The Frege Reader*, 325–345. Oxford: Blackwell, 1918. Translated by Geach P. and Stoothoff R.
[7] Kirkham, R. L. *Theories of Truth: A Critical Introduction*. The MIT Press, 1992.
[8] Russell, B. The problems of philosophy. Oxford University Press, 1912.

A Liar's Thorn in the Minimalist's Side[1]

Hou, Richard Wei Tzu
National Chung Cheng University
hou190506@ccu.edu.tw

ABSTRACT. Horwich (2005)[3] seeks a minimalist compatible solution to the Liar propositions. He proposes an account employing the combination of the Tarskian idea of hierarchical truth predicates and the Kripkean idea of groundedness. Let us call this henceforth the TK account. Horwich seems to suggest that, based on both Tarski's and Kripke's essential accounts, there is a minimal way to dispose of the Liar's threat with no substantive resources needed.

For Horwich, Tarski's hierarchisation can be seen as taken all sub-languages as well as all sub-truth-predicates constitute one single language and one single truth predicate. Instances of the Liar are excluded in terms of the warrant of the truth ascribable propositions' *groundedness*. The notion of groundedness in turn is explained as follows:

The intuitive idea is that an instance of the equivalence schema will be acceptable, even if it governs a proposition concerning truth ... as long as that proposition (or its negation) is grounded – i.e. is entailed either by the non-truth-theoretic facts, or by those facts together with whichever truth-theoretic facts are 'immediately' entailed by them (via already legitimized instances of the equivalence schema), or ... and so on.[3, p. 81]

It is not quite clear what Horwich means by saying that a proposition can be *entailed* by a fact or facts.

Similar claim is made by the following:

To put the [groundedness] constrains ... more precisely: the acceptable instances are those that concern grounded propositions – where every proposition of L_0 is grounded, and (for k ¿ 0) a proposition $< p_k >$ of L_k is grounded if and only if either it or its negation is entailed by the grounded facts of L_{k-1} and L_{k-2} and ..., in conjunction with the instances of the equivalence schema that are legitimized by these facts.[3, p. 82]

The groundedness of all propositions of all sub-languages results from the groundedness of the L_0 propositions. And their groundedness is supposed to do with non-truth-theoretic facts, and they are *entailed* by these facts. Apparently not every theory of a proposition is compatible with this sort of entailment relation, for instance, Fregean propositions.[2] For TK account to work, this is the

---

[1] The author is with assistant professor of Department of Philosophy, National Chung Cheng University. Email: hou190506@ccu.edu.tw This paper is the partial outcome of the research project "The Minimal Theory of Truth and the Liar Paradox" (95-2411-H-194-037-) and of the project "The Trinity Research of Dealftionary Semantics–Truth, reference, and Meaning (III-I)" (96-2411-H-194-010-) sponsored by NSC, Taiwan.

[2] This contravenes Horwich's assumption that minimalism is compatible with both Fregean and Russellian propositions, and even compatible with a hybrid kind of these two. [2, p. 17]

A Liar's Thorn in the Minimalist's Side

first problem Horwich has to answer. It is actually one of the major issues of Horwich's theory, but the problems of this theory will be brought close to the best understanding if we can start from a thorough analysis about how the threat of the Liar is neutralised.

By means of the liar instance of that $< D$ is not true $>= D$, Horwich further demonstrates how the potential contradiction can be explained away in the following manner:

Thus the potentially paradoxical $< D$ is not true $>$ will not be a grounded proposition of L_1 because there are no facts of L_0 which (given the equivalence schema) will entail either it or its negation; and similarly, it will not be a grounded proposition of any of the other sub-languages; so there will be no axiom governing it; so the contradiction will not be derived.[3, pp. 81–82]

Regarding the truth status of liar propositions, Horwich is so saying:

Note that there is no contradiction in supposing that D is true (or in supposing that it is false): the problems arise only if the equivalence schema were to be applied. Therefore, we can and should preserve the full generality of the Law of Excluded Middle and the Principle of Bivalence, by maintaining that D is *either true or false*. Of course we cannot come to know which of these truth values it has ...so ...it is indeterminate whether D is true or whether D is false. [3, p. 82, emphases added]

It is clear then that Horwich accepts semantic epistemicism, by which a given liar proposition is either true or false but the truth value of it is not knowable. And the solution to the Liar threat admits of no compromise of the principle of bivalence and the law of excluded middle. The reasoning behind is that Tarski's solution of the Liar paradox, according to Horwich, is quite independent from his compositional strategy.[3, p. 81] Horwich then claims that his TK account does not invoke Tarskian compositional principle, providing a solution that squares with minimalism.[3, p. 81]

In [1, p. 93], Armour-Garb and Beall raise a very compelling argument against Horwich's reasoning. The elaboration of why and how Tarskian solution can be adopted by the TK account and that of the notion of groundedness are supposed to give us reason to believe two things: that, as far as the equivalence schema is *not applicable* to Liar propositions, paradoxes do not occur and that Liar propositions such as '$< D >$' are either true or false but their truth values are in principle unknowable. So the principle of bivalence can be held without paying the price of allowing any instance of the Liar paradox. The rationale behind the account just briefed is that by holding the general *applicability* of the equivalence schema and the corresponding schema of the false predicate to every proposition plus the maintenance of classic logic, which gives us the law of excluded middle, the principle of bivalence can be derived. Armour-Garb and Beall's argument rests on the inconsistency between Horwich's account and the rationale behind it. The principle of bivalence is needed to explain why a Liar proposition, as an ungrounded proposition, is not neither true nor false but true or false, though its truth value is conceptually impossible to know. However, to explain why the principle of bivalence holds one needs the applicability of the equivalence schema and the false schema to Liar propositions, and this causes inconsistency.

To be successful, TK account has to neutralise the threat caused by the Liar to the minimal theory of both the property and the concept of truth. However, it is not free from producing its own parasitic Liar mirrors, such as (1),

(1) The proposition expressed by (1) is not grounded.

By 'parasitic Liars' I mean the Liar mirrors that parasitise a particular theory of truth. For instance, (2),

(2) does not correspond to a state of affairs,

is a parasitic Liar mirror of a sophisticated correspondence theory. The first question to ask is whether or not the proposition expressed by (1), $<(1)>$, that is, is grounded.[3] Although the intuition leads to the negative answer, TK does not seem to have enough resources to say it. Regarding the groundedness of the propositions of L_0, Horwich only says that, first, they are without the truth predicate and, second, they are entailed by non-truth-theoretic facts.[4] And since (1) does not contain the truth predicate, that proposition is grounded as far as it or its negation is entailed by a non-truth-theoretic fact. Further, if there is a fact for it to ground in, it is a non-truth-theoretic fact anyway. Therefore, the above-quoted criterion is not sufficient to show the ungroundedness of $<(1)>$. TK account has to say more about what a *fact* is. But without invoking Tarskian compositionality principle, it is difficult to see how to establish such a minimal account. This is the second problem.

One possible way to explain the ungroundedness of $<(1)>$ without invoking the compositionality principle is to appeal to the utilisation of semantic notions in (1), namely, the notion of groundedness. The reasoning is that, if one can decide whether or not a proposition contains the truth predicate without being accused of bringing up the compositionality principle, so do other semantic notions. How shaky this reasoning is, is not at issue here, but the fact that it overkills is. In many cases, to say that a proposition is grounded or not grounded is legitimate and reasonable. Consequently this strategy is not workable. [5]

For Horwich to maintain the Principle of Bivalence and the Law of Excluded Middle, the maneuvre is not to group ungrounded propositions according to their lacking truth value but according to the non-applicability of the equivalence schema to them. Ungrounded propositions are thus either true or false, but there is no way to know their truth values. The beauty of this maneuvre rests on that the equivalence schema is not applicable to any Liar proposition; it also rests on that because whether or not that it is true or that it is false is indeterminate, one can only maintain that any given Liar proposition is either true or false, and this is directly implied by the theory without employing the equivalence schema. Therefore the conception and the property of truth can still be explained in virtue of the equivalence schema and all its instances.[6] Liar propositions simply do not enter the explicans.

However, this does not apply to $<(1)>$. Let us put the second problem aside and assume that $<(1)>$ is ungrounded according to TK. The problem is that, if it is ungrounded, it says something true.[7] But the use of 'true' in "$<(1)>$ is true" cannot be explained by the equivalence schema and all its instances, for $<(1)>$ is ungrounded but with determinate truth value. Therefore, TK cannot

---

[3] I follow Horwich's angle brackets notation for propositions.
[4] Let's ignore the dubious entailment relation here.
[5] This criticism emerges from the discussion between me and my colleague Wang, Wen-Fang.
[6] The price to pay is a much more complicated account about truth generalisations. This, however, will not be discussed here.
[7] It cannot be claimed that, if it is false and is thus grounded, then this leads to a contradiction. For one needs to apply the equivalence schema to $<(1)>$ to get this contradiction, and this is not doable because of the ungroundedness of $<(1)>$

get away from the difficulty generated by its own parasitic Liar mirrors. This is the third problem.[8]

Consider the second and the third problems, perhaps one solution is simply to admit that $<(1)>$ is rather grounded than ungrounded. It follows that $<(1)>$ is grounded and false, for otherwise it leads to a contradiction – that is, it is both grounded and ungrounded. However, this bold concession is more undesirable because instances like $<(1)>$ are precisely similar to instances of the so-called truth teller – such as, this proposition is true. All of them are pathological in the sense that they are saying something about their either being true or being false or being none. This is why the intuition is that $<(1)>$ is not grounded. So we are back to the starting point.

Horwich explains that this TK account is very sketchy and is in need of further exploration. However, according to the analysis above, TK does not fail because of its sketchiness. A clearer picture can be given when comparing Kripke's idea of groundedness and Horwich's. In terms of a substantive account of facthood – that is, by means of the compositionality principle – Kripke uses the idea of groundedness to explain why all Liar propositions, and parasitic mirrors too, are ungrounded and therefore are expelled from the truth talk. Horwich's idea of groundedness, on the other hand, does not have the substantive resource to do this outright. Of course, he does not want to expel Liar propositions from the truth talk because of his holding of the principle of bivalence and the law of excluded middle. For this, three points can be made. First, the unclearness of what a fact is cripples TK account in the way that it cannot explain away the parasitic Liar mirrors, such as $<(1)>$, of TK, and the lack of resources makes his notion of groundedness not as clear as he would like it to be. Second, even if propositions like $<(1)>$ can be taken as ungrounded, the ascription of truth to it cannot be explained in terms of the equivalence schema and all its instances. Finally, moving toward Kripkean side of explanation by abandoning the principle of bivalence is not even doable, for his TK account still owes us a good explanation of what a fact is and it cannot deal with its own parasitic Liar mirrors, either. These points show that the minimal way proposed by Horwich, that TK account utilises Tarski's and Kripke's ideas freely without getting into the undesired substantive mud, cannot even enjoy the benefit of the doubt.

Bibliography

[1] Armour-Garb, B., Beall, J. Minimalism, epistemicism, and paradox. In Beall, J., Armour-Garb, B. (Eds.), *Deflationism and Paradoxes*, 85–96. Oxford: Oxford University Press, 2005.

[2] Horwich, P. *Truth*. Oxford: Clarendon Press, 2nd edn., 1998.

[3] Horwich, P. A minimalist critique of tarski on truth. In Beall, J., Armour-Garb, B. (Eds.), *Deflationism and Paradoxes*, 75–84. Oxford: Oxford University Press, 2005.

[4] Tarski, A. The concept of truth in formalised languages. In Tarski, A. (Ed.), *Logic, Semantics, and Metamathmatics*. London: Oxford University Press. Translated by Woodger J. H.

---

[8]This also emerges from the discussion between I and Wang (Wen-Fang).

A Brief Introduction to *Para-consistent logic and AI*[1]

Zhang Hansheng

Yanshan University

zhanghandsomennk@163.com

ABSTRACT. Para-consistent logic is a new thought current in international logic circles which arose in the last thirty years. Being a kind of "revolutionary" non-classical one, para-consistent logic allows "meaningful contradiction" to enter formal axiomatic systems; therefore, it attracts a great number of logicians, philosophers and mathematicians from Brazil, Australia, America, China and Eastern Europe. In Qiquan Gui, Zili Chen and Fuxi Ju's newly published *Para-consistent Logic and Artificial Intelligence* (Wuhan University Press, 2002) the authors probe the basic principles of para-consistent logic and introduce the reader to theories of para-consistent logic from abroad. They introduce, for example, N.C.A. da Costa's para-consistent logic, para-consistent deontic logic, and para-consistent dialectical logic. The authors' main contribution to the field, however, is that they also fully reform the systematics of para-consistent logic. In short, they make a significant attempt to establish a para-consistent relevant logic, a para-consistent relevant fuzzy logic and other new axiomatic systems.

1 Background and Foundation

Classified by the conclusions and the aims, there are two kinds of inferences about generic sentences. Conclusions of one kind of inferences are usually factual propositions, conclusions of the other kind are usually generic sentences. The second kind of inferences reason from some propositions and get the general conclusions which are represented as generic sentences. This kind of reasoning involves both deduction and induction. This paper discusses the second kind of reasoning about generics, and only about deduction part. For example, from 'sparrows are birds' and 'birds have feather' we get 'sparrows have feather'[2] .This is an inference getting generic sentences by deduction.

Yi Mao and Beihai Zhou have given a formal semantics for generics and some formal system [2, 3]. They interpret generic sentence with SP structure as '(normal S) (normally P)'. In semantics, the 'normal' in 'normal S' is a binary function $N(?_1, ?_2)$, $?_1, ?_2$ are used for representing subject sense and predicate sense respectively. They point out that the normal function has two restrict conditions

---

[1] The author is with College of Humanities and Laws, Yanshan University, Qinhuangdao, China, 066004. Email: zhanghandsomennk@163.com

[2] This example looks like syllogism' AAA, but the inference involves generic sentences, so it's incorrect to interpret it with the extension semantics and we should try to interpret it with the intension semantics.

A Brief Introduction to *Para-consistent logic* and *AI*   43

(see definition 2.4). Based on the points above, generic sentence SP' formalization is $\forall x(N(\lambda xSx, \lambda xPx)x > Px)$. They also construct some formal systems. Among these systems, M[6] and G[2] are most important. The proposition logic M contains the default implication MP: $(\alpha \wedge (\alpha > \beta)) > \beta$, and M's quantifier expansion G introduces axiom GU: $Gx(\alpha; \beta) > \forall x(\alpha > \beta)$. By GU, they transform the generic quantifier into universal quantifier. These systems are constructed for dealing with the inferences which get the factual conclusions from general premises.

This paper is based on the semantic analysis and formal semantics introduced above. The difference is that now we study the inferences leading to generic sentences, so we dont consider the transformation from generic quantifier to universal quantifier. Coming down to the formal system, we dont add the axiom GU, instead we investigate the relationship among the generic sentences by studying the normal function N further, and get some inference rules about generic sentences. First, we give some different semantic models by adding some different restrict conditions about N, and then we construct the corresponding formal systems.

2 Language and Semantics

The language L_G is based on the first order language, and we add new symbols: $>, \alpha, N$. L_G contains a denumerable set of individual variables (Var), a denumerable set of individual constants, and a denumerable set of n-ary predicate variables $(n > 0)$.

$$\alpha ::= \bot | \alpha \to \beta | \forall x\alpha | \alpha > \beta | (\lambda x\alpha)t | N(\lambda x\alpha, \lambda x\beta)t$$

We introduce symbols in L_G such as T, $\neg, \wedge, \vee, \leftrightarrow, \exists$, the definitions are as usual. $\lambda \vec{x} \alpha$ is a short-hand notation for $\lambda x_1 \lambda x_2 \cdots \lambda x_n \alpha$, where $x_1, x_2, \cdots, x_n$ are free individual variables in α. G and; are introduced symbols: $G\vec{x}(\alpha; \beta) = df \forall \vec{x}(N(\lambda \vec{x}\alpha, \lambda \vec{x}\beta)\vec{x} > \beta)$. $\forall \vec{x}$ is the abbreviations for $\forall x_1 \forall x_2 \cdots \forall x_n$, $\exists \vec{x}$ is the abbreviations for $\exists x_1 \exists x_2 \cdots \forall x_n$.

Definition 1 *W is a non-empty set. Function $\diamond_* : P(W) \times P(W) \to P(W)$ is a set selection function on W, if for any $X, Y, Z, X', Y' \subseteq W$, $\diamond_*$ satisfies:*

1. *If $X \subseteq X'$, then $\diamond_*(X, Y) \subseteq \diamond_*(X', Y)$;*
2. *If $\diamond_*(\{w\}, Y) \subseteq Z$ for every $w \in X$, then $\diamond_*(XY) \subseteq Z$;*
3. *If $\diamond_*(X, Y) \subseteq Z$, then $\diamond_*(W, X \cap Y) \subseteq Z$*
4. *If $\diamond_*(X, Y) \subseteq Z$ and $\diamond_*(X \cap Y') \subseteq Z$, then $\diamond_*(X, Y \cup Y') \subseteq Z$;*
5. *If $\diamond_*(X, Y) \subseteq Y'$ and $\diamond_*(X \cap Y') \subseteq Z$, then $\diamond_*(X, Y) \subseteq Z$*

Definition 2 *W and D are non-empty sets. $INT(W, D^n) = \{\cdots$ is a mapping from W to $P(D^n), n > 0\}$. $I(W, D) = \bigcup_{n=1}^{\infty} INT(W, D^n)$.*

Definition 3 *W and D are non-empty sets. For any ??, ??? $I(W, D)$,*

1. *$\tau_1 \subseteq \tau_2$ iff. $\tau_1(w) \subseteq \tau_2(w)$ for all $w \in W$;*
2. *$\tau_1 = \tau_2$ iff. $\tau_1(w) \subseteq \tau_2(w)$ and $\tau_2(w) \subseteq \tau_1(w)$ for all $w \in W$;*
3. *$\tau_1 = \tau_2 \sim$ iff. $\tau_1(w) = (\tau_2(w)) \sim$ for all $w \in W^3$;*

---
[3] $A \sim$ is the supplementary set of A.

4. For each $w \in W$, $(\tau_1 \cup \tau_2)(w) = \tau_1(w) \cup \tau_2(w)$, $(\tau_1 \cap \tau_2)(w) = \tau_1(w) \cap \tau_2(w)$;
5. Vacant sense $\tau_\top$, $\tau_\top(w) = D^n$ for all $w \in W$;
6. Full sense $\tau_\bot$, $\tau_\bot(w) = \emptyset$ for all $w \in W$.

Definition 4 *W and D are non-empty sets. $N : I(W, D) \times I(W, D) \to I(W, D)$ is a normal object selection function iff. for all ??, ???$\in I(W, D)$, N satisfies:*

1. $N(\tau_1, \tau_2) \subseteq \tau_1$ and;
2. $N(\tau_1, \tau_2) = N(\tau_1, \tau_2 \sim)$.

Definition 5 *W and D are non-empty sets. $N : I(W, D) \times I(W, D) \to I(W, D)$ is a normal object selection function iff. for all ??, ????$I(W, D)$, N satisfies:*

1. $N(\tau_1, \tau_2) \subseteq \tau_1$, and;
2. $N(\tau_1, \tau_2) = N(\tau_1, \tau_2 \sim)$

Definition 6 *W and D are non-empty sets. $[,] : D^n \times INT(W, D^n) \to P(W)$ is a function which satisfies: for all $\vec{d} \in D^n, \tau \in INT(W, D^n), [\vec{d}, \tau] = \{w \in W \mid \vec{d} \in \tau(w)\}$.*

Definition 7 *Quadruple $F = < W, D, N, \circledast >$ is a frame, if W and D are non-empty sets, N is a normal object selection function on $I(W, D)$, and $\circledast$ is a set selection function on W, and for all $w \in W$ any $\tau_1, \tau_2 \in I(W, D)$, if $\tau_1(w) \subseteq \tau_2(w)$, then for all $d \in D$, $\circledast(\{w\}, [\vec{d}, \tau_1]) \subseteq [\vec{d}, \tau_2]$.*

Definition 8 *$F = < W, D, N, \circledast >$ is a frame. ?? is interpretation on F, iff.*

1. For each individual constant $c \in C$, $c) \in D$;
2. For each n-ary predicate symbol $P^n (n > 0)$, ??. $P^n, w) \in D^n$.

Definition 9 *An ordered pair $S = < F, ???$ is a structure, if F is a frame and ?? is a interpretation function on F.*

Definition 10 *An ordered pair $M = < S, ????$ is a model, if S is a structure, ???named assignment??is a mapping function from Var to D. We use $\vec{t}$ to denote $< t_1, t_2, \cdots, t_n >$.*

Definition 11 *For each formula ??, symbolism $||?||^M$ is used to stand for the set of worlds in M in which ? is true, satisfying:*

1. $||\bot||^M = \emptyset$
2. $||P^n(t_1, t_2, \cdots, t_n)||^M = \{w \in W < t_1^M, t_2^M, \cdots, t_n^M >\in^M (P_n, w)\}$
3. $||\alpha \to \beta||^M = (W - ||\alpha||^M) \cup ||\beta||^M$
4. $||\alpha > \beta||^M = \cup\{X \subseteq W : \circledast(X, ||\alpha||^M) \subseteq ||\beta||^M\}$
5. $||\forall \vec{x}\alpha||^M = \{w \in W : \text{foreach } \vec{d} D^{M^n}, w \in ||\alpha||^{M(\vec{d}/\vec{x})}\}$
6. $||N(\lambda \vec{x}\alpha, \alpha \vec{x}\beta)\vec{t}||^M = \{w \in W : \vec{t}^M \in N((\lambda \vec{x}\alpha)^M, (\lambda \vec{x}\beta)^M)(w)\}((\lambda \vec{x}\alpha)^M \in I(W, D)$, is a mapping satisfying, for each $w \in W, (\lambda \vec{x}\alpha)^M(w) = \{\vec{d} D^{M^n} : w \in ||\alpha||^{M(\vec{d}/\vec{x})}\})$.

A Brief Introduction to *Para-consistent logic* and *AI*   45

Proposition 1 *For any variables' sequence $\vec{x}$, any formula α, β, any model M:*

1. $(\lambda \vec{x} \neg \alpha)^M = ((\lambda \vec{x} \alpha)^M) \sim$
2. $(\lambda \vec{x}(\alpha \vee \beta))^M = (\lambda \vec{x} \alpha)^M \cup (\lambda \vec{x} \beta)^M$
3. $(\lambda \vec{x}(\alpha \wedge \beta))^M = (\lambda \vec{x} \alpha)^M \cap (\lambda \vec{x} \beta)^M$
4. $(\lambda \vec{x}(\alpha \to \beta))^M = ((\lambda \vec{x} \alpha)^M) \sim \cup (\lambda \vec{x} \beta)^M$

Proposition 2 *For any M, $||\alpha(\vec{y}/\vec{x})||^{M(\vec{d}/\vec{y})} = [\vec{d}^M, (\lambda \vec{x} \alpha)^M]$.*

Proposition 3 *For any M, any $w \in W$, $w \in ||\forall \vec{y}(\beta \to \gamma)(\vec{y}/\vec{x})||^M$ iff. $(\lambda \vec{x} \beta)^M(w) \subseteq (\lambda \vec{x} \gamma)^M(w)$.*

Corollary 1 *For any M, any $w \in W$, $w \in ||\forall(\beta \leftrightarrow \gamma)(\vec{x}/\vec{x})||^M$ iff. $(\lambda \vec{x} \beta)^M(w) = (\lambda \vec{x} \gamma)^M(w)$.*

Definition 12 *Let M be a model and α be a formula, $X \subseteq W^M$, $X \neq \emptyset$. α is true at the set X (written as $M| =_X \alpha$) iff. $X \subseteq ||\alpha||^M$. When $X = \{w\}$, we also say that α is true at w. When $X = W^M$, we say that α is valid at M, we use $M| =_\alpha$ to denote it.*

Definition 13 *Given a formula α, α is valid ($| = \alpha$) iff. $M| = \alpha$ for all M.*

Theorem 1 *The formulas listed below are valid.*

1. $(\forall x(\alpha > \beta) \to (\alpha > \forall x \beta))$ *(x is not a free variable in α)*
2. $\forall \vec{y}(N(\lambda \vec{x} \alpha, \lambda \vec{x} \beta) \vec{y} \to \alpha(\vec{y}/\vec{x}))$
3. $\forall \vec{y}(N(\lambda \vec{x} \alpha, \lambda \vec{x} \beta) \vec{y} \to N(\lambda \vec{x} \alpha/\lambda \vec{x} \neg \beta))$
4. $\forall \vec{x}(\alpha \to \beta) \to \forall \vec{x}(\alpha > \beta)$

Theorem 2 1. *If $M| = \beta \leftrightarrow \gamma$ for all model M, then, $M| = \forall \vec{y}(N(\lambda \vec{x} \alpha, \lambda \vec{x} \beta) \vec{y} \leftrightarrow \forall \vec{y}(N(\lambda \vec{x} \alpha, \lambda \vec{x} \gamma) \vec{y})$ for all model M.*
2. *If $M| = \beta \leftrightarrow \gamma$ for all model M, then, $M| = \forall \vec{y}(N(\lambda \vec{x} \beta, \lambda \vec{x} \alpha) \vec{y} \leftrightarrow N(\lambda \vec{x} \gamma, \lambda \vec{x} \alpha) \vec{y})$ for all model M.*

Definition 14 *A frame $< W, D, N, o*.$ is a subject monotonic frame iff. it satisfies: for any $\tau_1, \tau_2, \tau_3 \in I(W, D)$, if $\tau_1(w) \subseteq \tau_2(w)$, then $N(\tau_1, \tau_3)(w) \subseteq N(\tau_2, \tau_3)(w)$. A model $< W, D, N, \circledast, ?,$ is a subject monotonic model iff. $< W, D, N, o*.$ is a subject monotonic model.*

Theorem 3 *For any subject monotonic model M, $M| = \forall \vec{y}(\alpha \to \gamma)(\vec{y}/\vec{x}) \to \forall \vec{y}(N(\lambda \vec{x} \alpha, \lambda \vec{x} \beta) \vec{y} \to N(\lambda \vec{x} \gamma, \lambda \vec{x} \beta) \vec{y})$.*

Definition 15 *A frame $< W, D, N, o*.$ is a full sense frame iff. it satisfies: for any $\tau_1 \in I(W, D)$, $\tau_1 \subseteq N(\tau_1, \tau_\perp)$. A model $< W, D, N, \circledast, ?,$ is a full sense model iff. $< W, D, N, o*.$ is a full sense frame.*

By Definition 4(1), $N(\tau_1, \tau_2) \subseteq \tau_1$ for any frame, so, we have $N(\tau_1, \tau_\perp) = \tau_1$ for any full sense frame. The intuition is that the normal subject sense selected from subject based on the full sense is the subject sense.

Theorem 4 *For any full sense model M*, $M| = \forall \vec{y}(\alpha(\vec{y}/\vec{x}) \to N(\lambda \vec{x} \alpha, \lambda \vec{x} \bot)\vec{y})$.

Definition 16 *A frame* $< W, D, N, o* .$ *is a selection-contained frame iff. it satisfies: for any* $\tau_1, \tau_2 \in I(W, D)$, *if* $\tau_1(w) \subseteq \tau_2(w)$, *then* $\tau_1(w) \subseteq N(\tau_1, \tau_2)(w)$. *A model* $< W, D, N, o*, .$ *is a selection-contained model iff.* $< W, D, N, o* .$ *is a selection-contained frame.*

By Definition 4(1) and Definition 16, for any selection-contained frame, if $\tau_1(w) \subseteq \tau_2(w)$, then $N(\tau_1, \tau_2)(w) = \tau_1(w)$. The intuition is that in a possible world w, if the object set corresponding to the subject sense is included in the object set corresponding to the predicate set, then, in this world, the object set corresponding to the normal subject sense which subject sense select according to predicate sense is the subject sense. For example, in the possible world we live, we have known 'sparrows are birds', then according to 'bird', the set corresponding to 'normal sparrows' is the set corresponding to 'sparrows'.

Theorem 5 *For any selection-contained model M*, $M| = \forall \vec{y}(\alpha \to \beta)(\vec{y}/\vec{x}) \to \forall \vec{y}(\alpha(\vec{y}/\vec{x}) \to N(\lambda \vec{x} \alpha, \lambda \vec{x} \beta)\vec{y})$.

Definition 17 *A frame* $< W, D, N, o* .$ *is a semi-degenerated frame, iff. it satisfies: for any* $\tau_1, \tau_2, \tau_3 \in I(W, D)$, $N(\tau_1, \tau_2 \cap \tau_3) \subseteq N(\tau_1, \tau_2)$. *A model* $< W, D, N, o* .$ *is a semi-degenerated model iff.* $< W, D, N, o* .$ *is a semi-degenerated frame.*

Theorem 6 *For any semi-degenerated frame, any* $\tau_1, \tau_2, \tau_3 \in I(W, D)$,

1. $N(\tau_1, \tau_2 \cup \tau_3) \subseteq N(\tau_1, \tau_2)$
2. *If* $\tau_2 \subseteq \tau_3$, *then* $N(\tau_1, \tau_2) = N(\tau_1, \tau_3)$.
3. $N(\tau_1, \tau_2) = N(\tau_1, \tau_3)$

By Theorem 6, semi-degenerated frame satisfies: for any $\tau_1, \tau_2, \tau_3 \in I(W, D)$, $N(\tau_1, \tau_2) = N(\tau_1, \tau_3)$. It means that according to any predicate sense, the normal subject sense selected is the same. It seems unreasonable, because that it makes the predicate sense useless when select normal subject sense. But, this principle can also be useful. In daily life, we always reason at some conditions or under relative context. At the same context, the normal subject sense selected from the same subject sense is usually the same. So we can attempt to use this principle as our base for the inference with order.

Theorem 7 *For any semi-degenerated model M*, $M| = \forall \vec{y}(N(\lambda \vec{x} \alpha, \lambda \vec{x}(\beta \wedge \gamma))\vec{y} \to N(\lambda \vec{x} \alpha, \lambda \vec{x} \beta)\vec{y})$.

We have given several different models based on the connection between subject sense and predicate sense. Next, we will give the corresponding logic systems.

3 Basis system G_0

Based on the system G[2], G_0 removed some axioms(e.g. GU: $G\vec{x}(\alpha;\beta) > \forall \vec{x}(\alpha > \beta))$[4], and added some axiom and rules. G_0 is a basic logic for generic reasoning.

Axiom schemata:
T(all tautologies); $\forall^-(\forall x\alpha \to \alpha(x/t))$;
$\forall_\to$ $\forall x(\alpha \to \beta) \to (\forall x\alpha \to \forall x\beta)$
$>_{BF}$ $\forall x(\alpha > \beta) \to (\alpha > \forall x\beta)$ (x is not a free variable in α)
C_K $(\alpha > (\beta \to \gamma)) \to ((\alpha > \beta) \to (\alpha > \gamma))$
$>_{MP}$ $(\alpha \wedge (\alpha > \beta)) > \beta$
T_{RAN} $(\alpha > \beta) \to ((\beta < \gamma) \to (\alpha > \gamma))$
A_D $(\alpha > \gamma) \wedge ((\beta < \gamma) \to (\alpha \vee \beta > \gamma))$
I_C $\forall \vec{x}(\alpha \to \beta) \to \forall \vec{x}(\alpha > \beta)$
N $\forall \vec{y}(N(\lambda \vec{x}\alpha, \lambda \vec{x}\beta)\vec{y} \to \alpha(\vec{y}/\vec{x}))$
$N_\neg$ $\forall \vec{y}(N(\lambda \vec{x}\alpha, \lambda \vec{x}\beta)\vec{y} \to N(\lambda \vec{x}\alpha, \lambda \vec{x}\neg\beta)\vec{y})$

Rules of inference:
MP; $\forall^+$
R_{CEA} From $\beta \leftrightarrow \gamma$, infer $(\beta > \alpha) \leftrightarrow (\gamma > \alpha)$;
R_N From β, infer $\alpha > \beta$;
R_M From $\alpha > \beta$, infer $(\alpha \wedge \gamma) > \beta$;
R_{NEA} From $\beta \leftrightarrow \gamma$, infer $\forall \vec{y}(N(\lambda \vec{x}\beta, \lambda \vec{x}\alpha)\vec{y} \to N(\lambda \vec{x}\gamma, \lambda \vec{x}\alpha)\vec{y})$
R_{NEC} From $\beta \leftrightarrow \gamma$, infer $\forall \vec{y}(N(\lambda \vec{x}\alpha, \lambda \vec{x}\beta)\vec{y} \to N(\lambda \vec{x}\alpha, \lambda \vec{x}\gamma)\vec{y})$

Some theorems and derived rules about generic sentences:
ThG_01 $G\vec{x}(\alpha;\alpha)$
ThG_02 $G\vec{x}(\alpha \wedge \beta;\alpha)$
ThG_03 $\forall \vec{y}(N(\lambda \vec{x}\alpha, \lambda \vec{x}\beta)\vec{y} \leftrightarrow (N(\lambda \vec{x}\alpha, \lambda \vec{x}\neg\beta)\vec{y})$
ThG_04 $\forall \vec{x}(\alpha > \beta) \to G\vec{x}(\alpha;\beta)$
ThG_05 $\forall \vec{x}(\alpha \to \beta) \to G\vec{x}(\alpha;\beta)$
R_{GN} From β, infer $G\vec{x}(\alpha;\beta)$

By Theorem 1, 2, the system G_0 is sound.

4 The expanded systems of G_0

4.1 System G_1

Based on the system G_0, we add an axiom in G_1:
N_{AM} $\forall \vec{y}(\alpha \to \gamma)(\vec{y}/\vec{x}) \to \forall \vec{y}(N(\lambda \vec{x}\alpha, \lambda \vec{x}\beta)\vec{y} \to N(\lambda \vec{x}\gamma, \lambda \vec{x}\beta)\vec{y})$

Some theorems and derived rules about generic sentences:
ThG_11 $\forall \vec{x}(\alpha \to \beta) \to (G\vec{x}(\beta;\gamma) \to G\vec{x}(\alpha;\gamma))$
ThG_12 $G\vec{x}(\alpha;\beta) \to G\vec{x}(\alpha \wedge \gamma;\beta)$
ThG_13 $G\vec{x}(\alpha \vee \beta;\gamma) \to G\vec{x}(\alpha;\gamma)$
R_{GIC} From $\alpha \to G\vec{x}(\beta;\gamma)$, infer $G\vec{x}(\alpha \wedge \beta;\gamma)$.

By Theorem 3, system G_1 is sound for all subject monotonic models.

---

[4] The '>' in this paper is denoted by '$\geq$' in [2], we will not talk about the detailed differences here.

4.2 System G_2

Based on the system G_0, we add an axiom in G_2:
$AG_\perp \quad \forall \overrightarrow{y}(\alpha(\overrightarrow{y}/\overrightarrow{x})) \to N(\lambda \overrightarrow{x}\alpha, \lambda \overrightarrow{x}\perp)\overrightarrow{y})$

Some theorems and derived rules:
$ThG_21 \quad \forall \overrightarrow{y}(N(\lambda \overrightarrow{x}\alpha, \lambda \overrightarrow{x}\perp)\overrightarrow{y} \leftrightarrow \alpha(\overrightarrow{y}/\overrightarrow{x}))(G_\perp)$
$ThG_22 \quad \forall \overrightarrow{y}(N(\lambda \overrightarrow{x}\alpha, \lambda \overrightarrow{x}\top)\overrightarrow{y} \leftrightarrow \alpha(\overrightarrow{y}/\overrightarrow{x}))(G_\top)$
$ThG_23 \quad G\overrightarrow{x}(\alpha;\perp) \leftrightarrow \forall \overrightarrow{x}(\alpha > \perp)$
$ThG_24 \quad G\overrightarrow{x}(\alpha;\perp) \to \forall \overrightarrow{x}(\alpha;\beta)$
$ThG_25 \quad G\overrightarrow{x}(\alpha \wedge \gamma;\perp) \to \forall \overrightarrow{x}(\alpha;\neg\gamma)$

By Theorem 4, system G_2 is sound for all full sense models.

4.3 System G_3

Based on the system G_0, we add an axiom in G_3:
$R_{NT} \quad \forall \overrightarrow{y}(\alpha \to \beta)(\overrightarrow{y}/\overrightarrow{x}) \to \forall \overrightarrow{y}(\alpha(\overrightarrow{y}/\overrightarrow{x}) \to N(\lambda \overrightarrow{x}\alpha, \lambda \overrightarrow{x}\beta)\overrightarrow{y})$

Some theorems:
$ThG_31 \quad \forall \overrightarrow{y}(\alpha \to \beta)(\overrightarrow{y}/\overrightarrow{x}) \to \forall \overrightarrow{y}(N(\lambda \overrightarrow{x}\alpha, \lambda \overrightarrow{x}\beta) \leftrightarrow \alpha(\overrightarrow{y}/\overrightarrow{x}))$
$ThG_32 \quad \forall \overrightarrow{y}(\alpha \to \neg\beta)(\overrightarrow{y}/\overrightarrow{x}) \to \forall \overrightarrow{y}(N(\lambda \overrightarrow{x}\alpha, \lambda \overrightarrow{x}\beta) \leftrightarrow \alpha(\overrightarrow{y}/\overrightarrow{x}))$

By Theorem 5, system G_3 is sound for all selection-contained models.

4.4 System G_4

Based on the system G_0, we add an axiom in G_4:
$AG_4 \quad \forall \overrightarrow{y}(N(\lambda \overrightarrow{x}\alpha, \lambda \overrightarrow{x}(\beta \wedge \gamma))\overrightarrow{y} \to N(\lambda \overrightarrow{x}\alpha, \lambda \overrightarrow{x}\beta)\overrightarrow{y})$

Some theorems and derived rules about generic sentences:
$ThG_41 \quad \forall \overrightarrow{y}(N(\lambda \overrightarrow{x}\alpha, \lambda \overrightarrow{x}(\beta \vee \gamma))\overrightarrow{y} \to N(\lambda \overrightarrow{x}\alpha, \lambda \overrightarrow{x}\beta)\overrightarrow{y})$
$R_{NCM} \quad$ From $\beta \to \gamma$, infer $\forall \overrightarrow{y}(N(\lambda \overrightarrow{x}\alpha, \lambda \overrightarrow{x}\beta)\overrightarrow{y} \to N(\lambda \overrightarrow{x}\alpha, \lambda \overrightarrow{x}\gamma)\overrightarrow{y})$
$R_{NCM+} \quad$ From $\beta \to \gamma$, infer $\forall \overrightarrow{y}(N(\lambda \overrightarrow{x}\alpha, \lambda \overrightarrow{x}\beta)\overrightarrow{y} \leftrightarrow N(\lambda \overrightarrow{x}\alpha, \lambda \overrightarrow{x}\gamma)\overrightarrow{y})$
$ThG_42 \quad \forall \overrightarrow{y}(N(\lambda \overrightarrow{x}\alpha, \lambda \overrightarrow{x}\beta)\overrightarrow{y} \leftrightarrow N(\lambda \overrightarrow{x}\alpha, \lambda \overrightarrow{x}\gamma)\overrightarrow{y})$
$ThG_43 \quad G\overrightarrow{x}(\alpha;\beta) \wedge G\overrightarrow{x}(\alpha;\gamma) \to G\overrightarrow{x}(\alpha;\beta \wedge \gamma)$
$ThG_44 \quad G\overrightarrow{x}(\alpha;(\beta \to \gamma)) \wedge G\overrightarrow{x}(\alpha;\beta) \to G\overrightarrow{x}(\alpha;\gamma)$
$ThG_45 \quad G\overrightarrow{x}(\alpha;(\beta \wedge \gamma)) \to G\overrightarrow{x}(\alpha;\beta) \wedge G\overrightarrow{x}(\alpha;\gamma)$
$ThG_46 \quad G\overrightarrow{x}(\alpha;\beta)) \wedge \forall \overrightarrow{x}(\beta \to \gamma) \to G\overrightarrow{x}(\alpha;\gamma)$

By Theorem 7, system G_4 is sound for all semi-degenerated models.

If we add $AG_\perp$ to G_4, then we can get a new system, in the new system, we have theorems: $\forall \overrightarrow{y}(N(\lambda \overrightarrow{x}\alpha, \lambda \overrightarrow{x}\beta) \leftrightarrow \alpha(\overrightarrow{y}/\overrightarrow{x}))$ and $G\overrightarrow{x}(\alpha;\beta) \leftrightarrow \forall \overrightarrow{x}(\alpha > \beta)$. That means the new system is full-degenerated, operator N have no use in it.

The completeness of G_0-G_4 is proved in [4].

Bibliography

[1] Eckardt, R. Normal objects, normal worlds and the meaning of generic sentences. *Journal of Semantics*, (16).

[2] Mao, Y. *A Formalism for Nonmonotonic Reasoning Encoded Generics*. Ph.D. thesis, The University of Texas, Austin, 2003.

[3] Mao, Y., Zhou, B. An analysis of the meaning of generics. *Social Sciences in China*, XXIV(3):126–133, 2003.

[4] Zhang, L. The completeness of the logic system $G_0 - G_4$ for generics inference. *Beida Journal of Philosophy*, 7(1):97–120, 2006.

[5] Zhou, B. On the essence of generic sentence and concept. *Journal of Peking University (Humanities and social sciences)*, 41(4):20–29, 2004.
[6] Zhou, B., Mao, Y. A basic logic for default reasoning. *Philosophical Researches*, supplement:1–10, 2003.

The Rise and Meaning of the Paraconsistent Model of the Philosophy of Science from the Perspective of Logic[1]

Ren Xiaoming

Nankai University

renxiaoming@nankai.edu.cn

Gui Qiquan

Wuhan University

qqgui@sina.com

ABSTRACT. After Laudan, the philosophy of science seems uneventful for a while, but a new model of philosophy of science has now arisen. From the perspective of para-consistent logic rationality, progress and truth as central concepts of the philosophy of science can now be united in harmony. The direct objective of science is to enhance the harmonious force of theory. Science develops the applicative process of conflict and harmony between theories. Progress means the gradual enforcement of the harmonious force of the theory. Rationality means harmony, and harmony is the ideal state of progress. A theory is called true because it has better harmonious force. So the criteria of rationality, theoretical progress and truth can be united in the criterion of harmonious force. This is the philosophical spirit of para-consistent logic and the soul and nature of the harmonious rationality mode.

Issues of scientific rationality have been controversial in scientific philosophy for nearly 40 years. It has become an important issue that has engaged the interest of scientists and philosophers. What is scientific rationality? What is scientific progress? How can we understand the relationship between rationality and non-rationality?

Scientists and philosophers have persistently searched for thousands years, and put forward many answers. Even though they promote scientific development and social progress to some extent, their attempts were not altogether successful. Therefore, after Laudan, the philosophy of science seems uneventful for a while, but now Chinese scientific philosophers, such as Lei Ma, have collected the merits of every philosopher and proposed a new model of scientific philosophy.

In scientific philosophy, nearly every theory and method has its merits and shortcomings. Logical positivists maintained that the progress of science, having the rationality of logic, should regard "truth" as the goal of science. They not only required new theories to explain the success of older theories, but also required new theories to explain more empirical facts than older theory did, and rightly so. However, while they distinguished between theoretical terms and observed terms, their cleavage went to extremes. While they regarded observed statements as the reliable foundation of scientific inquiry, their viewpoint was criticized and

---

[1] The authors are with Department of Philosophy, Nankai University, Tianjin 300071 China, and Institute of Philosophy, Wuhan University, Wuhan 430072 China, respectively. Email: renxiaoming@nankai.edu.cn, qqgui@sina.com.

could not be justified.

The viewpoint of logical empiricism that regarded the universal truth of logic as the criterion of rationality was also criticized by historicism. Proceeding from realities of scientific history, Kuhn and Feyerabend criticized the accumulated viewpoint of scientific progress. They believed that development of scientific is both continuous and non-continuous and thought that we cannot rely on the experience to judge the truth and false of theory, since observation penetrate the theory and observation get the pollution of theory. They denied that there is a criterion of rationality that is objective, neutral and beyond theory and believed that there is incommensurability between mutually competitive or consecutive theories. They maintained that the competition between theories required non-logical modes such as propagate, persuade. They rejected the logic criterion of rationality of scientific progress and showed that the rationality of scientific progress is a historical rationality and group rationality.

Laudan not only opposed the model of logical rationality but also criticized the mode of historical rationality about scientific progress and advanced a unique model of scientific rationality. He believed that the model of logical rationality deviated from reality of scientific history on the one hand; the mode of historical rationality excessively emphasized a factor of nonrationality in scientific decision on the other. Devoting himself to combine logic with history, Laudan raised a model of scientific progress – the model of problem solving. He maintained that even though there is incommensurability between different theories, we can make rational evaluations provided we can talk about different theories dealing with same problem meaningfully. Laudan maintained that the scientific objective is to search for a theory that can solve problems very well. That is, it can extend the range of problem solving as far as possible and reduce the number of abnormalities as far as possible. Laudan advocated a rationality of problem solving that is different from both logical rationality and history rationality.

Even though he presented unique opinion about scientific progress and scientific rationality, Laudan's theory met with difficulties. Fortunately, Chinese scholar Lei Ma raised a new theory that both inherit Laudan and surpass Laudan (c.f. [4]). In fact, Lei Ma attempted to give a new model of scientific rationality on the basis of Laudan's model, that is the model of harmonious force. The model had a new understanding for the force of problem solving.This model is the model of harmonious force. Thereby it is the extent of the range of scientific rationality, and deepens the research thereof.

According to the model of harmonious force, overall objective of science is to search for harmony in conflict. Similarly, the central ideal of paraconsistent logic is "searching for harmony in conflict." Lei Ma thought that overall objective of science, for instance, the harmonious force of experience, the harmonious force of conception, the harmonious force of background,are the sum of individual objectives of science, for instance, the harmonious force of empirical simplicity, the harmonious force of conceptual unity, the harmonious force of background technique, since individual objectives of science are the minimal unit of the overall objective of science . According to his opinion, any single objectives of science are equally significant – no objectives possess superiority.

The model of harmonious force maintained that between logic and history, between formal and informal, between stable and change there should be kept a necessary tension. Therefore, the model embodies the ideal of dialectics from the beginning to the end. The characteristic of the model of harmonious force is to find a more stable balance between two extremes. This model regarded harmony as the central concept and thought that truth consists of harmonization – harmonious is rational, and progress consists of augmenting the theoretic force of harmony. Thus this model puts forth new opinions different from current viewpoint about the issues of scientific philosophy. In fact, the scientific rationality embodied in this model may be call harmonious rationality.

It is obvious that from the logical rationality, the historical rationality, the problem-solving rationality to the harmonious rationality, every model tries to overcome the limits of the previous model, and make for more universality. The model of the harmonious rationality revealed a new system of scientific philosophy which critically assimilated the advantages of theories of scientific rationality, especially the theory of Laudan's scientific rationality, and provided a more open theory beyond logical and historical models.

From the perspective of paraconsistent logic, the rise of the paraconsistent model of the philosophy of science deepened certainly our viewpoint of the model of scientific rationality. Paraconsistent logic is a new thought current in the international logic circle, which arose in the late thirty years. Being a kind of "revolutionary" non-classical logic, paraconsistent logic allows "meaningful contradiction" to enter formal axiomatic systems. Therefore it attracts a great number of logicians, philosophers and mathematicians who come from Brazil, Australia, America, China and Eastern Europe.

In logical philosophy, non-class logic divided into two categories: one is extended logic, another is deviant logic. Extended logics are not rivals of classical logic, they do not removed the basis principles of classical logic, just add some new axiom and principles, but deviant logics are genuinely rivals of classical logic, because it replaced the basis principles of classical logic. Paraconsistent logic is a deviant logic, because it replaced or revised the basis principles of classical logic, that is, law of contradictory. It is common that law of contradictory is unmoved basis of classical logic, however, from the perspective of non-classical logic, the law of contradictory is revisable.

The concept of paraconsistent logic is not only applicable to logic but also scientific philosophy. From the perspective of the concept of paraconsistent logic, it is not difficult that we deal with dialectical contradictions regarding logic and history, formal and informal, stable and changing, in scientific philosophy.

Some successful examples of the application of paraconsistent logic is to Computer Science, the research of Artificial Intelligences showed that the concept of paraconsistent logic applied to all subjects, including scientific philosophy. The application of paraconsistent logic to a paraconsistent knowledge basis is also successful. One of the most important achievements of Artificial Intelligence is Expert Systems, which was established on the basis of classical logic. The problem is that classical logic admits Scott's Rule (which allows that a contradictory proposition can be followed by any proposition, therefore even if there are ten thousand consistent propositions they can be of no use only because there is one contradictory proposition which can be followed by any proposition). It is natural that there are disagreements between human experts (who are considered the prototypes of Expert Systems) such as traditional Chinese medicine experts and western medicine experts when they diagnose a disease. In order to solve this problem, we must solve the problem of the inconsistent structure of Knowledge

Base at first. Taking paraconsistent logic as its basis, each expert procedure just does its own business. Even if there is a contradiction in some respect, it can be "suspended" so that the whole system can work as usual.

Some philosophers and scientists advance a similar viewpoint. For instance, Niels Bohr advocated the principle of mutual supplementary in the "search for harmony in conflict". Feyerabend tried to introduce a new viewpoint, a new conception where rivals deviating from accepted background theory can finally reach new harmony. These principles percolate with the spirit of paraconsistent logic – the search for harmony in conflict.

Returning to the field of scientific philosophy, according to the spirit of paraconsistent logic, it was not difficult to see that these viewpoints were justified: rationality, progress, truth, as central concepts of the philosophy of science, can be united in harmony. The direct objective of science is to enhance the harmonious force of theory. Science is developed in the applied process of conflict and harmony between theories. Progress means the gradual enforcement of the harmonious force of the theory and rationality means harmony and harmony is the ideal state of progress. A theory is called true because it has better harmonious force. This is the philosophical spirit of paraconsistent logic and is also the essence and nature of the model of harmonious rationality in scientific philosophy.

Bibliography

[1] Gui, Q.-Q. Paraconsistent formal system: The logic of harmony under conflict. *Journal of Wuhan University*, (4), 1992.

[2] Gui, Q.-Q., Chen, Z.-L., Zhu, F.-X. *Paraconsistent Logic and Artificial Intelligence*. Wuhan University Press, 2002.

[3] Laudan, L. *Progress and its Problems*. Berkeley: University of California Press, 1977.

[4] Ma, L. *Conflict and Harmony, New Theory of Scientific Rationality*. The Commercial Press, 2006.

[5] Ren, X.-M., Cui, Q.-T. *The Pluralism of Logic and Logic Applied to the Humanities*. Beijing: Philosophical Researches, 2003. Supplement.

Second-level Hypothetical Inference and its Applications[1]

Zhou Xunwei

Beijing Union University

zhouxunwei@263.net

ABSTRACT. In rule-based systems general laws are regarded as inference rules used to infer theorems. Theorem proving in these systems can only be formalized, not automated. This paper, based on mutually-inversistic logic, proposes second-level hypothetical inference-based systems which transform rules in rule-based systems into logical theorems, taking second-level hypothetical inference as the sole inference rule. Theorem proving in second-level hypothetical inference-based systems can be automated.

1 Introduction

Rule-based systems can be divided into first-level rule-based systems and second-level rule- based systems. In first-level rule-based systems such as a geometrical theorem-proving system, parser, rules can be transformed into empirical or mathematical theorems, taking first-level hypothetical inference as the sole inference rule.The system can be studied in the first-level single quasi-predicate calculus of mutually-inversistic logic. In second-level rule-based systems such as Hoare rules[1] in program verification, Carroll Morgan's rule refinement in program refinement [2], operational semantics in formal semantics, and Armstrong's axiomatic system[3] in relational databases, rules can be transformed into logical theorems, taking second-level hypothetical inference as the sole inference rule. The system can be studied in the second-level single quasi-predicate calculus of mutually-inversistic logic. This paper studies a second-level rule-based system and its relationship to a second-level hypothetical inference-based system. In a second-level rule-based system, a general law is regarded as a rule used to infer a theorem. This tradition can be traced back to inference in Aristotelian logic. This type of inference is concise, skillful, and suitable for human inference. Since there are many rules, the system cannot choose from them automatically, so inference in this system cannot be made automatically. This type of inference is not suitable for a computer. A computer is only capable of making an inference that is mechanical, repetitive, or does not need a great deal of skill. This paper, based on mutually-inversistic logic, proposes a second-level hypothetical inference-based system that is suitable for a computer-generated inference. In a second-level hypothetical inference-based system, rules in the rule-based system

[1]The author is with Institute of Information Technology, Beijing Union University, Beijing 100101 China. Email: zhouxunwei@263.net

are transformed into logical theorems, with second-level hypothetical inference being the sole inference rule. Because there is only one inference rule, every time the system wants to make an inference, the system employs it. In the case of numerous theorems, the system searches them top-down, from left to right, depth first, plus backtracking in the way that Prolog does. In this way, an inference can be made automatically, and it is suitable for a computer to make the inference. The rest of this paper is organized as follows. Section 2 briefly introduces mutually-inversistic logic. Section 3 shows how inference is made in a rule-based system and in a second-level hypothetical inference-based system. Section 4 introduces some applications of a second-level hypothetical inference-based system. Section 5 presents the concluding remarks.

2 Brief introduction to mutually-inversistic logic

In mutually-inversistic logic, $int(x), rat(x), real(x)$ (corresponding to the naming of a predicate in classical logic) are first-order fact propositions. $P(x), q(x), r(x)$, where p, q, r are predicate variables ranging over predicate constants, int, rat, real, are second-order fact propositions. P, Q, R are the abbreviations of $p(x), q(x), r(x)$, respectively, ranging over first-order fact propositions. The logical operator is divided into compounder and connective. $\neg, \wedge, \vee$ are compounders. $\leq^{-1}$ (mutually inverse implication) is a connective. A connective connecting two fact propositions forms an empirical or mathematical connection proposition. $int(x) \leq^{-1} rat(x)$ (corresponding to $\forall x(int(x) \to rat(x))$ in classical logic, mutually-inversistic logic is quantifier-free; x is bound so long as it occurs on both sides of $\leq^{-1}$) is a first-order single empirical or mathematical connection proposition. A true first-order single empirical or mathematical connection proposition is an empirical or mathematical theorem. $P \leq^{-1} Q$ is a second-order single empirical or mathematical connection proposition. A compounder composing two empirical or mathematical connection propositions still forms an empirical or mathematical connection proposition; e.g. $\{P \leq^{-1} Q\} \wedge \{Q \leq^{-1} R\}$ is still a second-order single empirical or mathematical connection proposition. A connective connecting two second-order single empirical or mathematical connection propositions forms a second-order single logical connection proposition. A true second-order single logical connection proposition is a logical theorem. For example, $\{P \leq^{-1} Q\} \wedge \{Q \leq^{-1} R\} \leq^{-1} \{P \leq^{-1} R\}$ is a logical theorem.

3 Rule-based system v. second-level hypothetical inference-based system

Take "all integers are rationals, all rationals are real numbers; therefore, all integers are real numbers" as an example. In a rule-based system, hypothetical syllogism is an inference rule used to infer "all integers are real numbers" from "all integers are rationals" and "all rationals are real numbers," as follows:

$Int(x) \to rat(x)$
$Rat(x) \to real(x)$
───────────────
$\therefore int(x) \to real(x)$.

In a second-level hypothetical inference-based system, hypothetical syllogism is transformed into a logical theorem: $\{P \leq^{-1} Q\} \wedge \{Q \leq^{-1} R\} \leq^{-1} \{P \leq^{-1} R\}$. Second-level hypothetical inference is the inference rule: taking $\{P \leq^{-1} Q\} \wedge \{Q \leq^{-1} R\} \leq^{-1} \{P \leq^{-1} R\}$ as the major premise, taking $\{int(x) \leq^{-1} rat(x)\} \wedge \{rat(x) \leq^{-1} real(x)\}$ as the minor premise, infers $int(x) \leq^{-1} real(x)$. This inference is depicted as follows:

$\{P \leq^{-1} Q\} \wedge \{Q \leq^{-1} R\} \leq^{-1} \{P \leq^{-1} R\}$
$\{int(x) \leq^{-1} rat(x)\} \wedge \{rat(x) \leq^{-1} real(x)\}$
―――――――――――――――――――
$\therefore int(x) \leq^{-1} real(x)$

This example is small and is given only to demonstrate second-level hypothetical inference. Now consider a larger example: Hoare rules in program verification. In the Hoare rule system, there are 5 rules. The system cannot choose from them automatically, but the human experts can choose from them using skills. Therefore, this system is suitable for human inference, not for computer inference. In a second-level hypothetical inference-based system, the 5 rules are transformed into 5 logical theorems, with second-level hypothetical inference being the sole inference rule. Because there is only one inference rule, every time the system wants to make an inference, the system employs it. In the case of numerous theorems, the system searches them top-down, from left to right, depth first, plus backtracking in the way that Prolog does. In this way inference can be made automatically and it is suitable for a computer to make an inference.

4 Applications of a second-level hypothetical inference-based system

The author has applied second-level hypothetical inference-based systems in 18 fields of computer science, including resolution principle, logic programming, expert systems, recursion, iteration, formal semantics, database, automatic planning, semantic network, multi-agent planning, ontology, information flow, program verification, and parser.

5 Concluding remarks

A second-level rule-based system is suitable for human inference not for computer inference. It can be transformed into a second-level hypothetical inference-based system, which is suitable for computer inference and is very useful in computer science.

Bibliography

[1] Hoare, C. A. R. An axiomatic basis for computer programming. *Communications of the ACM*, 12(10):576–593, 1969.

[2] Morgan, C. *Programming from Specifications*. 1998.

[3] Silberschatz, A., Korth, H. F., Sudarshan, S. *Database System Concepts*. Beijing: McGraw-Hill Companies, Higher Education Press, 4th edn., 2002.
[4] Winskel, G. *The Formal Semantics of Programming Languages: An Introduction*. Massachusetts Institute of Technology Press, 1993.
[5] Zhou, X. Brief introduction to mutually-inversistic logic. European Summer Meeting of the Association for Symbolic Logic, The Netherlands: Utrecht, 1999.
[6] Zhou, X. Second level logical calculus of mutually-inversistic logic. 2005 Annual Meeting of the Association for Symbolic Logic, Stanford, CA, USA: Stanford University, 2005.
[7] Zhou, X., Bao, H. Second level hypothetical inference. Athens, Greece: 2005 Summer Meeting of the Association for Symbolic Logic, 2005.
[8] Zhou, X., Bao, H. Second level hypothetical inference based automated theorem proving. In *Proceedings of 2006 International Conference on Artificial Intelligence – 50 Years' Achievements, Future Directions and Social Impacts*. Beijing, China, 2006.

The Importance and Research Methodology of Chinese Logic[1]

Zhai Jincheng

Nankai University

zhaijc@nankai.edu.cn

1 Introduction

The study of Chinese logic as an independent field of logical inquiry dates to the beginning of the 20th Century. Qichao Liang published the book *The Thought of Mozi* (1904) in which there is a special chapter. This is the starting point for the study of Mozi's logic and logical thought in Chinese traditional philosophy. During the period between 1915 and 1917, Shi Hu wrote his PhD dissertation on Chinese logic *The Development of Logical Method in Ancient China*. It is the first book to study systematically the logical thought of different thinkers from the Pre-Qin period. It also is the first book to introduce Chinese logical thought to the world.

At the beginning of the 20th Century, there were very rich achievements in the study of Chinese logic. The study played an important role in transforming the traditional Chinese academic into the modern academic.[2] Chinese logic is unique for several reasons. It is the only traditional logic which is not based on an Indo-European language. Also, it focuses on ideas of demonstration during the development of ancient Chinese thought.

2 The Importance of Chinese Logic

In recent years, especially in the past decade, the study of Chinese logic has become more and more important both in China and abroad. By using new logical notions and research methods, the new achievements in Chinese have been scored. These achievements, expressing new features and new tendencies in the study of Chinese logic, are different from the ones discovered at the beginning of the 20th Century.

In China, since the beginning of the 21st Century, the study of Chinese has entered into a new stage. The latest achievements in display such new features as the following:

1. Studying Chinese logic with new methods - historical analyses and cultural interpretation. The best book is *Mingxue and Bianxue* (Naming and

---

[1]The author is with Faculty of Philosophy, Nankai University, Tianjin 300071 China. Email: zhaijc@nankai.edu.cn Supported by the MOE Project of Key Research Institute of Humanities and Social Sciences at UniversitiesProject No. 08JJD720044.

Debating Theory, 1997) edited by Qingtian Cui. It distinguishes between name theory and debate theory from the perspective of logical thought in ancient China. Other relevant books include: *A New Study on Chinese Logic* by Lin Mingjun and Zeng Xiangyun (2000) and *A Study of Naming Theory in Ancient China* by Jincheng Zhai (2004)

2. Summarizing systematically and analyzing comprehensively the research and development of logic in the last century in China. The best book is *The Spreading and Study on Logic in China* by Song Wenjian (2005). The author introduces the development of logic in China from 1900 to 2000 and makes a general analysis of the study of Chinese logic.

3. Continued study of the general history of Chinese logic. The study, which is based on achievements from the past, uses new materials to explain unique features of Chinese logic. In this area see *The Textbook of the History of Chinese Logic* edited by Wen Gongyi and Cui Qingtian (2001) and *The History of Chinese Logic* edited by Zhou Yunzhi (2004).[4]

4. Studies of some special topics. Centering on Chinese logic, some major achievements have been achieved through discussions from different angles using new methods. These achievements hopefully will promote the study of Chinese logic to new levels.

5. Chinese logic is being included in the study of the history of Chinese thought, which has been rare in the past. This shows that logical thought is an important part of the history of Chinese thought, which would not be comprehensive if it did not include the study of Chinese logical thought. Since the middle of the 20th century, the study of Chinese logic has become an important field for both sinology and logical study abroad. Many international scholars pay more attention to the study of Chinese logic. In 1977, A. Dumitriu published *The History of Logic*, including a chapter devoted to the study of Chinese logic as an example of "non-Indo-European Logic". In 1978, A.C. Graham published *Later Mohist Logic Ethics and Science*. In 1983, Chad Hanson published Language and Logic in China. "Language and Logic in Traditional China" is the topic of discussion in volume 7 of *Science and Civilization* in China by J. Needham and C. Harbsmeier (1998). Since 2000, a large number of papers have been published in the international sinology and logic journals. The common conclusion is that Chinese logic is the only logic which is based on a non-Indo-European language.[1]

There are several new ideas about the importance of Chinese logic. For example, Harbsmeier writes in Science and Civilization in China: "The history of logic reflection in China is therefore [because of its being based on a non-Indo-European language; K.G.] of extraordinary interest for any global history of logic and hence for any global history of the foundation of science."[1] Meanwhile, he enumerates other books to show the popularity abroad of Chinese logic.

We can quote Klaus Glashoff's review of the chapter about Indian logic in the Handbook of the History of Logic to explain the importance of Chinese logic in the system of world logic: "One should be aware of the fact that the Indian rooted logic provides us with formal logic systems which are not based on the Greek tradition, and this should be considered as an opportunity to critically reflect on our own tradition and the self-conception of our logic. However, this should require a careful discussion of the historical and philosophical background of Indian logic."[1] Chinese logic is a unique logic which is based on a non-Indo-European language; naturally it has a different base, character and form of expression from Indian and Greek logics. It is also important like Indian logic to critical reflection on Western logical tradition and the self-conception of western logic. This is one of the many values of Chinese logic in the world logic system.

3 Tendencies in the Study of Chinese Logic

After analyzing the study of Chinese logic both in China and abroad, we summarize its tendencies as "an important field with two returns":

The study on Chinese logic as an important field both in China and abroad:

Recent achievements in China show that it has become the focus of research. Different voices and discussion from various angles show that the study of Chinese logic has begun to influence other fields. The international interest in the study of Chinese logic shows that Chinese logic holds an important position in the world logic system. This opinion has become a common consensus. We also should research why Chinese logic is the only logic that is based on a non-Indo-European language and what features that gives it. This research calls for the guidance of correct logical notions and effective methods.

Two tendencies of the study on Chinese logic:

First, a return to Chinese culture.

As the study on Indian logic needs "a careful discussion of the historical and philosophical background of Indian logic", the same is true with Chinese logic.

It is becoming a common opinion that the study of the different logic must combine different cultural backgrounds. As an important part of Chinese culture, Chinese logic needs to be restored to its unique Chinese cultural and philosophical context in the analysis of its development, content and Chinese cultural features. In China, some scholars began to do this kind of research and made some new achievements. See for example Cui Qingtian's The Comparative Study on Moist and Aristotle's Logic (2004). In this book, the author discusses the relationship between logic and culture.

Second, the return to logic itself.

We should make clear that Chinese logic is a kind of logical thought, so we must use logical notions and logical methods to study Chinese logic. At the beginning of the 20th Century in China, the knowledge of logic was very elementary and basic. The logic that we understood at that time was not complete enough for us to study Chinese logic. Today, we have relatively enough materials to understand what logic is. For example, the works of Aristotle and Plato are now available in Chinese. This provides us with more logical tools to study Chinese than before.

4 Research Method of Chinese Logic

The research methods of Chinese logic include logical analyses, cultural analyses, dynamics analyses and comparative analyses.

The method of logical analyses: Chinese logic is logical thought, so we should analyze it using logical notions and methods, find out and analyze systematically the Chinese logical thought among the Chinese traditional thoughts.

The method of cultural analyses: since Chinese logic is an important part of Chinese culture, it requires a discussion of the historical and philosophical background to analyses the cultural character of Chinese logic. We should explain the basic meaning of the important concepts in Chinese logic according to the features of Chinese culture.

The method of dynamics analyses: The development of Chinese logic is changing in line with Chinese history. It also interacts with different thoughts in other fields, causing them to influence each other. In different historical stages, the subjects of Chinese culture and philosophy showed different features. The same is true for logical thought.

The method of comparative analyses: Chinese logic is in parallel with the development of Indian logic and Greek logic in the world logic system. We should conduct comparative research to find out why Chinese logic is the only logic which is based on a non-Indo-European language.

Bibliography

[1] Glashoff, K. Review. *the Bulletin of Symbolic Logic*, 10(4), 2004.

[2] Zhai, J. The cultural significance and value of the modern study on pre-Qin Ming-theory. *Nankai Journal*, (5), 2004.

[3] Zhai, J. Several issues on study of the history of China logic. *Nankai Journal*, (4), 2007.

[4] Zhou, Y. *History of Chinese Logic*. Shanxi Education Press, 2004.

On the Nature of Logical Truth based on Quine's Critique[1]

Long Xiaoping

University of Electronic Science and Technology of China

longxiaoping.jx@126.com

Since Quine criticized the distinctions between analytical truth and synthetical truth, many people think that we can not clearly and strictly give the distinctions between analytical truth and synthetical truth. Therefore, logical truth is not analytical truth without experience. We insist that, though Quine broke the distinctions between analytical truth and synthetical truth, logical truth can still be considered a kind of analytical truth independent of experience. The distinctions between logical truth and synthetical truth are still very clear.

Part One

It was Kant who first made a clear distinction between analytical proposition and synthetical proposition in the history of philosophy. He pointed out that all propositions can be divided into two kinds: analytical propositions and synthetical propositions. The definition of the analytical proposition is as follows: "express nothing in the predicate but what has been already actually thought in the concept of the subject." The synthetical proposition "contains in its predicate something not actually thought in the general concept of the body; it amplifies my knowledge by adding something to my concept."[4] The analytical proposition is necessarily true, and its truth is prior, namely, we can prove its truth with no experience. Therefore, Kant thought analytical proposition was endowed with necessity and priority, which is then named as prior and necessary proposition by him, while the synthetical proposition was endowed with accidentality and postpriority, namely, posterior and accidental proposition. But Kant admitted the exception – the prior and synthetical proposition. Afterwards, the logical experientialists disagreed with the exception and made an absolute distinction between Kant's two kinds of propositions. Because Kant's distinction between analytical proposition and synthetical proposition only fits propositions with a subject-predicate structure, and the subject's inclusion to the predicate with the ingredient of metaphor, therefore, the analytical proposition and the synthetical proposition are clearly expressed as follows: a proposition is analytical when and only when its truth only depends on semantic analytic, instead of experience; and a proposition is synthetical when and only when its truth only depends on experience. According to this definition, the division of prior necessary proposition and posterior accidental proposition still corresponds to analytical proposition and synthetical proposition. For example, the sentence "all bodies are extensible" is both an analytical proposition and a prior necessary proposition. "Some

---

[1]The author is with The School of Marxism Education, University of Electronic Science and Technology of China, Chengdu 610054 China. Email: longxiaoping.jx@126.com

bodies have weight", is both a synthetical proposition and a posterior accidental proposition. Hence analysis, priority and necessity are identical; synthesis, postpriority and accidentality are identical too.

The modern analytical philosophers have long considered Kant's differentiation of analytical proposition and synthetical proposition as an un-doubtable axiom. Russell, Wittgenstein, language philosophers, especially the logical positivists, all hold it as perfect truth. The logical positivists excluded Kant's prior synthetical propositions and categorized them into analytical propositions. Alfred Jules Ayer stated, "all genuine propositions can be divided into two sorts ...the former sort includes logical and pure mathematical prior propositions, which I consider necessary and definite only because they are analytical propositions. ...On the other hand, I think those propositions that are related to experience and matter are some hypotheses, which are only probable, but can never be definite."[1]

Quine thoroughly criticized the differentiation between the analytical proposition and the synthetical proposition in his book, *two dogmas of empiricism*.[6] He divided analytical propositions into two sorts, for example,(1) An unmarried man is not married.(2) A bachelor is not married. The two propositions are analytical, but they are different in that (1) is logical truth while (2) is not. But we can change (2) into (1) by a synonym. The proposition expressing the synonym is (3) a bachelor is an unmarried man. The use of (3) is to ascertain the two words, bachelor and unmarried man are synonymous. Hence, substitute "bachelor" in (2) for "unmarried man", (2) is changed into (1).

In natural language such a proposition as (3) is generally considered as a definition. Nevertheless, a new problem arises here. That is how to define a bachelor as an unmarried man. Because according to dictionaries, whose compilers are in fact experiential scientists, these words are defined according to experience. The relation of synonyms has existed in people's brain before he compile the dictionary, so the definition is only the report of synonymous experiences which he has observed., not the grounds of synonyms. Hence a vicious circle appear here: the synonymy of "bachelor" and "unmarried man" is dependent on the definition in the dictionary, while the dictionary depends on the experiences, and the people's experiences have something to do with the definition of other words in dictionary, and these definition in the dictionary still depend on the other experiences, and so on with the rest. Therefore, Quine insist that the foundation that supports the definition of such sorts of synonymous proposition as (3) is vague. Consequently the analyticality of (2) resulting from (3) is vague too, thus make the distinctions between analytical proposition and synthetical proposition vague too. This is the main reason that Quine object to the point that there is essential distinction between analytical proposition and synthetical proposition.

The absolute distinction between the analytical proposition and the synthetical proposition is a dogma of empiricism, especially the foundation of modern empiricism – logical positivism. Quine's critique led to a debate for decades in Euro-American analytical philosophy which completely revealed the limitations in the philosophy of logical positivism and resulted in its gradual decline since the late 1960s.

Part Two

Quite a few people consider that Quine's critique broke the distinction between the analytical proposition and the synthetical proposition. It also broke the division between logical truth and synthetical truth. They consider that "although Quine let pass the logical truth temporarily when he criticized the *two dogmas of empiricism*, his final conclusion involved it. This is because: since it is impossible to make off the division between analysis proposition and synthesis proposition, there is no analytical proposition without any experiences or facts, therefore the truth of logic and mathematics is not such an analytical proposition as mentioned above in that they have not the generally-supposed character of analysis, and that it can be adapted according to experiences and evidences."[2] "The logical truth also has the character of experience."[7]

I don't agree to the above-mentioned viewpoints. In my opinion, they have misapprehended Quine's dissertation and misconceived the essentiality of truth. Now let us take a glimpse at some former statements of logical truth.

There are a great many statements about logic truth. Wittgenstein thought a logical truth is a tautology. He said, "tautology has no truth conditions, because it is unconditionally true." "tautology and a contradiction can not be combined together because they don't describe any possible circumstance." "In tautology, the relation of condition − description (die darstellenden Beziehungen) − is retractile with each other, so that it doesn't have any description with the reality."[9] Carnap pointed out, "A sentence whether is true or not can be judged according to the syntax rule."[10] Proposition variables and their experiential signs and such syntax are accordant with logic rules. So logical truth doesn't express experiences and facts, doesn't need experiential evidence, and has been pre-determined by its relevant semantical expressions and syntax rules.

Housel thought that logic only seek such essential things as common truth, current argumentation relation or proposition itself etc, whose studying objects are idealistic, prevalent things which are not limited by time, and such prevalent essential things are invariable and possess the character of objective truth. So logic rule, namely logical truth, is not a rule about causality, but a certain truth between precondition and conclusion, which is neither dependent on facts nor interested in the existence of facts, including nothing of consciousness and its activity of empiricism. Quine pointed out, "a logical truth is a statement which is true and remains true under all reinterpretations of its components other than the logical particles."[6]

Thus it can be seen that, logical truth doesn't reflect subjective facts and experiential facts, but reflects the relation of logic and of thought. Logic truth is essentially different from synthetical truth or fact truth which reflects experiences and facts. Synthetical truth can form corresponding relation between cognition and the world, but logical truth can't. Logical truth lacks any fact content, and transfers no information in any sphere of experiential topic. Logical truth may have something to do with experience, but this kind of truth is entirely independent of experiential facts. Even Quine himself also thinks that logic truth is independent of experiential facts. So, "If some experiential facts have to be singled out from logical truth, it is a misunderstanding not only to logic, but to the science of modern logic."[8]

Quine thinks that a logical truth is true in all interpretations given to its components except its logical constants. He distinguishes logical truth into two sorts, namely, the first sort proposition of logical truth and the second analytical proposition based on synonymous notions. In two dogmas of empiricism, what Quine criticized is the second sort of analytical proposition, not logical truth.

Susan Haack pointed out, "The fact that Quine's skepticism about analyticity doesn't extend to logical truth is pertinent here."[3] In *two dogmas of empiricism*, "the main problem that Quine concerns is analytical proposition, not logical truth." I approve of this viewpoint that Quine's critique is aimed for the second sort analytical proposition instead of logical truth, and that what he breaks up is the distinction between analytical proposition and synthetical proposition, not the distinction between logical truth and synthetical truth. Consequently, there is an obvious distinction between the analytical and the synthetical proposition, and any denial of this viewpoint is but a misunderstanding of the nature of logical truth.

Part Three

In a word, logical truth is a sort of analytical truth independent of experience, which is different from synthetical truth. Generally, logical truth has the following features:

First, logical truth is formalized and independent of experience. The symbolizing and formalizing of modern logic makes the validity of logical truth dependent on its form. The formalizing gives logical truth the feature of great abstraction. In such circumstance, the propositions expressing logical truth aren't words or sentences, but the symbols extracted from original sentences or words, which makes logical truth independent of experiences. Whether a logical truth is sound is completely decided by its logical expression, namely, relies on the relation of the symbols which compose the expression of logical truth is in accordance with logical rules, which completely has nothing to do with experience.

Second, logical truth contains no logical contradiction. The law of contradiction is the most primary logical law (at least in classical logic). Whether a logical truth is sound relies on its having contradiction. It is impossible that a logical truth contains "$p \wedge -p$", and vice verse. The negation of logical truth contains contradiction, so it can't be possible.

Third, logical truth is a universally valid formula. The key that a logical truth is true lies in its universality. Take the logical truth "$A \to (B \to A)$" as an example, no matter what is the content of "A" and "B", namely, no matter how to give a semantic interpretation to the formulas, the nature of it as a logical truth is not changed.

Fourth, whether a logical truth is sound depends on its relevant formal system. A logical truth is valid just in its formalizing system. For example, "$(x)F(x) \to F(y)$" is an axiom of first-order predicate logic in its system, but it is not a logical truth beyond first-order predicate logic.

Last, logical truth is sound in every possible world. Any logical truth is true not only in the actual world but also in every possible world. Possible world means the possible condition (or history) of actual world, or the unreal situation of possible world.[5] Just as Mr. Chen-Bo stated, "logical truth is relevant not only to the real world, but to all possible worlds. Logical truth means it is true in every possible world." "Logical truth roughly corresponds to the inferential truth, namely, it is a truth which is true in every possible world."[2, pp. 207, 131, 137] As necessary proposition, logical truth is certainly true in all possible worlds.

According to Kripke, logical truth is just analytic truth that he approves, that is to say, logical truths are those propositions that are prior and necessary. In Naming and Necessity, he pointed out, "An analytic truth is one that which depends on meanings in the strict sense and therefore is necessary and a priori."[5]

Whereas synthetical truth is experiential, its truth is based on experience rather than on the form or structure of itself. According to Kripke's viewpoint, some synthetical truths are necessary. For example, the two proper names of "Hesperus is Phosphorus" are rigid designators, which designate the same individual in all possible worlds. Kripke regarded such propositions as posteriori necessary propositions. Some synthetical truths are contingent, e.g., "Aristotle is a philosopher".

Bibliography

[1] Ayer, A. J. *Language, Truth and Logic*. Shanghai Translation Publishing House, Chinese edn., 1981.

[2] Chen, B. *Introduction to Philosophy of Logic*. Chinese People University Publishing House, Chinese edn., 2000.

[3] Haack, S. *Philosophy lo Logics*. Cambridge University Press, 1978.

[4] Kant, I. *Prolegomena to Any Future Metaphysics*. The Commercial press, Chinese edn., 1981.

[5] Kripke, S. A. *Naming and Necessity*. Basil Blackwell Publisher, 1980.

[6] Quine, W. V. *Two dogmas of empiricism, From a Logical Point of View*. Shanghai Translation Publishing House, Chinese edn., 1987.

[7] Tang, Y. On logical truth and factual truth. *Journal of Social Science*, (1), 2004.

[8] Wang, J. Truth, logical truth and factual truth. *Logic*, (4), 2000. China Renda Sciences Information Center.

[9] Wittgenstein. *Tractatus Logico-philosophicus*. The Commercial press, Chinese edn., 1962.

[10] Zhang, L. Logical truth and its strict relativity to logical system. *Logic*, (3), 2001.

On the Philosophical Foundations of the Evolution of Logical Methods[1]

Ning Lina

Heilongjiang University

ninglinanln@yahoo.com.cn

> As effective inference and argumentation, logical methods are closely connected with philosophy. Philosophy not only promotes logic but also helps logical method to evolve at every stage. New logical methods also change the directions of the development of philosophy. In this article the writer tries to reveal the philosophical foundations of logic from the internal needs of philosophical research on logic methods and tries to understand the relationship between logic and philosophy in several different dimensions. The paper explores the value of logic methods in the development of philosophy and makes a prediction about the trend in logical methods for the future.

1 Aristotle's logical methods emerged from the womb of philosophy

Logic forms are based on philosophical foundations. The existent value and meaning of logical methods in different human history period, different cultural background often depend on different philosophical ideas. Born over the perception of nature and reflection of inference, Aristotle's logical system has been conceived in ontological philosophy. In Ancient Greek philosophy it carried the mission of philosophy to seek the truth, and distinguish falsehood.[1]

1.1 To embody the spirit of seeking truth

Aristotle was the first to systematically study logical problems in western philosophical history. He studied the inevitable outcome of inference, which just met the requirement of philosophical argument. Aristotle set up logic methods intentionally in the process of analysis and assessment of philosophical categoryies He found evidence to establish his logic when forming his idea of ontology. His philosophical ideas displayed the characteristics of his logical analysis while his logical theory permeated philosophical connotation. It was just because his idea of ontology provided the basic categories for a logical system which settled the philosophical foundations that his logical theory was also called ontological logic.

Influenced by philosophy, Aristotle regarded the concept as the most important part and explored it before developing rules for judgment and inference. To

---

[1]The author is with The College of Philosophy and Public Administration, Heilongjiang University, Harbin 150080 China. Email: ninglinanln@yahoo.com.cn

him, understanding usually came from the concept, while inference was based on the concept. He found, in the process of establishing an ontological philosophy system, effective ways were needed to solve such problems as avoiding self-contradiction when presenting an argument, uniting ideas to the same concept and the same judgment, enriching and empowering an argument. He also found a set of scientific ways to guide inference and argumentation.

1.2 The ideal of seeking truth

Aristotle clearly understood that seeking and holding truth was the task of logic. There's no difference between the truth on discussion and the truth discussed by philosophers. He thought the important principle to distinguish truth and falsehood of the subject is to see whether the ideas and reality are conformed. Logic must reflect objective reality. The truth of subject lies in the consistence of objective object and certain attribute.

Understanding true reasonableness involves predictability. Such philosophical ideas set demands on logical methods. Free argument in ancient Greece helped develop philosophy. Objects of philosophy have extended from the natural world to human knowledge. Problems concerning the tools of understanding truth have been particularly attached importance and have greatly influenced the use of Aristotle's logic, they also enlightened his view on logical study, and thus ensuring its status as a tool to understand the truth.

To Aristotle, the task of logic was to use effective inference to obtain truth. Such truth reflected correct thinking rules that human beings should follow, and the essential characteristic of holding the truth. In reality, whether this effect is direct or indirect and whether there are any other subjects searching truth, to study actual truth in logic had a methodological importance.

1.3 To present wisdom and distinguish it from falsehood

Philosophy originated from human beings' questioning the form and meaning of the existence of the world and human beings. In the discussion of these problems, how to obtain the truth, how to distinguish falseness, how to argue against sophism, how to correctly express oneself, how to set the necessary connection between ideas, and how to make arguments which were non-contradictory were the concerns of philosophers at the time. Aristotle displayed his ancient Greek spirit and love of wisdom in seeking knowledge. He explored effective forms of inference and argumentation and helped people to know the necessity of inference in the process of perception. Logical methods become a powerful tool for philosophical argument and appear to meet the requirement of philosophical study.

Aristotle initiated the law of contradiction and the law of the excluded middle to avoid falsehood in recognition of inference. He believed that the law of contradiction is not only ontology and the law of logic but the law of inference which possesses unarguable justice property. He set forth the requirement to the law of contradiction that is, opposite depictions of the same thing at the same time can not be true at one time. If one goes against this law he will be self-contradictory and mix up truth and falsehood. In his understanding, when you have to approve or disapprove something you cannot consider either side false. There must be one truth. There is no middle part in front of truth and falsehood. We cannot take an ambiguous attitude to it. Aristotle's logic, logical methods, effective logical tools, became proven methods for distinguishing truth from falsehood.

2 Symbolic logic methods from philosophy reflected in recent times

In recent times, Bacon directed his criticism at Aristotle's logic. He maintained that recognition started from one particular thing, went through a middle stage and finally reached generally acknowledged truth the reliability of which must be proved by experience. Based on this he advanced his own inductive theory of logic. Although Bacon's criticism was prejudiced against Aristotle's logic, there is no doubt that his epistemology is the philosophical foundation of logical inference, logical law and rule. He closely linked induction to observation, analysis, experiment and sought relations between cause and effect to provide logical methods necessary to the support and confirmation of scientific perception and discovery.[2]

Descartes insisted on replacing belief with knowledge. He worshipped authorities with logical proof. He brought forward his general doubt theory and advocated the search for truth by correctly using the rational. He opposed scholasticism because it only explained the known and only made tedious arguments with no penetrating judgments. Even though Descartes did not establish his own logical system, he made the necessary preparations for the solution to the philosophical problem when he demonstrated his skepticismdirect perception principle and perfect deductive method. His philosophical ideas such as the repulsion of the suspicious subject and the search for definite truth had great impact on the transition of logical methods in this period.

With the fantastic advancement of science, philosophy was correspondingly required to give an accurate and precise summary and depiction of the progress. It set a new requirement for logical methods. William Leibnitz first introduced mathematical methods into Aristotle's logical thought and replaced the natural language of logic with an artificial language which became the basis for symbolic logic. His system provided a new tool for the clarity in the definition of philosophical problems. Leibnitz highly regarded Aristotle's argument method and tried to turn inference language into a kind of symbolic art in order to express oneself clearly and briefly. When he stated his ontological idea, he emphasized that the contradiction principle is the essential principle of logic. He believed if one idea is true, there should be no conflicts in it. An important reason for him to replace natural language with artificial language was to avoid the ambiguity and contradiction in natural language. Although he failed to complete his symbolic logic system, his idea became the precursor for modern symbolic logic.

Since Leibnitz transformed Aristotle's logic in a symbolic way, there have been two tendencies in research into innovations in logical methods. First, Kant advanced a priori logic to study innate and a priori formation. Hegel demonstrated his dialectical logic and put the development of thinking and every stage in human recognition together to study, linked thinking form with each other and made them mutually convertible. Hegel's logic mainly expressed his philosophic ideas. Mill's empirical induction logic brought forth systematically induction theory. Although these logic theories of philosophical methods did not help directly

Leibnitz's mathematical logic concept, the need for change and novelty in modern philosophy indicated a new breakthrough in the development of modern logic methods. Secondly, some logicians tried to set up a kind of expressive form of accurate language for logic to make up for the flaws in traditional logic. Among them, famous English logician Augustus De Morgan transplanted the methods from algebra into traditional logic and designed a symbolic logic which facilitates accurate calculation. Another famous English logician George Boole first realized Leibnitz's assumption about algebraic calculation in logic which could be used to solve tough problems in traditional logic. The German logician Frege proposed a logic calculus system and tried to prove that logic and mathematics are the same. One can deduce the whole of mathematics with logic. To guarantee accurate inference, he created a "concept language" as a symbolic language. Modern logician Russell followed and developed the logic idea and thought in which there's no strict line between logic and mathematics but they are closely connected. He also proposed a strict and complete subject calculating system and discussed the theory of description fully and systematically. [6]

3 Modern blended logic methods transformed from philosophical reform

At the beginning of the 20th Century, language diversion began to take place in western philosophies. This was not only the unavoidable result when philosophy was seeking reform but it was also an inevitable choice when it was trying to escape the crisis in which it found itself trapped at the end of the 19th Century. Philosophy needed updating and a breakthrough in its system and content if it was going to continue to develop itself. The priority was to upgrade philosophical methods and discard the traditional metaphysical methods, replacing them with new methods that could be analyzed and be more accurate. Leibniz started to renovate traditional logic using mathematical methods while Frege and Russell introduced logical methods into philosophy. Modern logical methods were turned to the study of such philosophical issues as meaning, truth, existence, possible world and certainty. The philosophical reform itself brought about the diversion of logic methods. Modern logic provides a system and measures of language analysis for the development of philosophies, thus breaking through the contemplative pattern of the traditional philosophical ideas and methods with speculative nature and greatly pushing forward research on and solution to philosophical problems. A new revolution in philosophies became possible which was based on language analysis. Traditional philosophy used abstract speculative measures to study issues concerning ontology. However, after logical analysis came into use, people realized that issues concerning ontology could be discussed in the perspective of language analysis and that some abstract subjects could be obtained through language analysis.[3]

Whenever philosophies experience reforms, there will be a transition in logic methods as well. Introspection on ontology and epistemology brought about branches of logic such as language logic, modal logic, temporal logic and epistemic logic. Just as analytical philosophers and positivism have had a great effect on modern logics, the boom of non-formalized logic is heavily influenced by philosophy of daily language. At present, the study of the philosophy of mind and of artificial intelligence in Britain and the US is directing the use of the corresponding logical methods into new arenas.

Ontology and epistemology took a dominant role in western philosophies for quite a long period. Since the 20th Century, philosophical thought has been divided and diversified. Nowadays, however, they are on the way to blending again. Philosophy needs logic, which in turn satisfies the demands of philosophy. It is believed that the combination of logical methods will surely bring fresh ideas and methods into philosophy.

In reviewing the origin and development of logical methods, we can conclude that the methods of effective inference have proven philosophical issues. The method of language analysis has clarified philosophical issues and the modern non-formalized method has formed a close tie between philosophy and everyday life. Just as Professor Jaakko Hintikka has commented, almost all logical branches have direct or indirect philosophical connotations. Each time when logic method is updated, it helps to solve major philosophical problems and it is significant for the world, human beings and the relations between them. If philosophy originated out of a love for wisdom and is aimed at studying ways of living wisely, logic must be an important method for studying philosophy and interpreting wisdom. Philosophical reform calls for renovation in logical methods. It may be true, in different periods of history, that philosophy constitutes the foundation of logic, directly or indirectly. Nevertheless, in the development of logic methods, philosophical expectation is always in sight from which we can see its value in satisfying the demands of philosophy.

Bibliography

[1] Aristoteles. *Aristoteles Collected Edition*, vol. 1. Renmin University of China, 1990.

[2] Bacon, F. *New Tools*. Commercial Press, 1982. Trans. by Kui, Hsu Po.

[3] Frege. *Frege Philosophy of the Selected*. Commercial Press, 2006. Trans. by Wang, L.

[4] Haack, S. *Philosophy of Logics*. Commercial Press, 2003. Trans. by Yi, Luo.

[5] Heidegger, M. *Being and Time*. Joint Publishing Co, 1987. Trans. by Chen, J.

[6] Leibniz, G. W. *On Human Reason*. Commercial Press, 1982. Trans. by Chen, X.

[7] Wang, L. *Logic and Philosophy*. People's Publishing House, 2007.

[8] Wittgenstein, L. *Philosophy of Logic*. Commercial Press Ltd, 1962. Trans. by Ying Guo.

Kripke's Research on the Theory of Proper Names – About Essentialism[1]

Liu Shanshan

Yanshan University

szmcz@yahoo.com.cn

Song Hang

Yanshan University

ABSTRACT. Kripke proposed a new theory of proper names on the basis of his theory of essentialism. By distinguishing Kripke's contemporary essentialism from Aristotelian classical essentialism and refuting Quine's anti-essentialism, he clarified the progress of his essentialism of individuals and further illustrated a rational basis for the causal-historical theory of proper names. If we correctly understand Kripke's theory of proper names, we must correctly grasp first the theory of essence. In our account, the main reason why the appearance of Kripke's theory of proper names has triggered a long-lasting debate in the philosophy of logic lies in the fact that scholars hold different understandings of what Kripke means by "essence".

1 Kripke's Theory of Proper Names and the Essence of Individuals

Aristotle originated Essentialism. Aristotelian essentialism is well represented in his theory of four predicates, namely property, definition, genus and accident. Aristotle defines "definition" as "a phrase signifying a thing's essence."[2] Thus it is not difficult to see that the summary of the theory of Aristotelian terms is that the property of objects has essential attribute and accidental property. What essential attribute is that some objects have, and the others do not have, in addition, objects which have some attributes necessarily have these attributes, however, accidental property is that the property of certain objects can be shared with other things' which can change, or even have dispensable property.

Aristotelian essentialism was called classical essentialism, which was to be compared with contemporary essentialism. With the development of modern philosophy, essentialism did not disappear, but has been a new development. Kripke was the first advocate for essentialism in modern philosophy. He said, the so-called essential attribute was bound to have the attribute in the relating possible worlds, and remained unchanged through changing of the world, making things was still the thing, namely to maintain their own identity. For the relationship between the essential attributes and the non-essential attributes, He thought that, as long as things did not lose their essential nature, they could maintain their own identity; however, non-essential attributes were dispensable. If things

---

[1]The author is with College of Humanities and Law, Yanshan University, Qinhuangdao 066004 China. Email: szmcz@yahoo.com.cn

lost their essential attribute, and kept non-essential attributes unchanged, the things were no longer their own.

Kripke's contemporary essentialism and Aristotelian classical essentialism have similarities and differences. The differences can be attributed to that Aristotle cares of the real world, and his essentialism is the essentialism of the real world; however, and Kripke's essentialism is the essentialism of the possible worlds based on semantics of possible worlds as a tool. It can be said that the mainly difference is the difference of their theory tools. The same is that essential attribute is both defined necessary property. Ultimately, Kripke analyzed the essence of objects by using semantics of possible worlds in modal logic, and as a result, made Aristotelian classical essentialism develop a new stage.

Kripke presented his theory of proper names based on his essentialism and semantics of possible worlds. Kripke views that the proper names are the rigid designators, which only have reference but no sense. The support of proper names is determined by causal-historical chains. The references of proper names are initially identified through the following ways: either through initial naming baptism or through the use of definite descriptions to illustrate the references of proper names which are to satisfy their objects. Once the references of proper names are identified, they will pass on from generation to generation in a causal-historical chains, through series of social activities, so Kripke is the progenitor of the theory of causal-historical references.

"Rigid designator" is the core of concept in Kripke's theory of names. Let's call something a rigid designator if in every possible world it designates the same object, a non-rigid or accident designator if that is not the case.[1, p. 48] The concept of rigid designator closely contacts with Kripke's sense of essence. About the essence of individuals, Kripke offered two principles: first, the origin of an object is essential to it; second, the substance of which it is made is essential to it. He thinks, first of all, the essence of individuals can only resort to their origins which decide the individuals' self-identity; secondly, the materials composing individuals are the essence of individuals. So the proper names of individuals are just names.

How can we correctly understand the essence of individuals, so as to correctly refer individuals? Kripke said: "It was with the help of the other speakers in our social life that we referred a particular person by tracing the object itself." One of the important contributions in Kripke's causal-historical theory of proper names was that he combined social factors with theoretical system, which revealed its decisive role in the decision of reference.

2 A Discussion of Quine's Anti-Essentialism

Quine is a strong opponent of essentialism. In order to refute it, he had once constructed a paradox about a cyclist who loves math. To read the whole quote from Quine:

> "Perhaps I can evoke the appropriate sense of bewilderment as follows. Mathematicians may conceivably be said to be necessarily rational and not necessarily two-legged. And cyclists necessarily two-legged and not necessarily rational But what of an individual who counts

among his eccentricities both mathematics and cycling? Is this concrete individual necessary rational and contingently two-legged or vice vesa? Just insofar as we are talking referentially of the inst cyclists, or vice versa, there is no semblance of sense in rating some of his attributes as necessary and others as contingent. Some of his attributes count as important and others as unimportant, yes, some as enduring and some as fleeting; but none as necessary or contingent."[4]

Quine thought essentialism undesirable, because an attribute of the object which can be either necessary, or contingent, obviously constituted a "paradox".

Kripke animadverted Quine on the criticisms of quantified modal logic. He thought Quine was not aware of the distinction between rigid designator and non - rigid designator. We say, it is inevitable that nine are more than seven, but not necessarily the number of planets in solar system is more than seven. Why? The number of the solar system's planets could be different from what it is in fact, but nine could not be different from itself.[3] In other words, the number of solar system's planets is not always nine in any possible world; however, nine are always nine in any possible world. In the real world, Pluto no longer belongs to the solar system, so nine doesn't equate the number of planets in the solar system. We could imagine if the universe in future happens to change, it hard to say that the planets in the solar system will not be reduced to six. In terms of the possible worlds to see this problem, it is easier to understand.

In my view, Quine's ill-considered problem is because he changes a discussion of groups into a discussion of individuals, and changes a discussion of class attributes into a discussion of individual attributes, which in fact are not the equivalent to class attributes. The conclusion is "mathematicians", as a group, must be necessarily rational and not necessarily two-legged, so "cyclists" should conceivably be said to be necessarily two-legged and not necessarily rational in the group. If W, as an individual, is a mathematician, he will be necessarily rational and not necessarily two-legged; that is to say, he could be two-legged or no two-legged. Similarly, if W is a cyclist, he will be two-legged, not necessarily rational, which means that he can not also be rational. If W is both a mathematician and a cyclist, obviously, he is bound to be necessarily rational and to be necessarily two-legged; therefore Quine says this is a paradox. But do not forget to discuss the objects from class to individuals. As both a mathematician and a cyclist, the individual W, who is necessarily rational and necessarily two-legged, doesn't conflict with mathematicians who must be necessarily rational and not necessarily two-leggedand cyclists who must be necessarily two-legged and not necessarily rational.

In addition, "mathematicians" and "cyclists" are regarded as common names; however, an individual, who counts among his eccentricities both mathematics and cycling, belongs to description. In *Naming and Necessity*, Kripke does not give sufficient proofs to prove common names, which like proper names, are the rigid designators, so it exits two situations: first, although Kripke offers full proofs, common names are the rigid designators; second, common names are not the rigid designators.

In the first case, in Quine's reasoning, the mathematicians and cyclists are the rigid designators, but an individual, who counts among his eccentricities both mathematics and cycling, isn't the rigid designator. In the possible worlds, what's the problem for discussing the necessity of the non-rigid designators? Also, is it feasible to convert the discussion of the attributes into the discussion of the proper names? In the second case, is there a necessary attribute for mathematicians and cyclists? These are questions to be worthy of considering.

If Quine's reasoning is correct, it is not reasonable that we totally ignore essentialism. Based on "Watergate Affair", Kripke had constructed a paradox in Outline of a Theory of Truth. He certified that even a paradox can easily be made in the natural language that we humans have been using for several thousand years; can we assume this natural language is undesirable? Of course, you can say that it is undesirable, but how can we communicate with each other without natural language? Is it possible for Quine to conduct his reasoning without using natural language? A system has flaw, and then it negates completely. It is not a good way to solve the problem.

Evidently, the essence of individuals held by Kripke is closely relative to Kripke's theory of proper names. Depending on his theory of proper names, he refers to the essence of individuals; therefore, the essence of individuals serves for Kripke's theory, and proves his theory to be reasonable in the construction.

Bibliography

[1] Kripke, S. A. *Naming and Necessity*. Basil Blackwell Publisher, 1980.//
[2] LiTian, M. *the Complete Works of Aristotle (Vol. I)*. China Renmin University Press, 1990.//
[3] Liu, T. *Study on Kripke's Theory of Names*. The People Daily Press, 2006.//
[4] Quine, W. V. *Word and Object*. The MIT Press, 1960.

A Theory of Causative Constructions in Chinese Based on Type-Logical Grammar[1]

Zou Chongli   Xia Nianxi
CASS   Capital Normal University
chongli@263.net xia54321@sina.com

1 Causative Constructions in Chinese

Chinese, the language spoken by the largest population in the world, has been developed by the Chinese people over the course of several thousand years. It contains many special constructions such as the collocation of verbs and prepositional phrases, verb-copying constructions, "ba" constructions, constructions with verbs and their complements, and pivotal constructions with commands. From a certain point of view, the common characteristic of these constructions is that they all represent causative meanings and relationships between two events. We give the following examples:

(1) ta fang yiben shu zai zhuozi shang   ta fang shu fang zai zhuozi shang
She put a book on table top   She put book put on table top
She put a book on the table.   She put a book on the table.
(3) ta ba shu fang zai zhuozi shang   (4) Zhangsan he jiu he zuile
He take book put on table top   Zhangsan drank wine drank drunk
He put a book on the table.   Zhangsan drank and got drunk.
(5) Zhangsan jiao Lisi lai
Zhangsan call Lisi come
Zhangsan asked Lisi to come.

What is the causative meaning? In short, it means that the two events described by the sentence stand in such a relation that one event causes the other. In (5) the event "Zhangsan called Lisi" caused the event "Lisi came". There are a lot of literatures on causative constructions such as Dowty's analysis of a causative construction [2, pp. 217-218].

2 Inadequacy of Current Type-Logical Grammar

Type-logical grammar declares that cognition is equal to computation, grammar is equal to logic and analysis is equal to deduction. Type-logical grammar establishes syntax and semantics of natural language from the perspective of deduction and computation. In general, the operations in type-logical grammar are just for those current components of a sentence without reducing or adding any of them.

---

[1] The authors are with Institute of Philosophy, CASS and Department of Philosophy, Capital Normal University, respectively.

And the higher-order logic, which is regarded as the semantic representation of natural language, doesnt have appropriate means to describe causative meaning. Consider (5), its usual derivation according to type-logical grammar is the following :

The conclusion "jiao (Zhangsan, lai (Lisi))" means there is a relationship "jiao" between the individual corresponding to the individual constant "Zhangsan" and the truth value of the logical formula "lai (Lisi)". This reading is obviously unsatisfactory, because the relationship "jiao" exists between the individual corresponding to "Zhangsan" and the individual corresponding to "Lisi", and this event causes the second individual to have the property "lai". So the next logical formula, which emphasizes the meaning of causativeness, may be more acceptable as the semantic representation of (5).

jiao (zhangsan, lisi) CAUSE **lai (lisi)**

In addition to the lack of logical means describing the causative meaning, the current type-logical grammar has another weakness, namely the semantic representations of causative constructions are not intuitive enough. According to Carpentor's method [1, p. 224], the sentence with multi-prepositional-phrase such as "John put the key into put the box on the table" can be analyzed into the final result:

(6a)s: **put(i(on(i(table))(into(i(box))(key))))(j)**

The semantic references of the higher-order logic forms "**on(i(table))**" and "**into(i(box))**", embedded in (6a), are both functions from sets to sets, and their semantic types are $<< e, t >, < e, t >>$. Though "**put(i(on(i(table)) (into(i (box))(key))))(j)**", as a whole, is a first-order logical formula, we can find that the first argument "**i(on(i(table))(into(i(box))(key)))**" resulted from computation of higher-order logical forms. (6a) has the following intuitive reading: John put the key, and the key was selected from those keys which result from the restriction on "the key put in the box" by "on the table". The logical formula in (6a) seems complicated and non-intuitive for human cognitive ability.

It is difficult for people to have a perception on too complex logical expressions. In general, the expressions in first-order logic would be more natural and intuitive for logical analysis on natural language. The semantic representations of natural language, provided by first-order logic, are relatively easily accessible for humanity's cognitive ability. If the semantic type of prepositions gets adjusted on the base of representing causative meaning, we can derive the following logical formula from the English sentence with multi-prepositional-phrase.

(6b)s :

put(i(key))(j)CAUSE [into(i(box))(i(key)) $\wedge$ on(i(table))(i(box))]

(6b) is very easy to be cognized. It means that the event "John placed the key" caused another event "the key is in the box" and "the box is on the table". The more difficult problem is that the current type-logical grammar has nothing to do with analyzing causative constructions with verb-coyping and causative constructions with a verb and its complement. Take (4) for an example. Here the verb "he" appeared twice, if the predicate corresponding to the verb also appears twice in the logical expression, and the first argument of the first appearance is the individual constant involving subject "Zhangsan", then what is the argument of the second appearance? If the second appearance of the verb "he" is not removed in (4)'s type-logical semantic analysis, the arguments of the predicate corresponding to that appearance would be out of the question. Furthermore, where is the argument of the predicate corresponding to the adjective "zuile"? Obviously the type-logical analysis on the second half of (4) will have no solution. The relationship between the events "Zhangsan drank" and "Zhangsan was drunk" would not be established without the matchmaking of "causative meaning".

3 The Improvement of Type-Logical Grammar

In order to reveal the logical semantics of causative constructions in Chinese, we should improve the current type-logical grammar. Firstly, following Dowty we add a logical constant "CAUSE" to high-order logic as the semantic representation of causative meaning in natural language [2, pp. 353], it would enhance logical tools' efficacy. Secondly, some components of causative constructions in Chinese need movements and changes such as adding or reducing some of them, the aim is to derive type-logical semantic value representing causative meanings. In other words, the following specific structure rules are added to the system on type-logical grammar [3, pp. 120–121] :

Besides, we should add a specific inference rule about the introduction of the constant CAUSE.

If we can derive their causative meanings respectively from certain permutations which are needed in analyzing (1) - (5) from the perspective of type-logical grammar, then we can separately deduce such causative meanings from their original sequences. Namely causative constructions can lead to their own causative meaning as their alterations can do by means of SR1-SR5.

Moreover, we also should adjust the type-logical semantic values assigned to common nouns and prepositions in Chinese. The Chinese preposition "zai" involving space originated from verb; therefore it is reasonable to assign the same category to "zai" as to a transitive verb. On the other hand, a bare common noun sometime refers to an entity specially. We use "shu" to represent "naben shu". Hence we define the following entries in the Lexicon of type-logical grammar.

According to current inference rules [3, pp. 120–121] in type-logical grammar with our supplements, we can get more natural and intuitive analysis on causative constructions in Chinese. For example,

Above inference shows that we can derive that the causative construction "ta fang shu zai zhuo shang" has the same causative meaning as its permutation "ta fang shu daozhi shu zai zhuo", which is needed in analyzing (1a) from the perspective of type-logical grammar.

Our treatment on causative constructions in Chinese belongs to formal semantics for natural language. On the other hand, we need to make an abstraction from these results, that is, just to concentrate on categorical deduction process, and replace natural deduction rules of type-logical grammar with their axiomatic representations, The related structure rules would become structure postulates of multi-modal system in categorical-type logic. It would be shown after the usual configuration of NL. Those structure postulates are the following:

A Theory of Causative Constructions in Chinese ...   79

$$A \cdot (B \cdot (C \cdot (D \cdot (E \cdot F)))) \rightarrow (A \cdot (B \cdot C)) \cdot_c (C \cdot (D \cdot E))$$
$$(A \cdot (B \cdot C)) \cdot (B \cdot (D \cdot (E \cdot F))) \rightarrow (A \cdot (B \cdot C)) \cdot_c (C \cdot (D \cdot E))$$
$$A \cdot ((G \cdot C) \cdot (B \cdot (D \cdot (E \cdot F)))) \rightarrow (A \cdot (B \cdot C)) \cdot_c (C \cdot (D \cdot E))$$
$$(A \cdot (B \cdot C)) \cdot (B \cdot D) \rightarrow (A \cdot (B \cdot C)) \cdot_c (A \cdot D)$$
$$A \cdot (B \cdot (C \cdot D)) \rightarrow (A \cdot (B \cdot C)) \cdot_c (C \cdot D)$$

Adding these structure postulates to the multimodal system NL [3, p. 102] of categorical type logic, we would establish a system which can represent causative constructions in Chinese. If we construct a frame class based on certain corresponding constraints for the accessibility relationships R and R_c, we will be able to prove that the system is sound and complete with regard to such frame class. The following are these constraints.

$\forall xyzuvwm(\exists pqst(Rpvw \& Rqup \& Rszq \& Rtys \& Rmxt)$
$\Rightarrow \exists abcd(Rauv \& Rbza \& Rcyz \& Rdxc \& R_c mdb))$
$\forall xyzuvwm(\exists pqrst(Rpvw \& Rqup \& Rryq \& Rsyz \& Rtxs \& Rmtr)$
$\Rightarrow \exists abcd(Rauv \& Rbza \& Rcyz \& Rdxc \& R_c mdb))$
$\forall xyzuvwmn(\exists pqrst(Rpvw \& Rqup \& Rryq \& Rsmz \& Rtxs \& Rntr)$
$\Rightarrow \exists abcd(Rauv \& Rbza \& Rcyz \& Rdxc \& R_c mdb))$
$\forall xyzuv(\exists nst(Rnyu \& Rsyz \& Rtxs \& Rvtn) \Rightarrow \exists pqr(Rpxu \& Rqyz \& Rrxq \& R_c vrp))$
$\forall xyzuv(\exists st(Rsuv \& Rtzs \& Rxyt) \Rightarrow \exists pqr(Rpuv \& Rqzu \& Rryq \& R_c xqp))$

The proof process is as follows: define the canonical model based on the ternary frame $< W, R, R_c >$, which is the counter-model of non-theorems of the above expanded multi-modal system. It is easy to prove the related truth lemma of this canonical model, as well as that its frame satisfies those constraints for the accessibility relationships R and R_c [4].

4 Summary

The scheme proposed in this paper has the following significance: (a) Introducing Dowty's CAUSE, representing causative constructions in natural language, to type-logical grammar. (b) Giving an intuitive and simple treatment to five kinds of causative constructions in Chinese. Such as constructing a semantic analysis of verb-copying constructions, and describing the fact that the Chinese preposition "ba" has a causative meaning. (c) Adding structure postulates, which are used to represent causative constructions in Chinese, to NL and extending a multimodal system of categorical type logic with its completeness.

Bibliography

[1] Carpentor, B. *Type-logical grammar*. Cambridge/London: MIT Press, 1997.
[2] Dowty, D. *Word Meaning and Montague Grammar*. Reidel: Dordrecht, 1979.
[3] Moortgat, M. Categorial type logics. In van Benthem (Ed.), **??**, 93–178. 1997.
[4] Zou, C. Study on multi-modal categorical logic. *Philosophy Research*, 9, 2006.

B

General Philosophy of Science

Researches on Deduction in Chinese Logic[1]

Zhang Zhongyi
Yanshan University
zh-y-zhang@126.com

Logic in the narrow sense examines the deductive relationship between propositions, with deduction as the major focus of research. From Aristotle to modern times, the study of logic has always been permeated with the concept of deduction. We believe that logic is science providing apodeictic inference. Whether or not China has such logic has been the focus of academic debate. The Physics Nobel Prize winner Yang Zhenning, citing the I Ching: Book of Changes, said, "Chinese traditional culture has a major characteristic of induction, but no inference". He drew the conclusion, "Chinese culture did not develop inference. How can we prove this 'lacking'? Look at the example of Guangqi Xu. Here 'not having' is a universal negative, and is the negative of 'having'; however, 'lacking' is 'having', which means, the amount is not enough, so both are self-contradictory." We recognize that Mr. Yang's inferential "lacking" is a fact, but it is only an issue of how much rather than whether having or not, so we may draw the substantive conclusion: Chinese logic has deductive inference, and to some extent deductive inference has been explored, which means we have studied deduction.

1 Researches on "Bi" and "Gu" in Chinese logic

"Bi", which appears 70 times in the Mo Jing, is a major term in Mohist logic, equal to a necessary proposition. The Jing Shuo Shang claims that "Bi" is an undoubted apodeictic propositions used to emphasize necessity between propositions. "Bi" in Da Qu should be translated as "must", and here it emphasizes the importance of these three things, expressing conclusions on the premise of the requirements of necessity, that is, all prerequisites are owned, "Ci" (Conclusion), is "enough", that is, inevitable to draw conclusions from the premise, which stresses that the logical necessity of the problem is said here. As Wittgenstein said, "As another one was incident, an incident was bound to happen, which is not mandatory, only a logical necessity".[2]

Later Mohist logic explicitly defined "Gu": the reasons for the things to come into existence. Later Mohists divide "Gu" into "Xiao Gu" and "Da Gu". "Xiao Gu" reflects the logical property of the necessary conditions, that is to say, the drawing of the conclusion is the prerequisite for the necessary conditions. Conversely, the conclusion will not be inevitably drawn without the prerequisite, which reflects their relation of necessity. For "Da Gu", Mohists give a stricter requirement–on one hand, they stress that "it is not necessary without it", on

---

[1]The author is with College of Humanities and Law, Yanshan University, Qinhuangdao 066004 China. Email: zh-y-zhang@126.com

the other hand, they stress that "it is necessary if it exists", which shows that the deduction of a conclusion can be inspected in two ways: First, it would not necessarily draw the conclusion in the negative premise; secondly, "Da Gu" can draw the conclusion, provided a precondition exists. Here, having "Da Gu" certainly gives the desired results. In fact, "Da Gu" is a necessary and sufficient hypothetical proposition. Jin Yuelin said, "symmetric including", "inevitable" is "non-symmetric including". This is the problem of inferential inevitability.

2   Research on Logical Law in Chinese Logic

Apodeictic inference must follow logical laws. As Shen Youding said, "The logical laws and the logical forms of human thinking have no nationality and class character."[1] Cui Qingtian also said that "all humans must follow these laws, even though their geography, ethnicity, and cultures are different. Logical theory or thinking in different cultures reflects and encapsulates these laws." Deductive inference must also follow these "inferential rules". The Mohists also studied logical laws.

The *Jing Shang* says, "any one contradictory proposition is neither true at the same time, nor false at the same time." When Shen Youding delivered some lectures to us in 1985, he said that the middle of previous "no" dropped the word "both". "You" expressed one pair of contradictory propositions, "A is not non-A" expressed law of contradictions, and "A or non-A" expressed the law of the excluded middle. I think Mr. Shen's addition of the word "both" is thoughtful and follows the principles set by the collation. The *Mo Jing* did propose thoughts on the law of contradiction. We try to analyze the possibilities of "both". According to Liu Peiyu, Mr. Shen mentioned two reasons:

1. in such a tight Mohist system, these two "may" had different meanings, and the addition of "both" would enable the two senses of the word "may" to have the same meaning.
2. In a debate, one cannot say "both can not be" first, but should say "both can be" first, and then argue what is not.

We believe that the Mohist logic not only skillfully uses the law of contradiction, but also reveals the thoughts of the law of contradiction. Shuo: "(Debate) There is an object, someone says, 'It is a cow'. But another man says, 'It isn't a cow'. This is a controversy over a couple of contradictory propositions. A couple of contradictory propositions cannot be true at the same time, and one of the couple must be false." Since "debate" is "debate between each other" and "Bi" is a contradictory proposition, so this contradictory proposition cannot be true at the same time, that is, not consistent with reality, there must be a false one. This latter sentence essentially also reveals the basic content of the law of contradiction, where "not true at the same time" is the best explanation of "A or non-A".

As previously stated, in the *Mo Jing* "both can be" concludes two contradictory propositions both of which are true. This is a contradiction, that is to conclude that "the non-cattle and the cattle", and "not both can be" denies contradictive propositions. Expressed in modern logical symbols, "both can be" is to conclude (P and non-P), "both cannot be" denies that "both can be" is true at the same time, using logical symbols to express, not $(P \wedge notP)$, which is the formula of the law of contradictions in modern logic. The meaning of "both can be" respectively negates the contradictory propositions, namely, $(not(P) \wedge not(notP))$, "not both can not be" namely: $not(not(P) \wedge not(notP))$ is equivalent to $P \vee notP$. This is the law of the excluded middle in modern logic. It shows that the expressions "not both can be" and "not both can not be" are fully consistent with the formulation of the law of contradiction and the law of the excluded middle in modern logic. The two formulas are the negative of the contradictions, we can see that this is a deduction from negative contradiction to derive a tautology. This is also the reference of necessity.

Mohist logic indeed contains three abstracts for logical rules:

1. "(debate) There is an object, someone says, 'It is a cow'. But another man says, 'It isn't a cow'. This is a controversy over a couple of contradictory propositions." In the Jing Shuo Shang cites some specific examples, strictly speaking, the "cow" and "non-cow" in fact also play the role of variables, which are completely changed as we well known into the "horse" and "non-horse". Here "cow" is used to replace a similar thing which is the embryonic form of the law of the excluded middle.
2. "debator said that yes, or said that no"(Jing Shuo Xia). Concrete things are abstracted into "yes" and "no". This is more abstract than the previous one, because this is not a debate about concrete things; it is a tautology about "yes" and "no".
3. "For any one contradictory proposition, it is neither true at the same time nor false at the same time."(Jing Shang) This is fully consistent with the expression of the laws of modern logic: "not both can not be" is the negative of "both can not be", equivalent to $P \vee notP$, which is the expression of the law of the excluded middle (or tautology), and it is an abstract for "or said that yes" and "or said that no".

3 Jin Yuelin's Necessary Condition Hypothetical Inference

Chinese logic has studied Mou, Xiao, Zhi, equal inferences, and so on. Due to limitations of space, we focus only on Jin Yuelin's necessary condition hypothetical inference. In the 1930s, Jin Yuelin published his Logic, in which he argues that "unless ... no" is a logical connective of the necessary condition hypothetical proposition. Maybe it absorbed the Mohist thoughts about "Xiao Gu". Moreover, like the Mohist logic, they regard natural language as variables which in Mohist terms only use "this" and "that". Jin thought that the number of "this" or "that" is too little and used instead A, B, C and D as variables. Jin used these familiar variables to construct necessary conditions for hypothetical inference.

Unless A is B, A is not C;

A is not B,

So A is not C.

Jin also proposed the rules of these hypothetical inferences, namely, to negate the former is to negate the latter, but to affirm the former is not to affirm the latter; to affirm the latter is to affirm the former, but to negate the latter is not to negate the former.

Jin's researches on necessary condition hypothetical inference supplemented and developed the researches of inference in Chinese logic and Western traditional logic. Although the Stoics found five basic patterns which can be used to derive countless arguments–almost the entire propositional logic–these five patterns don't contain necessary condition hypothetical inference. Neither does traditional logic. also Clearly, Jin Yuelin's formulation of "unless ... no" as a logical connective of the necessary condition hypothetical proposition and its inference supplements the Stoic's logic.

4 The Value of Research in Deduction in Chinese Logic

Logical theory with deductive reasoning as an essential characteristic has a rich tradition in ancient China, an important reason why Chinese logic is comparable to Western logic and Indian Hetuvidy and one of the three major sources for modern world logic. As Shen Youding said, "an attempt to prove that China has no logic, or the idea that the Chinese thinking mode follows a specific logic which strays away from the regular track of human science" is a fallacy. The viewpoint that "China has no logic" is no less absurd than the viewpoints of "China has no science", which is a conclusion which many serious scholars have tried to argue and defend.

If the result of our preliminary research is to show the characteristics of Chinese logic, the summary we made about deductive reasoning in Chinese logic implies that we arrived at the same conclusion although we explore different logical theories in different ways. This paper seeks to promote "ontological integration" between Chinese logic and Western logic by exploring this issue. We stress that Chinese logic is neither traditional Ming Bian nor a vassal of Western logic. It converges with the world logical system. So far, we established the same status for Chinese logic and Western logic in the history of logic through research on the legitimacy of Chinese logic. From this we claim that the origins of world logic can be found in Chinese logic, Western logic and Indian Hetuvidy.

Bibliography

[1] Shen, Y. *Moist Logic*. China Social SciencesPress, 1980.
[2] Wittgenstein, L. *Tractatus Logico-Philosophicus*. The Commercial Press, 1999.

Cognitive Context and Scientific Explanation[1]

Yan Kunru

South China University of Technology

kunruyan@163.com

Duan Xiufang

Guangzhou Administration Institute

duanxiufang@tom.com

Cognitive context, which developed from the definition of traditional context, mainly includes propositions, knowledge scripts, psychological schemata and socio-psychological representations. The study of cognitive context, which is different from the traditional study of context, lays great emphasis on the analysis of the communication participants' mental states and cognitive processes. It structures and cognizes the context of traditional context, containing not only the study content of the traditional context, but the language speakers' hypotheses about the world and their cognitive abilities. Thus a psychological construct is formed and a platform is offered for pragmatic reasoning.

Relevance Theory is a cognitive pragmatic theory accounting for language communication. A context is considered as "a subset of the hearers assumptions about the world[3, p. 141]". According to it, context is a cognitive process; its elements are not stated, yet they are a set of changeable propositions; therefore, being dynamic is one of its most prominent characteristics. It is a new way for the study of scientific explanation. There are a lot of models of scientific explanation. We can divide them into two parts. One is semantic models of scientific explanation. The other is pragmatic models of scientific explanation. C.G. Hempel's theory belongs to the first type and is not based on context. Hempel thinks explanatory relevance is logical relevance. Some scholars don't agree with Hempel on this point. Peter Achinstein gives us an example to disprove it:

> Jones ate a pound of arsenic at time t
>
> who eats a pound of arsenic dies within 24 hours
>
> Therefore,
>
> Jones died within 24 hours of t.[1, p. 170]

But this is truly not Jone's day. Shortly before succumbing to the poison, in his hurry to get to a hospital, Jones steps off the curb without looking and is hit by a bus, which kills him. Suppose he did die from being hit by a bus within 24 hours of t, then the explanations above do not correctly explain the explanandum,

---

[1]The authors are with School of Marxism and Philosophy, South China University of Technology, Guangzhou 510640 China and Guangzhou Administration Institute, Guangzhou 510070 China, respectively. Email: kunruyan@163.com, duanxiufang@tom.com Supported by "the Fundamental Research Funds for the Central Universities, SCUT" and National Fund of Social science

even though all the conditions of Hempel's model are satisfied. It is his being hit by the bus that explains Jone's untimely demise, but assuming that all the premises in the original argument are true, it still stands as an explanation of Jone's death. Insofar as conformity to Hempel's model, the explanantion is taken to be a sufficient explanation.

The pragmatic model of van Fraassen and Achinstein differs from the semantic model. Van Fraassen's pragmatic theory of explanation takes explanations to be answers to why-questions. The why-question Q expressed by an interrogative in a given context will be determined by three factors:

The topic P_k

The contrast-class $X = \{P_1, \cdots P_k \cdots\}$

The relevant relation R.[4, p. 142]

Every Why-question contains a topic, with a contrast-class of alternatives, and the relevance relation for which any answer A must stand in relation to the topic to count as a good explanation. Most of van Fraassen's energy is directed to a discussion about the contextual factors surrounding the explanation. He says "the question arises in this context[4, p. 146]". "It might be thought that when we request a scientific explanation, the relevance of possible hypotheses, and also the contrast-class are automatically determined."[4, p. 129]

Every explanation acquires its explanation due to its context relation to the antecedent explanation-request. Explanations are just descriptions of a relevant part of the empirical information contained in theory. Explanatory power is not distinct from descriptive power; it is up to the explainer to provide the pertinent description in response to the context. A why-question arises in a context K provided that K entails that the topic is the only true member of the contrast class and does not entail that there is no answer to the question. We reject the why-question Q in context K if the question does not arise in this context.

In a given context, several questions agreeing in topic but differing in contrast-class, or conversely, conceivably differ further in what counts as explanatorily relevant. Hence we cannot properly ask what is relevant to this topic or what is relevant to this contract-class. Instead, we must say of a given proposition that it is or is not relevant (in the context) to the topic with respect to that contrast-class. For example, in the same context one might be curious about the circumstances that led Adam to eat the apple rather than the pear (Eve offered him an apple) and also about the motives that led him to eat it rather than refuse it.

According to Van Fraassen, not only the topic of why-question, the contrast-class and the answer of why-question are based on context, but also the evaluation of the answer of why-question is relied on context.

Van Fraassen maintained that the two chief problems inherited from the traditional accounts of explanation that he wants to solve are the problem of rejections and the problem of asymmetries. He thought he has solved the two problems which exist in the semantic explanation model, but I don't agree with his.

Salmon thought the fundamental problem van Fraassen's theory encounters with respect to the relevance relation. Van Fraassen's contextual relevance is not the explanatory relevance.

Such as:

Horace is a member of the Greenbury School Board
All members of the Greenbury School Board are bald
_____
Horace is bald[2]

Van Fraassen thought explanation is the answer to a why-question. It is an answer to "why is Horace is bald?" Van Fraassen looked on it as a explanation to "why is Horace is bald?" but according to Hempel's model, it is not a explanation to "why Horace is bald?" because there is no law in the explanans. So we don't think van Fraassen's model is better than Hempel's model.

The context determines which the appropriate contrast class is. The context in which a question is posed involves a body of background knowledge K. If the question Q does not arise in a given context, we should reject it rather than trying to providing a corrective answer to why-questions.

There is a profound difficulty with van Fraassen's theory centering on the relevance relation R. The theory of the pragmatics of explanation given by van Fraassen in The Scientific Image is highly illuminating the best, but he has not succeeded in showing that all the traditional problems of explanation can be solved by appealing to pragmatics.

Another important representative of pragmatic explanation is Peter Achinstein's theory, whose view is articulated in great detail in The Nature of Explanation (1983) Explaining is what Austin calls an illocutionary act, like warning and promising.

Achinstein denies that all explanations are answers to why-question. Achinstein's illocutionary theory of explanation is a new pragmatic explanation model. Achinstein said "If S repeats the sentence above on different occasions when he explains this phenomenon, he has engaged in several explaining acts (he has explained several times), even though the product – his explanation – is the same on each occasion"[1, p. 3]

In general, there are three aspects in an Achinstein's explanation, the explaining act, the product of an explaining act and the evaluation of an explanation. He distinguished the explaining act from the product of an explaining act. Achinstein pays more attention to the explaining act. The explaining act is a necessary condition of the product and the evaluation of an explaining act. He also distinguished explaining act out from general illocutionary act. With the brand-new eyes, he has advanced the research of the scientific explanation.

Achinstein points out that "explanation" may refer to either a process or a product. A physician might explain John's malaise by saying "He drank too much last night." John's wife might use the same words to criticize his behavior, and then his wife's speech act is not an explanation. What she produced is not an explanation.

Achinstein pays more attention to the explaining act, but there are different explanations in the light of different people. Different people have different explanations in accordance with the same phenomenon. Every explaining subject has different background knowledge. It is not a unified standard of evaluation. It means pragmatic explanation has no explicit style.

Cognitive context is not given but chosen and the psychological deductive process of the choice of context is linked with human cognition. Context is a cognitive process; its elements are not stated, yet they are a set of changeable propositions, therefore, being dynamic is its most prominent characteristics. It is a new way for the study of pragmatic inference and scientific explanation.

I want to expand on the relationship of cognitive context, knowledge situation and contextual relevance. Scientific explanation must always be evaluated relative to cognitive context and a knowledge situation. The central criterion on an explanation is the explanans to increases the belief value of the explanandum, where the belief value of a sentence is determined from the given knowledge situation. My intention is to push ahead the research of contextual relevance through theory of cognitive context and knowledge situation.

An explanation is an act to answer one question. We must analysis explanatory act about the answer. It is very important of speaker's cognitive condition in an explaining process. We must pay more attention to the speaker's cognitive condition. Furthermore, the appropriate explanation must eliminate hearers knowledge status.

Cognitive context includes traditional context and the hypothesis about the world in the users' mind. Scientific relevance relies on the explanatory background knowledge and interests. It is very difficult for scientists to avoid affecting by cognitive context. The scientists accept certain scientific explanations which are related to the style what they look at the world. Different scientists have different explanations which they are right. They depend on their cognitive knowledge.

In all, not only the questions of explanation and the answers to the questions of explanation, but also the evaluation of the answers to the explanation bear a close relevance to the cognitive context.

Bibliography

[1] Achinstein, P. *The Nature of Explanation.* New York: Oxford University press, 1983.

[2] Kicher, P.and Salmon, W. *Scientific Explanatio.* Minnesota: Unversity of Minnesota Pres, 1989.

[3] Sperber, D., Wilson, D. *Relevance: Communication and Cognition.* Blackwell Publishers Ltd, 1995.

[4] Van Fraassen, B. C. *The Scientific Image.* Clarendon Press, 1980.

Presupposition, Evidence and Model-based Reasoning in Science[1]

Wei Yidong

Shanxi University

weiyidong@sxu.edu.cn

In this brief paper, the author will try to answer the following four questions:

1. is it necessary for presuppositions in science to be preconditions of scientific reasoning?
2. how can presuppositions in science be determined?
3. what is the relation between the presuppositions and the hypotheses of a specific question in science?
4. are presuppositions in science real and rational?

1 Different Names for Presupposition in the History of Philosophy

Every conclusion of science requires presuppositions, just as every conclusion of science requires evidence. Without appropriate presuppositions, evidence loses its evidential role, and that will undo science. Just as O'Hear once said, "Our presuppositions are always with us, never more so than when we think we are doing without them[? , p. 54]. Therefore, presuppositions are not only useful, but also important to science.

In the history of philosophy, presupposition has different names, such as assumption, supposition, starting point, *a priori* belief, axiom, premise, first philosophy, first principle and first truth. I name a few philosophers' ideas.

Aristotle claimed that in every systematic inquiry where there are first principles, or causes, or elements, knowledge and science result from acquiring knowledge of those. Albertus Magnus built on the concept of suppositional reasoning and handled science's presuppositions by an appeal to conditional necessity. Thomas Aquinas substituted conditional certainty for conditional necessity. For John Duns Scotus, evidence is considered a sufficient foundation for true and certain knowledge, regardless of whether the latter is of a necessary nature. Francis Bacon had two key ideas – the belief that science ought to proceed by means of observation without presupposition and the idea that scientific research can be conducted by means of the systematic tabulation of data. He held that science does not need any presuppositions at all.

---

[1]The author is with Research Center for the Philosophy of Science and Technology, Shanxi University, Taiyuan, China. Email: weiyidong@sxu.edu.cn

Rene Descartes' starting point was the famous "I think, therefore I exist". He said that each of us can become indubitably convinced first of his own existence, then of the existence of God, and finally of the essence of material things and the true nature of the human mind. David Hume, a skeptic, thought that science's ambitions must be limited to describing our perceptions, avoiding philosophical speculations about some external physical world.

Thomas Reid regarded common sense as the only secure foundation for philosophy and science. His conception of science based on common sense had five elements – the symmetry thesis, harmonious faculties, parity among presuppositions, reason's double office (belief and action should match), and asking twice. He was one of the few philosophers who regard presuppositions as the base of science.

Immanuel Kant had a subtle and complex philosophy of science that is not easily summarized. He held that we see the world through colored glasses, or rather that our minds impress a certain pattern on the physical world. Consequently, scientific propositions express factual propositions, since they tell us something about our experiences, but they can be known by pure reason, since it is the mind that stamps them on reality.

2 Gauch's Definitions of Various Presuppositions

In Gauch's point of view, presupposition is something that is taken to be true without a premise for some conclusion. In this sense, if A presupposes B, B is derivable from A. As a semantic notion, presupposition is a relation between two statements A and B such that A presupposes B if the truth of B is a necessary condition of A of being either true or false. But the relation of presupposition differs from the relation of entailment, for if A entails B, the truth of B is a necessary condition of the truth of A, rather than a necessary condition of A possessing a truth value at all.

Gauch gives the following general definition of a presupposition: "presupposition P is a presupposition of a statement S if and only if P must be true for S to have a truth-value, either true or false"[6]. His definition of a question is that, "presupposition P is a presupposition of a question Q expressed by the hypothesis set H1 to Hn if and only if (a) P must be true for every hypothesis H1 to Hn to be possibly true and (b) P makes no hypothesis more or less credible than another." He also gives the following definition of science: "a presupposition of science is a belief that is necessary for an ordinary realist implementation of science if and only if that belief cannot possibly be proved by any evidence or reasoning whatsoever but rather must be accepted by common sense and faith" In this way, there are two kinds of presuppositions, philosophical and scientific. In my opinion, a presupposition in philosophy has traditionally been thought as a starting point of reasoning. In science, a presupposition is not looked at as a starting point but as a promise of inference.

3 An Integrated Model of Scientific Reasoning

As we know, the philosopher of science Popper proposed a model of scientific inquiry: $P_1 \to TT \to EE \to P_2(PTE)$. This means that a scientific inquiry begins at a question that makes a scientist boldly imagine and bring forward various hypotheses. Then he designs an experiment eliminating errors and testing those hypotheses. Gauch also put forward his own PEL model (Presupposition+Evidence+Logic). The Meaning of PEL Model is that presuppositions are beliefs that absolutely necessary in order for any of hypotheses under consideration to be meaningful and true but that are completely non-differential regarding the credibilities of the individual hypotheses. Evidence is data that bear differentially on the credibilities of the hypotheses under consideration. Logic combines the presuppositional and evidential premises, using valid reasoning to reach a conclusion. The author thinks that the above two models have their shortcomings, and should be integrated into a complete model.

According to Popper's model and Gauch's model, the author poses a new integrative model: $Q \to HS \to [PEL] \to EEp \to C$. Among these, Q is a question, Hs stands for Hypothesis set, PEL is Gauch's model, EEp stands for eliminating errors, C is stands for conclusion. The new model shows that presuppositions, hypotheses, evidences and logic are included in the process of reasoning which begins at a question, and that model-based reasoning is based on presuppositions. An Example of the new model is as follows:

Q: is there a coin in the cup? (we have seen that there are a cup and a coin)

Hs: H1: there is a coin in the cup.

H2: there is not a coin in the cup.

P: Seeing implies existence (namely, seeing is believing).

E: We see a coin in the cup.

L: Modus ponens is a correct rule for deduction .

EE_p: there is indeed a coin in the cup (also seen by others).

C: There is a coin in the cup.

We can see, from this example, presupposition is determined by the given question that produces the hypotheses tested by experiments. From a contextualist point of view, the hypotheses in the new model are context-dependent. In the cup/coin test, hypothesis H_1 is the context of H_2, and H_2 is the context of H1. The two hypotheses are mutually negative. These hypotheses constitute a hypothesis set that contains all possible hypotheses. A question's presuppositions are those beliefs that are held in common by all of the hypotheses in the question's hypothesis set.

4 Reality Check of Presuppositions of Science

Presuppositions of science can be tested by common-sense reality check. The author thinks that it is rational, true, objective, realistic and certain that "Drawing fire against oneself is hazardous". This is a common-sense reality check. Philosophical analysis of the reality check reveals varieties of presuppositions, such as ontological, epistemological, and logical ones.

Reality check presupposes distinctions, that is, physical reality has multiple things that are not all the same, for instance, natural kinds such as tree, flower and sheep. This is Ontological presupposition.

Reality check presupposes that our sensory organs provide generally reliable information about the external world and our brain can process and comprehend these sensory inputs; human language is meaningful; humans have abilities of language and communication. These are epistemological presuppositions.

Realty check presumes coherence. This is logical presupposition. For example, to assert that "drawing fire against oneself is hazardous" legitimately, coherence demands that we not also assert the negation that "drawing fire against oneself is not hazardous ". But "the logical premises of factuality are not known to us or believed by us before we start establishing facts, but are recognized on the contrary by reflecting on the way we establish facts."[7, p. 162]

Realty check also presumes truth. Coherence requires merely that the reality check and its negation are not both to be asserted, but it does not indicate which one to assert.

So to speak, reality check expresses a true belief that corresponds with the physical world in an orderly and comprehensible way. This is the premise of scientific research. Science's presuppositions supply a basic and indispensable picture of our real world and have their factuality and rationality. When asking a scientific question, these presuppositions focus a set of viable hypotheses to a limited set that can be sorted out with ordinary, attainable evidence.

5 Summary

In the above arguments, the author proves that presuppositions, together with hypotheses, evidence and logic, play an important role in the process of model-based reasoning. Within philosophy, presupposition has traditionally been thought as a starting point of reasoning, but in science, it is not looked at a starting point, but as a promise of inference. By comparing Popper's PTE model about the scientific inquiry with Gauch's PEL model, the author puts forward a new integrative model of $QH_S[PEL]EE_p$. The new model shows that there are presuppositions not only in philosophy, but also in science, and that presupposition, hypothesis, evidence and logic are included in the process of reasoning which begins at a question, and also shows that model-based reasoning is premised on presuppositions which are based on the model and come from the context by comparing different hypotheses originating from the question being asked.

From the scientific realist's point of view, a presupposition of science is regulated by the given question which produces the hypotheses which can be tested by experiment. Presuppositions can be defined by scientific realism, so the author embraces the definition given by Gauch and divided into two kinds – a general one or philosophical one which appears in every scientific inquiry and a special one or scientific one which appears in only some scientific inquiries.

According to the contextualist point of view, the hypotheses in the model of $QH_S[PEL]EE_p$ are context-dependent. That is, hypotheses interact with one another. Presuppositions are also included in the set. That is, a question's presuppositions are those beliefs held in common by all of the hypotheses in the question's hypothesis set. By contextual analysis of hypotheses and presuppositions in the model, the author believes that scientists can understand their discipline in greater depth.

Finally, the reality check that "drawing fire against oneself is hazardous" is used to analyze the reliability of presuppositions of science. Philosophical analysis of the reality check reveals various kinds of presuppositions, such as ontological, epistemological, and logical ones. All in all, an inquiry's presuppositions are those beliefs held in common by all of the hypotheses in the inquiry's hypothesis set. Common sense is the basis of science, and the reality check includes different presuppositions. Science's presuppositions supply a basic and indispensable picture of our actual world.

Bibliography

[1] AAAS (American Association for the Advancement of Science). *Science foe All Americans: A project 2061 Report on Literacy Goals in Science, Mathematics, and Technology.* Washington, DC, 1989.

[2] Caldin, E. F. *The Power and Limits of Science: a Philosophical Study.* London: Chapman & Hall, 1949.

[3] Davis, W. A. *An Introduction to Logic.* Englewood Cliffs, NJ: Prentice-Hall, 1986.

[4] Jr Gauch, H. G. *Scientific Methods in Practice.* Cambridge: Cambridge University Press, 2003.

[5] Magnani, L., J. Nersessian, N., Thagard, P. (Eds.). *Model-based Reasoning in scientific discovery.* New York: Kluwer Academic/Plenum Publisher, 1999.

[6] O'Hear, A. *An Introduction to the Philosophy of Science.* Oxford University Press, 1989.

[7] Polanyi, M. *Personal Knowledge: Towards a Post-Critical Philosophy.* University of Chicago Press, 1962.

Contextual Realism and Relativism[1]

Cheng Rui
Shanxi University
chengruier@126.com

In recent years, researches in contextual realism have become one of the new approaches in the philosophy of science. The method of contextual analysis has also become a promising methodological trend. Investigators try to take 'context' as the basis of the combination of syntax, semantics and pragmatics so that they can expand and construct the whole building of linguistic philosophy on this foundation (Guichun Guo) or even have an influence on the development of the philosophy of science, because the contextual analysis method has obvious advantages as well as profound philosophical roots. The concept of "context", however, is imprecise in its connotation and extension. There are, therefore, some problems with the concept of context one of which is relativism.

When it comes to relativism, It's very easy for us to recall Logical Positivism. Logical positivists failed in explaining the problem of scientific progress and scientific method etc., this led to the appearance of Kuhn's paradigm theory. The characteristics of paradigm such as incommensurability and no common evaluation criteria made science become irrational or subjective. The same concepts may have different meaning in different scientific community and the world view may change entirely, these all promoted the rise of relativism.

For contextual realists, they also need to think about the issues of scientific progress and scientific methods. In contextual realism there also exists the problems of the changes of concepts and relationship between the contexts. Can contextual realists make better answers than Kuhn's paradigm in this respect?

1 The transition of meaning and the relativism of truth

From the origin of 'context', we can find a specific character in both Frege and Wittgenstein that 'context' and 'meaning' related closely, it emphasized that meaning is rich and concrete. This is a very important change on meaning. In late Wittgenstein's concept of "the meaning of a word is how we use it" (Wittgenstein, 32) and interpretation of the concept of the Hermeneutics, "context" linked up with pragmatism. Furthermore, all the theoretical, social and historical macro-background were bound to infiltrate through the microscopic structure of the context. This made the connotation and extension of the concept of "Context" blurred and the elements of the context diversified. The scientific concept as the

---

[1]The author is with Research Center for Philosophy of Science and Technology, Shanxi University, Taiyuan 030006 China.

centre of scientific language maybe has different meanings in different contexts. We know that unsatisfactory explanations of meaning of changing is one of the critical reasons why Kuhn got a relativistic result, and it is also a focus that antirealists retort context realism: the meaning of scientific terms is changing with the development of scientific contexts, so the meaning of scientific terms are different in different theories or different contexts. Those who think that there is no comparability between different theories will lead to relativism on truth, and instrumentalism or pragmatism on the aim and usage of science.

For the transition of meaning, contextual realism pointed out that the dynamic changes of the meaning of words are the results of the dynamic changes of contexts. Context is a kind of reality, the semantic content of words will be bound to expand with spreading of experience. From the formal features of scientific theories, the meaning will be adjusted from time to time in order to maintain the internal logic of formal language system. From a pragmatic point, the transition of the meaning of words is related to the main background of epistemology, psychological factors and so on. But it's relationship with the referential objects is keeping stable. In contextual realism, a very important point is that the boundaries of contexts are continuous before and after and the new context is a context beyond the old one. This kind of continuity of context makes that "the basic, inherent and logic things turn into the new context in the development of history." (Sumei Cheng and Guichun Guo) Even if the same concept has different meanings in different contexts, these differences are not solitary existence, but conversion or deepen under complex Contexts. For example, the development of the concept of physics space-time has experienced the process of deepening of Newton space-time, relativity space-time and modern quantum gravitational space-time. Although the content of the concept changes, different space-time theory is not isolated structure, in fact it is embedded in a broader epistemological context. It's development inherently includes physicists' metaphysics consideration of space-time in different times, the development of mathematical theory and the depth of physical formal system, the semantic transformation and expansion of representative formulas and symbols, the change of physicists' thoughts from absolute to relative and other contextual factors, and these factors in the dynamic development of space-time theory interlocks, which is a complex organic whole. Every change of the space-time concept is based on our understanding of the space-time concept and the foundation of related theoretical development. The understanding of concepts requests the continuity of context, we need to admit the recontextualization function of contexts, admit that the changes of meaning are the adjustment under different contexts. These help to overcome the incommensurability difficulty of paradigm. For the different analysis of the meaning of different theories, we should recognize the same root of different meaning, that is to say, we should recognize the ontological coherence under the phenomena, and see that different meanings serve for the same ontology. In other words, contextual realism admits the relative changes of meaning, but in the ongoing development and beyond of context, the final purpose of the changes of meaning is to explain the nature of the world. This change reflects the relativity of epistemology, but will not lead to relativism.

2 The range of context and the pluralism of truth

In the study of the Philosophy of Science, the units of meaning change from "words" to "propositional meanings", but the range of contexts experienced the extension from "words" to "utterance", from the "utterance" to the "sentence system", which went through interpretative turn, the rhetorical turn, and affiliated the historical, psychological and social factors. It provides a whole background for us to understand the changes of "meaning" in the overall contexts. The significance of contextual analysis consists in that we can go on our research of philosophy of science on a holism position. However, if we blindly extend the context to the entire range of scientific activities, the range of the context will be too large, which may be possible to ablate the meaning of the words, the meaning of "meaning" in the endless chaos of the overall context, which will cause extreme holism. (Guichun Guo and Hongmei Wang) Everything is in context, it will be difficult to make a valuable analysis for a theoretical development. There is no uniform standard for judgment, and this will be very easy to lead to the pluralistic of the truth if one use the extreme holism to ablate the unreasonable aspects of the existence of every kind of theory.

In order to avoid dropping into "extreme holism", when emphasizing holism, contextual realists also pay attention to studying the microstructure of context. The "context" has a very specific content about the choice and description of the language system. For the levels between various contextual factors of some specific problem, it requires the specification of the statements about related behaviors, the statements about the background knowledge - even the statements about the time of the language etc. In a word, this is a entirely specific regulation, its has very strong realistic feature. For example, the very measurement context is such a completely concrete and realistic option. It gives specific content for every specific measurement, including the purpose of implementing the measurement, the relevant knowledge and the logical or mathematical rules which work as a background knowledge of the measurement, the accepted presumptions and ontology and the selected space-time reference system etc. In such a overall context of the structural system of its own, the meaning from "words" to "statement system" got adjusted, It provides the stability of the meaning and realizes its objectivity, and ultimately reveals the public nature of meaning. There are some differences with the extreme holism. In extreme holism, the carrier of the meaning is the whole language, different languages interconnected, if philosophers want to describe such a language, they will become omnipotent, it is thoroughly impossible. Therefore, the wholeness of the context is essentially holism in constrained choices on the basis of realistic context.

Another important advantage for contextual realism to avoid the relativism and pluralism of the truth is that it has a unified standard for judging the progress of science. The criterion is the similarity between the theory and the world under different context. (Sumei Cheng and Guichun Guo) it can overcome the shortcomings of logical positivists who wanted use the number of real propositions to measure the fidelity of a theory thus can explain the scientific progress reasonably as well as explain why different theories could confer a concept with different meaning.

In a word, contextual realism may propose a more reasonable answer in the study of scientific progress and the issue of truth relativism. In this sense, it may be demonstrated from one aspect to be a relatively mature choice during the development of scientific realism.

Bibliography

[1] Cheng, S., Guo, G. Contexual realism. *science, technology and dialectic*, 3, 2004.
[2] Guo, G. On contextual realism. *Philosophy Stuty*, 4, 1997.
[3] Guo, G., Wang, H. The significance of measurement context. *science, technology and dialectic*, 4, 1999.
[4] Wittgenstin. *Tractatus Logico-Philosophicus*. Commercial Press, 1992. Ying Guo translated.

Bayesian Test and Kuhn's Paradigm[1]

Chen Xiaoping
South China Normal University
chenxp@scnu.edu.cn

Wesley C. Salmon, the famous American philosopher of science, pointed out that in criticizing the soCcalled testing pattern of science, Kuhn focused all his attention on a single test model, namely the hypotheticoCdeductive (HCD) schema. As a matter of fact, however, many philosophers of science had already abandoned that schema and adopted the Bayesian schema as a more proper test model. Among them, H. Reichenbach and R. Carnap began to advocate Bayesian schema respectively in 1949 and 1950. In addition, L. J. Savage, a personalist, referred to himself directly as a Bayesianist beginning in 1954. (see [9, p. 325])

The main difference between Bayesian schema and the HCD schema lies in that the former is a testing model for more than one theory while the latter just for a single theory.[2] Since Kuhn, multi-theoretical testing model has become a consensus among experts, that is, a theory and its rivals should be faced with testing together, rather than a theory being tested in isolation. Kuhn was correct in finding the HCD schema not appropriate to scientific test, but didn't catch the propriety of Bayesian schema in this field. This led to his disapproval of the logic or method of scientific test. So Salmon argued that Kuhn's doubt on the HCD schema was untenable to Bayesian

I agree with Salmon's point of view largely, and plan to give a further argument for it in this paper.

Just when logical positivism and falsificationism were fairly popular, Kuhn achieved spectacular results in his *The Structure of Scientific Revolutions* (1962), bringing in a refreshing breeze of historicism. In a detailed investigation into the history of science, Kuhn put forward a pattern of scientific progress in which the phases of normal science and scientific revolution dominate alternately. Its key concepts are "scientific community" and "paradigm". A paradigm "stands for the entire constellation of beliefs, values, techniques, and so on shared by the members of a given community". Meanwhile, it offers some exemplars as puzzle-solutions that "can replace explicit rules as a basis for the solution of the remaining puzzles of normal science".[5, p. 175] In the period of normal science,

---

[1]The author is with School of Public Administration, South China Normal University, Guangzhou 510006 China. Email: chenxp@scnu.edu.cn

[2]The HCD schema can be expressed as the following two forms of inference:

1. Confirmative Inference: $h \Rightarrow e, e, \therefore$ very likely h.
2. Disconfirmative Inference: $h \Rightarrow \neg e, e, \therefore \neg h$.

The Bayes's theorem can be stated as follows(only take the two-hypotheses case into consideration):

$$P(h_1/e) = \frac{P(h_1)P(e/h_1)}{P(h_1)P(e/h_1) + P(h_2)P(e/h_2)}$$

Bayesian Test and Kuhn's Paradigm

the scientific community, engaged in scientific practice under the same paradigm, has not or does not give any consideration to other paradigms in opposition to the present one. When confronted with knotty problems and anomalies, scientists only put the blame on their ability to solve puzzles, but not on the paradigm that they follow. Only when the knotty problems or anomalies resist solutions long enough and increase rapidly will scientists suspect the paradigm itself. At that moment, all the previous knotty problems or anomalies turn out to be counterexamples directed against the paradigm. It is then that a crisis appears. As a result, a new paradigm comes out and scientists replace the old one with the new one eventually. This is called scientific revolution and a period of new normal science follows. Take for instance Copernicus's Heliocentric Theory, Darwin's Theory of Evolution, Einstein's Theory of Relativity, and so on, all of them became a paradigm of new normal sciences through scientific revolution. From the angle of Bayesian testing model, I shall give a further explanation about the structure of scientific revolutions depicted by Kuhn.

According to Kuhn, members of the scientific community act on the same paradigm in the period of normal science. It seems as if the H-D testing model for a single theory were well adapted at that time. However, as a matter of fact, when the sole existing paradigm encounters an anomaly, namely e, the scientific community will not take it as a counterexample of the paradigm, but as a puzzle remaining to be solved under the paradigm. For that reason, the H-D model of disconfirmation does not work in this case. So, Kuhn said: "Once it has achieved the status of paradigm, a scientific theory is declared invalid only if an alternate candidate is available to take its place. No process yet disclosed by the historical study of scientific development at all resembles the methodological stereotype of falsification by direct comparison with nature".[5, p. 77] It means that anomalies will not have a chance to turn into counterexamples playing the part of disconfirmation until the appearance of another paradigm competing with the existing one. Consequently, instead of one theory, the scientific testing has to involve more than one.

From Bayes's theorem we learn that the Bayesian schema is a multi-theoretical testing model. It involves two competing theories at least, namely, the number of the competing theories $n \leq 2$. In the course of normal science, however, there is one paradigm alone, namely $n = 1$, with the result that it doesn't meet the prerequisite for Bayesian test. It signifies that any evidence cannot pose testing to the existing paradigm, or the existing paradigm is exempt from test. With respect to this point, a further analysis is carried out below.

When the existing paradigm encounters an anomaly, the number of the paradigm in testing is: $n = 1$, i.e. the hypothesis to be tested, h_1, exists alone, without other hypotheses competing whit it. Under this condition, the Bayes's theorem changes into a simplified form:

$$P(h_1/e) = \frac{P(h_1)P(e/h_1)}{P(h_1)P(e/h_1)}$$

As can be seen from the above formula, if both $P(h_1)$ and $P(e/h_1)$ are not equal to zero, then $P(h_1/e) = 1$. We know that in the period of normal science, the scientific community is absolutely positive about the existing paradigm, or rather, about key theory h_1, therefore, the members consider the degree of h_1's prior belief (the prior probability of h_1), $P(h_1)$, to be 1; with the anomaly e, the degree of prediction of h_1 on e is: $P(e/h_1) > 0$, but not: $P(e/h_1) = 0$, because the scientific community doesnt believe there is any phenomenon the existing paradigm is unable to explain. Since $P(h_1)$ and $P(e/h_1)$ do not equal zero, $P(h_1/e) = 1 = P(h_1)$ according to the Bayes's theorem. And e is irrelevant to h_1, as stated in the Positive Relevance Criterion.[3] That is to say, confronted with the anomaly, the members of the scientific community still firmly believe the existing paradigm just as before, and the degree of their belief will not be affected by the anomaly e. In this sense, the existing paradigm is exempt from being tested.

In the period of normal science, the logical reason why the scientific community considers $P(e/h_1)$ greater than zero when the existing paradigm encounters the anomaly is as follows: The existing paradigm entails a prediction, $\neg e$, and it is manifested to be false. In other words, e is verified, so scientists will regard e as an anomaly of h_1. But the usual statement of "h_1 entails $\neg e$", namely "$h_1 \to \neg e$", is not accurate, because $h_1 \to \neg e$, actually, is just an elliptical expression of $h_1 \wedge A_1 \wedge \cdots \wedge A_n \to \neg e$, in which $A_1 \cdots A_n$ are auxiliary hypotheses. So, it is not clear whether this erroneous prediction about $\neg e$ is inferred from h_1 or from a certain auxiliary hypothesis A. The probability that h_1 entails $\neg e$ depends on the one that $A_1 \wedge \cdots \wedge A_n$ is true. If $A_1 \wedge \cdots \wedge A_n$ is firmly believed to be true, then it cannot entail the false preposition $\neg e$, thereby $\neg e$ must be inferred from h_1 alone, (of course, there is no conflict between $A_1 \wedge \cdots \wedge A_n$ and h_1) ; on the contrary, if $A_1 \wedge \cdots \wedge A_n$ is not firmly believed to be true, only with some probability of being true, then the probability that h_1 entails $\neg e$ is the same as it. In this sense, $h_1 \to \neg e$ is untenable, therefore $P(\neg e/h_1) < 1$, correspondingly $P(e/h_1) > 0$.

Strictly speaking, there is a process in which the scientific community makes adjustments in its belief system. Initially, the members of scientific community are convinced that the auxiliary hypotheses $A_1, A_2, \cdots A_n$ are true, thus $P(e/h_1) = 0$; but they don't believe that the existing paradigm is unable to explain e, so they make a skeptical attitude towards the truth of the auxiliary hypotheses, and in this way reduce the degree of their belief in those auxiliary hypotheses and their relevant prepositions, thereby leading to $P(e/h_1) > 0$. Luckily enough, such adjustment in the degree of belief in auxiliary hypotheses normally doesn't affect the general situation, because they are ordinarily located on the edge of a system of theories. The way of adjusting a belief system from the edge to the core follows none other than the famous Duhem-Quine Thesis, which is also known as the viewpoint of Hard Core-protective Belt posed afterwards by Lakatos.

Confronted with anomalies, scientists direct first their suspicions at auxiliary hypotheses. Scientists would not aim their arrow of doubt at the key theory until the anomalies remain untouched and increase unceasingly with successive appearance even after all the relevant auxiliary hypotheses are scrutinized. As a result, the period of crisis comes, as stated by Kuhn. In that period, scientists finally identify that those anomalies are really inferred from the key theory h_1 of the existing paradigm, i.e. identify that $h_1 \to \neg e$, and then $P(\neg e/h_1) = 1$, in other words, $P(e/h_1) = 0$. In this process of change, a new paradigm h_2 is

---

[3] Positive Relevance Criterion is as follows: If $P(h/e) > P(h)$, then e confirms h; if $P(h/e) < P(h)$, then e disconfirms h; if $P(h/e) = P(h)$, then e is irrelevant to h. (see [3, pp. 133–141], [4, chap. 7])

Bayesian Test and Kuhn's Paradigm

gradually taking shape and obtains a certain degree of prior belief – $P(h_2) > 0$. The new paradigm h_2 is raised to cope with those counterexamples confronting the old one h_2, thus h_2 holds a higher degree of prediction on the counterexample $e - -P(e/h_2)$ is rather close to 1, even $h_2 \to e$ being made tenable. At this moment, the state of a multi-theoretical testing turns up, correspondingly, Bayesian schema is able to be employed (only two competing paradigms are taken into account here):

$$\begin{aligned} P(h_1/e) &= \frac{p(h_1)P(e/h_1)}{P(h_1)P(e/h_1) + P(h_2)P(e/h_2)} \\ &= \frac{P(h_1) \times 0}{P(h_1) \times 0 + P(h_2)P(e/h_2)} = 0 \\ P(h_2/e) &= \frac{p(h_2)P(e/h_2)}{P(h_1)P(e/h_1) + P(h_2)P(e/h_2)} \\ &= \frac{P(h_1) \times 0}{P(h_2)P(e/h_2) + P(h_2)P(e/h_2)} = 1 \end{aligned}$$

These show that in the period of crisis, counterexamples play the role of crucial test between the new and the old paradigms, and make the degree of posterior belief of the new paradigm equal one and that of the old one equal zero. So far, the scientific community accepts the new paradigm and gives up the old one, thus the scientific revolution by Kuhn is accomplished. As can be found from the two equations above, the critical factor in this crucial test is $P(e/h_1) = 0$, that is to say, the degree of prediction of the older paradigm on the evidence is equal to zero.

In short, the process from normal science to scientific revolution and the function of the anomaly or the counterexample against a paradigm, which are described by Kuhn, can be interpreted using the Bayesian testing model. Specifically speaking, in the period of normal science, the paradigm cannot be tested by anomalies, which can be regarded as a special case within the Bayesian schema – it only deals with one theory under test. In the period of crisis in science, anomalies turn out to be counterexamples and play the role of Bayesian crucial test between the new and the old paradigms, thereby promoting the accomplishment of scientific revolution.

Bibliography

[1] Earman, J. *Bayes or Bust? A Critical Examination of Bayesian Confirmation Theory.* Cambridge, Massachusetts: The MIT Press, 1992.

[2] Gamow, G. *Biography of Physics,*. Beijing: Commercial Press, 1961. Quoted from Wulixue Fazhanshi (the Chinese edition of Biography of Physics), Gao Shiqi.

[3] Hesse, M. *The Structure of Scientific Inference.* Berkeley: University of California Press, 1974.

[4] Howson, C., Urbach, P. *Scientific Reasoning: The Bayesian Approach.* Chicago: Open Court Publishing Company, 1993.

[5] Kuhn, T. S. *The Structure of Scientific Revolutions.* Chicago: University of Chicago Press, 1970.

[6] Kuhn, T. S. *The Essential Tension*. Chicago: University of Chicago Press, 1977.

[7] Kuhn, T. S. Commensurability, comparability, communicability. In Conant, J., Haugeland, J. (Eds.), *The Road Since Structure*. Chicago: University of Chicago Press, 2000.

[8] Lakatos, I. *The Methodology of Scientific Research Programmes*. Cambridge: Cambridge University Press, 1978.

[9] Salmon, W. C. The appraisal of theories: Kuhn meets bayes. In *Proceedings of the Biennial Meeting of the Philosophy of Science Association*, vol. 2, 325–332. 1990.

[10] Xiaoping, C. *Inductive Logic and Inductive Paradox*. Wuhan: Wuhan University Press, 1994.

Does the "Grue" Paradox Have a Solution? How to Solve It?[1]

Lv Shirong

Henan University

lvshirong6@sina.com

Jin Zhichuang

Henan University

jizhichuang@sina.com

Since Goodman raised the New Riddle of Induction problem, the "Grue Paradox" has stimulated a wide-ranging response in philosophy, especially in the field of logic. Many varied schemes have been proposed to resolve the paradox using different approaches. These schemes, though many in number, however, tend to be invalid due to a persistent recourse to circular reasoning in demonstration.

1 Part I

The first scheme, named "predicate projectibility", mainly includes Barker and Achinstein's scheme for Time Orientation, Gardenfors' scheme for Conceptual Space and Goodman's scheme for Entrenchment. Barker and Achinstein's scheme is a non-symmetrical demonstration directed against Goodman's symmetrical statement. It discusses the non-symmetry in logical character between "green" and "Grue", through the demonstration that under the specially designated circumstances, Grue is also predicate of Time Orientation, as far as the Grue language users, Mr. Grue, are concerned.[1] But its result just depends on the fact that "green" and "blue" are daily language, however, Grue is not, that is depend on the entrenchment of "Green" and "blue", so can't constitute the refutation on their symmetry. The crux of the question is that Barker and Achinstein demonstrate their thought under the circumstance of Green language from beginning to end. Quine's Natural Kind tries to argue that Grue doesn't reflect the natural attribute, so it can't project. But Quine also can't give an exquisite statement on the fact that predicate Green reflects the natural attribute. He only takes Darwin's natural selection theory as consolation.[4] Gardenfors provides an operative standard with "Protruding regulation" for natural attribute, and he further raises the scheme of Conceptual Space. The attribute "green" and "blue" expresses in daily conceptual space presents protruding color field, so it can project. However, according this, Grue can't project. But we can't eliminate the possibility that we can constitute a conceptual space, in which the attribute "green" and "blue" express is not protruding, but Grue is contrary to it. Therefore, we can't negate the projectibility of Grue. Up to this point, the resolving schemes through time and space have fell into t a circle on receiving demonstration of the frame of belief conception, for the determination of "predicate projectibility" depend on the basement of conceptual frame which remains to be demonstrated.

[1] The authors are with Department of Philosophy, Henan University, Kaifeng 475001 Henan.

The second kind of resolving scheme carries out from scientific methodology to provide operative standards for the choice and acceptance of hypothesis, which mainly include simplicity and falsifiability. The scheme of simplicity determines projectibility of Grue through syntax and semantic simplicity. But syntax simplicity is relative to conceptual frame or signal expressing system, so we can't inevitably determine which is simpler. Though semantic simplicity rarely depends on the expressing system, the extent that it relies on the determination circumstance is still great. In other words, the determination on simplicity depends on the choice and acceptance of the frame of background cardinally, the Paradox still exists. For the same reason, determination of big or small on potential class which is falsified of the hypothesises that the standard of falsifiability depends on, and the confirmation of pre-verified probability which Bayesian scheme depends on to make determination, can only draw support from the hypothesis belief existing in advance and can't escape the circulatory demonstration in the preceding discussion.

2 Part II

Does it have no solution or are the resolving routes wrong? Is it one-sided as a simple logical paradox or do we ignore its philosophical background? How to resolve it? In fact, the same place in resolving schemes in preceding discussion lies in the reality that they only carry out themselves from pure logical angle. However, as a logical paradox, the Grue Paradox possesses a distinctive philosophical background but the neglect of its philosophical allegory only makes us diverge its philosophical meanings. This is the fundamental reason why it falls into the resolving dilemma.

In the "The New Riddle of Induction" chapter of Fact, Fiction and Forecast, after Goodman criticizes misunderstandings of Hume's Induction Question, Goodman argues that the demonstration task of induction inference as deduction inference is that it conform to effective inductive principles, and the effectiveness should depend on the conformity with accepted deduction or inductive practice. Though the proving process is a circle, Goodman thinks it is virtuous, in which they try to mutually adjust themselves to find the best demonstration both sides needs between principles and accepted inference. Here, Goodman expresses two meanings. The first is that induction breaking away from inductive practice can't be testified on the pure logical necessity and it must not be a logical question escaping from practice, that he sums up the induction question as the opposite of the demonstration of rationality or establishment of legitimacy can be taken as the illustration. The second is that Hume's answer of "habit" just can resolve the question and at least he provides a valid direction. So "His answer was incomplete and perhaps not entirely correct; But it was not beside the point, the problem of induction is not a problem of demonstration but a problem of defining the difference between valid and invalid predictions."[2, p. 65]

Now, our task is that we should establish a regulation which can give the definition, Confirmation Theory. Obviously, this kind of judgment and definition is tightly connected with the given confirmation and acceptation of hypothesis and belief, and the above-mentioned question has been naturally transformed into the definition question for confirmation of belief and hypothesis. The definition

Does the "Grue" Paradox Have a Solution? How to Solve It?   105

is going to define the hypothesis that can be and not be confirmed, or find and provide the standard for what we accept and refuse. But, he finds that the appropriate construction task of Confirmation Theory has not been completed. Though following with the fact that testifying question has been displaced by definition question, and the important question neglected long before has been announced and also has been given the answer, "But our satisfaction is short-lived. New and serious trouble begins to appear."[2, p. 72] Suppose before specified time t, all jadeite are green. According the definition of confirmation now available, we can give our proof statements that jadeite a is green, jadeite b is green etc, at time t, all of them support this hypothesis and also everyone confirms this universal hypothesis. Now, we draw the other predicate "Grue" and it can be applicable in entire objects which have been testified before time t and also at this time, they are green, and also all others are blue. So, at time t, the proof statements, such as jadeite a is green and jadeite b is green etc. all incline to confirm the universal hypothesis that all the jadeites are grue. But according to the definition of Grue that all grue jadeites testified at time t are green, the testified predicts including all the jadeites are green and all the jadeites are grue have been confirmed with the same proof statement through the same depicted observing result in the same method. But "if we simply choose an appropriate predicate, then on the basis of these same observations we shall have equal confirmation, by our definition, for any prediction whatever about other emeralds – or indeed about anything else. ...... We are left once again with the intolerable result that anything confirms anything."[2, p. 75] Obviously, the question of Grue Paradox is to find a standard of distinguishing "law-like or confirmable hypotheses from accidental or non-confirmable ones"[2, p. 80] and also to determine whether it can be accepted or not.

Obviously, Goodman makes induction question transform into confirmation question. But, it is worth noticing that Goodman doesn't continue his research under the academic circumstances that it generally forms the logical interpretation on Hume's question, but he continues the thinking of his critique in the process of misunderstanding Hume. However, it is the logical interpretation under the academic circumstances at that time that is prevalent. So it leads to the fracture between logical resolving method and philosophical resolving goal, because in the process of understanding Grue Paradox, it pays more attention to the meaning of logical interpretation in the logical resolving way. Therefore, we can see the circle which makes the resolving schemes incline to be invalid appearing in Section 1 of this dissertation.

3 Part III

Then, how does Goodman resolve the Grue Paradox? From Goodman's statement in his book *Fact, Fiction, and Forecast*, he holds that in fact, we have misunderstood our task, that is, we consider the statement for necessary result as a method statement which is over-restricted to make sure that we may get this result. To confirm an hypothesis, given proof and possessed hypothesis is indeed necessary, but it doesn't mean that they are all of our material to be utilized. In fact, "whenever we set about determining the validity of a given projection from a given base, we have and use a good deal of other relevant knowledge.

......the record of past predictions actually made and their outcome."[2, p.85] Actually, these recorded material is the record on the passed actual projection, in other words, we can define the projectivity according to the passed actual projection. But conflict often exists in the actual projection, the paradox in the projective process in hypothesises of Green and Grue is the case in point. Goodman's resolving method is "we must consult the record of past projections of the two predicates, Plainly green", as a veteran of earlier and many more projections than "Grue", has the more impressive biography, The predicate "green", we may say, is much better entrenched than the predicate "Grue".[2, p. 94] Therefore, one projection can be eliminated if it is in conflict with a more entrenched predicate "projection", certainly, the hypothesis which uses it is also abandoned. In fact, here Goodman has drawn the non-logical practical standard into the resolving process. Obviously, this resolving method briefly avoids the need of the predetermined conceptual frame and it also refrains from the circle that the demonstration of accepting belief hypothesis must depend on accepting belief hypothesis because it adopts practical standard.

Now, we can go back to the analytical circumstances of different schemes in the preceding discussion again to make further investigation and recognition on their erroneous origin. Obviously, these schemes carry out a logical interpretation basing on the academic circumstances of that time, and behind their interpreting process, they always potentially possess strong dependence and trust on the logic of scientific deduction and scientific conceptual frame, which is taken as the unique belief hypothesis that can be accepted. In fact, through drawing into the predicate "Grue", second to none in logic, Goodman suggests that any induction might be possible to sink into paralysis and in the use of induction, there is always a presupposition of projective determination, which expresses the use and practical essence of induction, and also we can't think that all the predicates are projectible and also it is the fact that there is a reason making determination between projectible and non-projectible, which shows that the projectibility of predicate can't be taken as a ultimate means of theory and the determined process can't summit itself to treating formally, that is to say, we are impossible to find a theoretical confirmation fundament and guaranty and the determination must depend on practice; at the same time, pluralism can't be avoidable. Then, which one of the hypotheses with plural existential forms can be accepted? It can only depend on the practical natural becoming, but can't come true through logical demonstration based on the unique acceptance of belief hypothesis. In this sense, the Grue Paradox is not only a logical paradox inherently, but also a philosophical allegory to great degree which eradicates the logical unique conviction of science essentialization.

Though Goodman's entrenchment scheme isn't the ultimate answer, it is important that he draws practice into logical demonstration, and therefore he has filled the vacancy the artificially isolation between pure theory and life practice leads to since long time before, which leads a perfectly right way to resolve Grue Paradox at this stage. In fact, Goodman holds that there doesn't have an absolute essential substance about the world, and the world is not found and also doesn't exist with plural forms but it is constructed.[3, pp. 1–7] So there doesn't have the unique correct cognition about essence and logic and science also doesn't possess the supreme necessity; our belief hypothesis, conceptual frame, knowledge and our world are all established under the system of the background naturally becoming in practice, which includes systematic conditions specially designated and our practical goal. That whether we accept every hypothesis or not is not because necessary determination of the cognition and law in essential meaning can be demonstrated by logic, but because it is generated in practice under much additional information additions outside the logic. Entrenchment is the description of practice, so entrenched judgment can only be carried out in the practice, which is not a pure theoretical demonstration. Therefore, only when we recognize that its substance isn't taken as a pure logical paradox but as a philosophical allegory does the Grue Paradox mean it has no solution; only when we transform the pure logical demonstration into introducing the practice and its demonstration into can we obtain its genuine solution and the new philosophical view it contains and His whole philosophical background provides us with the navigation guidelines which can guide us to the solution.

Bibliography

[1] Barker, S. F., Achinstein, P. On the new riddle of induction. In Catherine, Z. E. (Ed.), *Nelson's New Riddle of Induction*, 65–68. New York & London: Garland Publishing Inc, 1997.

[2] Goodman, N. *Fact, Fiction, and Forecast.* Cambridge: Harvard University Press, 1983.

[3] Goodman, N. *Ways of Worldmaking.* Indianapolis, IN: Hackett, 1985.

[4] Quine, W. V. Natural kinds. In Douglas, S. (Ed.), *Grue! The New Riddle of Induction*, 49. Chicago: Open Court, 1994.

A Defense of *Ceteris Paribus Laws*[1]

Wang Wei

Tsinghua University

wangwei@tsinghua.edu.cn

ABSTRACT. Earman *et. al.* raise several objections to *Ceteris Paribus* laws. In this paper, I argue that CP clauses could be ineliminable even with scientific terminology, and that it is also possible to test the contraposition of a CP law and therefore the law itself. Earman's account of differential equations may violate his MRL view of the laws of nature. Again, Earman's view of laws of nature may be inconsistent with his supervenience thesis.

    The John Earman, John Roberts and Sheldon Smith (hereafter ERS) paper "*Ceteris Paribus* Lost?" (hereafter ERS) raises the severest criticism of *Ceteris Paribus* (hereafter CP) laws. They criticize the concept of CP laws in the following six ways: (i) Appeal to examples from physics. ERS argue that CP clauses can be easily eliminated by known conditions if we properly use scientific language. (ii) Confusing Hempel's provisos with ceteris paribus clauses. ERS think the conditions of the provisos are conditions for the validity of the application, not conditions for the truth of the law statements of the theory. So they would accept Hempel's proviso but reject the CP clauses. (iii) Confusing laws with differential equations of the evolution type. ERS argue that those examples provided by CP law proponents are just differential equations of evolution type. But differential equations of evolution type depend on non-nomic assumptions, therefore are not laws. (iv) Early Cartwright on component forces. ERS raise two objections: in many cases component forces are measurable; it is not clear that it follows that something is not occurrent just because it is not measurable. (v) Cartwright's argument from Aristotelian natures and experimental method. ERS repeat the supervenience: "One can grant that there is a lot more to being a law of nature than just being a true behavioral regularity, and even grant that what laws state is helpfully understood in terms of capacities, while maintaining that laws (and capacities) must supervene on the behaviors of physical systems." (vi) The world as a messy place. The CP laws proponents would argue: "The. Therefore, we just have not good reason to." ERS acknowledge the world is an extremely complicated place, but believe that there are any non-trivial contingent regularities that are strictly true throughout space and time.[4, pp. 283 – 288]

    ERS also mention two objections to CP laws: (1) there seems to be no acceptable account of their semantics; (2) there seems to be no acceptable account how they can be tested. They think the first objection is "not fatal to CP laws", but they find the untestability of CP laws decisive.[4, p. 293]

---

[1]The author is with Institute of Science, Technology and Society, Tsinghua University, Beijing 100084 China.

So I summarize ERS' main arguments against CP laws follow three theses: (1) CP clauses can be easily eliminated if we properly use the scientific language, i.e. ERS' (i). (2) The CP laws can not be tested, if we can not substitute testable auxiliaries for the CP clauses. (3) So called "CP laws" are just differential equations of evolution type (which hedged on non-nomic assumption), but laws are strict, i.e. ERS' (ii), (iii), (iv) and, perhaps, (vi). I will argue against the first two theses in Section and the third in Section .

1 Eliminability and Untestability

Is a CP clause eliminable? Here is Lange's example. To state the law of thermal expansion [the change in length of an expanding metal bar is directly proportional to the change in temperature], "one would need to specify not only that no one is hammering the bar on one end, but also that the bar is not encased on four of its six sides in a rigid material that will not yield as the bar is heated, and so on". [6]

ERS think this example is expressed in a language that "purposely avoids terminology from physics". If we use technical terms from physics, the condition can be easily stated: "The 'law' of thermal expansion is rigorously true if there are no external boundary stresses on the bar throughout the process."[4, p. 284]

But how can we be sure any forces on the metal bar, say gravity by the earth or electric force by electric charges nearby, would not be a stress, which could influence the expansion of the metal bar? Even we agree with ERS' strict terminology, consider the temperature is raised higher than the melting point of the metal, would the length of the metal bar be still be proportional to its temperature? In fact, ERS do not mention the melting temperature at all in their strict or rigorous reconstruction of the thermal expansion law.

ERS give another example "... Kepler's 'law' that planets travel in ellipses is only rigorously true if there is no force on the orbiting body other than the force of gravity from the dominant body and vice versa."[4, p. 284]

But again, is "other than" terminology from physics? Even it is, would the ellipse law still hold if, say, the mass of the sun is increasing or decreasing because of certain chemical reaction, which is not "force" at all? If we consider all such interference, I am afraid that ERS' rigorous reformulation would finally have to expand infinitely.

Is a CP law testable? ERS mention two common views for testability of CP laws: (1) We can confirm the putative law that CP, all Fs are Gs by finding evidence that in a large and interesting population, F and G are highly positively statistically correlated; (2) We can confirm the hypothesis that "CP, all Fs are Gs" if we find an independent, non-ad-hoc way to explain away every apparent counter-instance, that is, every F that is not a G.

ERS think the former just lend confirmation to the stronger claim that in some broader class of populations, F and G are positively statistically correlated, that would not be a CP law. And the latter is not sufficient. Here is their counter-example: "CP, white substances (or compounds containing hydrogen) are safe for human consumption." Although we can explain away any white substance would be not safe for human by modern biology or medicine, it is not a law at all.

I can agree ERS' two objections are nice, especially the first, but I would like to raise another testability possibility for CP laws – their contrapositions are testable.[2] If we write a CP law "If CP, then L" as "CP→L", it is logically equivalent to "$\sim L \to CP$". We can easily get, say, "If not all Fs are Gs, then not CP".

Here I consider two interpretation of CP, "there are no interferences" and "other things being equal". I give the former the logical form $\sim (I_1 \vee I_2 \vee I_3 \cdots)$, Ii refers to different Interferences, which could be a infinite set. The latter can be written as $(E_1 \wedge E_2 \wedge E_3 \cdots)$, here E_i means various Equal conditions, which again could be infinite. So whenever F is not G, there must be at least one interference or one conditional unequal. So experimenters try to find the interference or unequal condition – when they finally do there is a disturbing factor or something unequal. I think that is a confirmation of the CP law!

So my testability argument can be summarized as the following logical steps.

For "there are no interferences" interpretation

1. $CP \to L \equiv \sim L \to \sim CP$
   A conditional equals to its contraposition.
2. $[\sim (I_1 \vee I_2 \vee I_3 \cdots) \to \forall x(Fx \to Gx)] \equiv [\sim \forall x(Fx \to Gx) \to (I_1 \vee I_2 \vee I_3 \cdots)]$
   We substitute the logical form for CP and L
3. $Fa \wedge \sim Ga$
   A counter-instance is found.
4. $\sim \forall x(Fx \to Gx)$
   L seems not hold.
5. $I_1 \vee I_2 \vee I_3$
   CP does not hold, i.e., there is inference.
6. I_n
   Scientists find the interference (or interferences), which confirms the CP law.

For "other things being equal" interpretation

1. $CP \to L \equiv \sim L \to \sim CP$
   A conditional equals to its contraposition.
2. $[(E_1 \vee E_2 \vee E_3 \cdots) \to \forall x(Fx \to Gx)] \equiv [\sim \forall x(Fx \to Gx) \to \sim (E_1 \vee E_2 \vee E_3 \cdots)]$
   We substitute the logical form for CP and L
3. $Fa \wedge \sim Ga$
   A counter-instance is found.
4. $\sim \forall x(Fx \to Gx)$
   L seems not hold.
5. $\sim E_1 \vee \sim E_2 \vee \sim E_3$
   CP does not hold, i.e., there is inference.
6. $\sim E_n$ Scientists find the unequal condition (or conditions), which confirms the CP law.

---

[2] I learned the idea from [5, pp. 441–450] They raise a contraposition argument against Nancy Caright's claim that the fundamental laws do not apply in the real world.

With regard to ERS's "white substances" counter-example, I think it involve the distinction of genuine laws and accidental generalizations. According to the best knowledge of modern science, we regard "CP, white substances (or compounds containing hydrogen) are safe for human consumption" as an accidental generalization rather than a genuine law, even if it is true. But generalizations of the strict form would face the same problem. "All gold are less than 10^6 Kg" would be true while not regarded as a law of nature. So I do not think ERS counter-example justify the untestability of CP laws.

2 Differential Equations and Supervenience

ERS argue that those so called "CP laws", say, thermal expansion law or Kepler's law, are just differential equations of evolution type. They are not laws at all. But I think, that claim would be inconsistent with Earman's so called "system approach"[1, p. 4] to the understanding of laws of nature.

In the discussion on laws of nature, there are mainly two camps in the philosophy of science. David Armstrong, Michael Tooley and Fred Drestke give a necessitarian view. They think a kind of physical or nomic necessity distinguishes the genuine laws from the accidental generalizations. But it is still difficult for them to work out an explicit definition of that necessity. So nowadays J.S. Mill, Frank Ramsey and David Lewis' view (therefore MRL) is more popular. John Earman is in this camp.[3]

MRL think laws are "consequences of those propositions which we should take as axioms if we knew everything and organized it as simply as possible in a deductive system".[8, p. 38] So "a contingent generalization is a law of nature if and only if it appears as a theorem (or axiom) in each of the true deductive systems that achieves a best combination of simplicity and strength".[7, p. 73]

Therefore, according to MRL, and Earman, "No sphere of uranium-235 has diameter greater than 100 meters" is a law of nature, while "No sphere of gold has diameter greater than 100 meters" is just an accidental generalization. Because the former belongs to our knowledge system of quantum physics, but the latter does not.

A system approach (MRL and Earman) would acknowledge even theorems (not only axioms) of our deductive system of the modern sciences are laws of nature. The thermal expansion law or Kepler's law are the consequences of modern physics, say, solid mechanics or Universal Gravitation Law. Why Earman insists they are just differential equations of evolution type, not laws of nature? I think his claim in the discussion of the CP law is inconsistent with his point concerning the law of nature.

ERS think there is a distinction between conditions for the truth of a law (CP) and conditions for the validity of its application (Provisos). They accept Hempels provisos but reject CP clauses. Conceptually, conditions for the truth of a statement may not equal to the conditions for the validity of its application. "The Indpendency Day of USA is July 4" is true; but is not necessary for me to apply it in my pursuing a lady. But the situation in fundamental physics is a little bit different.

The fundamental laws of physics are always abstract. It seems there is no direct way for us to justify their truth. Of course, it does not mean those fundamental laws are untestable. We can deduce something, usually with the help of bridge principles (or correspondence sentences), from the abstract laws, combined with non-nomic assumptions or initial conditions. From my point of view, that is a kind of application of the abstract laws to the real situations. Since these derivations are testable, we can confirm (or disconfirm) the truth of the fundamental laws. So the conditions for the truth of a law are closely related, if not logically equivalent, to their applications.

Consider ERS' supervenience thesis. They insist laws must supervene on the behaviors of physical systems. According to *Oxford English Dictionary*, "supervene" means "to come on or occur as something additional or extraneous; to come directly or shortly after something else, either as a consequence of it or in contrast with it; to follow closely upon some other occurrence or condition". So the supervenience thesis should remind us that laws always "come after" the real behaviors.

Here is Cartwright's nice analogy. She regards the relation between laws of nature and real situations as a kind of abstract-concrete relation, like morals and fables. She quotes Leesing's claim, "The general exists only in the particular and can only become graphic (anschauend) in the particular". Consider the moral: The weaker are always prey to the stronger.[2, pp. 37–43] We can find the real and concrete situation as described in Lessing's fable "A marten eats the grouse. A fox throttles the marten; the tooth of the wolf, the fox." I think her analogy gives a wonderful example of supervenience: the moral supervenes on the fable. In what sense is the moral true? It provides a nice idealization of the real situation, say, the relation between martens, grouses, foxes and wolves. It can be applied to the relation between other animals, perhaps humans and nations too. Suppose in a possible world, there are no animals, humans, nations or something like that. The moral need not be true any longer, since there is nothing it can supervene. Now suppose in another possible world, the electric charge of everything is removed because of a certain evolution of the universe, while other things (so laws) remain equal. Would the law of electromagnetic force, one of the four fundamental forces in modern physics, still hold? According to Earman's distinction among the conditions for truth and for application, one can argue the law of electromagnetic force still holds however it cannot apply any longer. But according to the supervenience thesis, the law cannot supervene any physical behavior since there is no electric charge at all. Again, I am afraid Earman's point on the CP laws is inconsistent with his understanding of the laws of nature.

3 Conclusion

Earman et. al. raise several objections to the *Ceteris Paribus* laws. In this paper, I argued that CP clauses could be ineliminable even with scientific terminology, and that it is also possible to test the contraposition of a CP law, therefore the law itself. Earman's account of differential equations may violate his MRL view of laws of nature. Again, Earman's view of laws of nature may be inconsistent with his supervenience thesis.

Bibliography

[1] Carroll, J. (Ed.). *Readings on Laws of Nature*. Pittsburgh: University of Pittsburgh Press, 2004.

[2] Cartwright, N. *The Dappled World*. Cambridge: Cambridge University Press, 1999.

[3] Earman, J. Laws of nature. In Balashov, Y., Rosenberg, A. (Eds.), *Philosophy of Science – Contemporary Readings*. London & New York: Routledge, 2002.

[4] Earman, J., Roberts, J., Smith, S. Ceteris paribus lost. *Erkenntnis*, 57:283 – 288, 2002.

[5] Elgin, M., Sober, M. Cartwright on explanation and idealization. *Erkenntnis*, 57:441 – 450, 2002.

[6] Lange, M. Natural laws and the problem of provisos. *Erkenntnis*, 38:234, 1993.

[7] Lewis, D. *Counterfactuals*. Cambridge, MA: Harvard University Press, 1973.

[8] Ramsey, F. *Foundations of Mathematics*. Altantic Highland, NJ: Humanities Press, 1978.

On the Evolutional Structure of Science Theory[1]

Hu Guang

Dalian University of Technology

huguang8111@163.com

The axiomatic structure of science theory is constructed completely during and as a result of scientific practice.

1 Scientific invention and the initiation of science theory

Scientific invention invents hypotheses to solve science problems which can be solved only within axiom systems. If one scientific problem can be solved by virtue of the original axiom structure, it is a process of discovery. If one scientific problem can't be solved with the original axiom structure because solving the problem using the original axiom structure will cause inconsistency or contradictions in the original axiom structure, then a new hypothesis must be invented and the original axioms or axiom structure must be denied. This denial establishes a new axiom structure. Although the axiom structure does not need to be perfect, the newly-established axiom structure must be effective for solving problems and guiding observation and experiment in future scientific practice to qualify as a new axiom structure. This is innovation and the process of invention.

One important question here is how the new hypothesis is raised. Since the new hypothesis denies the original axiom structure, it can be deduced from the original axiom structure through the logical process, which requires creative thinking, i.e. imagination, aspiration and instinct. Imagination, aspiration and instinct are not independent thinking methods. They are rather the establishment of the embryonic form of a new axiom structure based on the adoption of various pieces of knowledge, methods and practices which break the constraints of the original axiom structure under certain favorable conditions and through the mutation realized by rapid summarization in the cognition process. As a method of creative thinking, imagination, aspiration and instinct don't take the form of concept, judgment and logical deduction, but come and go suddenly. Also, this method of thinking cannot be exchanged between subjects and the achievements of science theory can't be provided directly. The formation of any science theory is always completed by logical thinking, must be expressed in logical form and established only through logical systems. The new logical system is not the old logical structure but a new one with breakthrough cognition.

---

[1] The author is with Dalian University of Technology, Dalian 116024 Liaoning.

2 Scientific interpretation and construction of science theory

Scientific interpretation transforms science problems or scientific facts into a special case. It is possible to perfect the axiom system by adding axioms or putting forward new principles. It is the three basic characteristics of structures raised by Piaget when expounding on their features, i.e. integrity, transformability and self-adaptability. Self-adaptability is the driving mechanism by which the axiom structure maintains its own stability and without which the axiom structure can't survive.

As an axiom system, the new theory is the revelation of general rules of certain mode of motion or material system in nature. Hence, when the interpretation and function of predictability achieve amazing success, it becomes the prospect for the scientific circle dominant in this epoch. Thus, people regard it as a kind of knowledge background or a method of cognition, as the rule of a scientific method or scientific activities to explore the unknown field or new scientific fields and guide the generation and formation of new scientific cognition. For example, contemporary Newton Mechanics was adopted by most scientists in the 18th Century as a kind of cognitive pattern and scientists were satisfied only after having included the new natural phenomena and form of material motion into the track of mechanics. Hence, various forces appeared. Why could materials be bonded together? It is because of a kind of affinity between materials? Why could organisms move? It is because of a force of life inside them. Electricity has electric force and magnetism has magnetic force. When Coulomb introduced Newton's mechanics into the science of electricity, he formulated Coulomb's Law in a manner similar to Newton's law for explaining mechanics. Dalton introduced it into physics and discovered the law of partial pressure and into chemistry and found the mass isochronization of chemical elements and atoms. As it was generally accepted over time, the axiom structure of Newton's mechanics became the standard pattern of science.

3 Scientific growth and isomorphism of science theory

Scientific growth has been an important topic in the study of the philosophy of science since it relates to the source and mechanism for knowledge generation, the criteria for knowledge appraisal and selection and also the link between new and old axiom systems. Any school of philosophy of science must have its own theory of scientific growth. To me, the evolutionist structure emphasizes that scientific growth is self-adapting and the perfection of the axiom system structure is fulfilled in the process of scientific practice. The axiom structure reveals the things' content, nature and relationship as a whole or from the angle of motion transformation. The axiom structure that applies to the entire system of scientific knowledge refers to the cognition that the overall state of objective things exceeds individualism and is of general significance. The axiom structure is absolutely not an empirical logical structure or rigid formal system but a motile adapting system that grows along with scientific practice. It plays an important role in stipulating the solution and digesting empirical materials. It is also the requisite methodologist principle for growth of scientific knowledge. The axiom system is the tool to solve problems and also rational confidence that consolidates the view of science and the world.The theory of electromagnetism in the 19th Century integrated achievements in electricity and magnetism from the 18th Century. Coulomb determined the law of acting forces between electric charges and magnetic poles and laid the foundation for electrostatics. Afterwards, Danish physicist H.C Oersted found that the live wire could cause deflection of nearby magnetic needles and revealed the relationship between magnetism and electricity. A.M. Ampere and J. Davy and others discovered convertibility between electric energy and magnetism. Then M. Faraday revealed the Law of Electromagnetic Induction and introduced the concept of field of force. J.C. Maxell inherited his predecessor's achievements, established the quantitative mathematical mode for Faraday's concept, and turned electromagnetism into an axiomatized system. He used only four partial differential equations to interpret all the known electromagnetic laws of that time.

The axiom system is a hierarchical system pattern and is a complicated, dynamic and self-adaptive system. The function of a scientific revolution is to accumulate experience and knowledge within the conceptual pattern comprised by the existing axiom systems. This newly added experience and knowledge results in adaptive changes in theory concept levels. The levels inside each axiom structure not only interact horizontally but also lead to new adjustments and organization patterns vertically which result in changes to the entire axiom structure so that knowledge, methods and principles of various disciplines interact, interpenetrate, integrate and transplant with each other to form the overall features of modern natural science.

4 Scientific logics and the reconstruction of science theory

Science theory has followed a long and windering path. When a certain number of concepts, categories, principles and theorems accumulate in a scietific discipline, they begin to get sorted out based on the logical relationship within the theory to determine systematically the status of various statements and make the axiom system more precise. One problem we need to clarify here is the relationship between axiomatization and formalization of scientific theories which make up an axiomatized structure. Although this axiomatization is of a different extent, some are quite apparent and complete from inception (as in mathematics and cosmology), some are not that apparent or perfect (as in geology and biology), and some become more apparent or perfect (as in physics and chemistry) after a period. In each case, the system starts with some inceptive axioms. These axiom structures could be added to or deleted from the whole but can't be denied absolutely. Piaget concluded in his "Structuralism" pamphlet that "the structure could be formalized. However, it could refer to soon after the structure is discovered or the inceptive period immediately after. It needs to explain here that formalizing structure is the theorists' task. The structure itself is independent to theorists. This formalization could be expressed by symbolic logic equations or regarded as the intermediate stage by the controlled mode. Hence, formalization may have different transitional stages and it depends on the decisions of the theorists. The existence of the structure he discovered shall be illustrated in each specific study field."[2, pp. 2–3] It is observed that the axiomatization of science theory is the essential feature of scientific theories and the formalization of scientific theories is the expression of axiom theories. Once science theory, formalized or not, becomes an axiomatized system, the formalization represents only its perfection to a certain degree. We know that extremes meet. The more perfect the axiom system becomes, the weaker becomes its ability to reflect accurately the outside world. The more stable it is, the more fixed and conservative its function becomes. It leads finally to its own collapse.

The perfection of an axiom system itself from axiom formalization to form axiomatization has strengthened the logic of science theory. The induction and conclusion are both established on rigorous logical rules. On the other hand, the compatibility of an axiom system can't be proven within the system. It must be within a stronger one or its axiom structure has to change. As scientific cognition goes deeper, the human practice with respect to objective nature provides objective material premises for the transformation of the axiom structure while the appearance of essential contradictions inside the science theory creates a solid base for transformation of the axiom structure. The development of grand scientific theories, i.e. scientific revolution must surely follow.

5 Scientific revolution and the reframing of science theory

In the history of scientific theories, nothing excites or upsets scientists more than scientific revolution. Since Thomas Kuhn published *The Structure of Scientific Revolutions*, the term "science revolution" has been agreed upon by most science philosophers and aroused many disputes. What is science revolution? When and how does science revolution take place? These questions haven't been satisfactorily solved. For example, according to Kuhn, science revolution is a paradigm shift, i.e. from an old axiom structure to a new axiom structure. Paradigms are incompatible and lead to cutthroat fights between competing axiom structures. The change from one paradigm to another reflects the reality that different science communities have different likes and dislikes without inherent and inevitable logical relationships. Kuhn thus wrote that, "When I wrote the book regarding revolution, I always said them as incidents that some scientific terms have meaning reforms. I also raised that as a result of the revolution, the incommensurability between different viewpoints and local interruption of exchanges between different theory supporters appear ... Now, I already believed that incommensurability and local exchange problems could be solved by another method. Supporters of different theories or paradigms, in accordance with the broad sense of this term, speak different languages, mainly to express different cognitions to adapt to this world. Hence, their ability in understanding each other's viewpoints will be constrained inevitably by the translation process and incompleteness of references determined."[3, p. 215]

I think that Kuhn's analysis of the inherent reasons for revolutions in science are not sufficient, since not all new theories are identified as science revolutions when they appear. How can we then claim that a new theory is a revolutionary theory but another one is not? How could we know to what extent that a theory develops when the revolutionary change takes place? Kuhn also realized that abnormality is a normal phenomenon in regular science. How could we know that abnormality A or B could become the start point of science revolution? Kuhn explained the extensive occurrence of abnormalities in crisis situations but doesn't reveal their fundamental causes. Regarding these questions, we need to review again from the angle of evolutionist structure of scientific theories to further understand the essence and significance of science revolution. I think that science revolution is the fundamental change of an axiom structure and the contradiction paradox within the axiom system has important significance in science revolution.

The new axiom structure could explain the scientific facts that could and those that couldn't be described by the original axiom structure. "On one hand, the new knowledge shall be consistent in logic and be extended from the existing knowledge at certain sense. On the other hand, the new knowledge shall be contradictory to previous knowledge and would be difficult to understand, unreasonable or absurd from the past viewpoint."[1] As the science develops, this new axiom structure adjusts continuously its own structure for further enrichment and perfection and transits from axiom formalization to formal axiomatization. We must be clear with one point that no matter how perfect the structure is, its contradiction, i.e. incompatible axiom system, can't be eliminated in this system. Once this contradiction is revealed by the paradox, a new round of science revolution commences. The new axiom structure announces the death of the old structure, but it also breeds the seeds of its own death. New theories are replacing old ones like this and the old structure fulfills its own revolution while adapting to the new structure. It is the evolutionary process of scientific cognition, scientific development and also scientific progress.

Bibliography

[1] Kopnin, P. V. *Dialectics Logic and Scienc.* East China Normal University Publishing House, 1981.

[2] Piaget. *Structuralism.* Commercial Press Printing Co., Ltd, 1986.

[3] Shu, W., Qiu, R. (Eds.). *Review on Modern Western Science Philosophy.* People's Publishing House, 1987.

The Extension of Vienna Circle Protocol Sentences Debates: A Comparative Study of W. Quine and P. Feyerabend [1]

Yuann Jeu-Jenq

National Taiwan University

jjyuann@ntu.edu.tw

    W. V. Quine (1908-2000) and P. Feyerabend (1924-1994) shared many essential ideas in both historical and methodological contexts. Both claimed to be descended from the Vienna Circle. Both inherited the result of the protocol statements debates taking place in the Circle and took it as a crucial part of their philosophies. Both resorted to something like conceptual schemes by which they interpreted all experiences, even the most commonplace ones (the idea of "ontology" for Quine and the idea of "theory" for Feyerabend). These similarities are not exhaustive, yet they offer a picture of Quine and Feyerabend as analogous in essentials and essences in their ideas.

    In what follows, I will first explore Quine's philosophy of science by distinguishing it from a relativistic stand without maintaining an observational core. This wavering position between relativism and positivism is best reformulated from the transition "from empirical stimulus to empirical content". Then, in contrast to Quine's exposition of the nature of science, I will examine Feyerabend's concerns in the same regard mainly considered from a methodological point of view. It demonstrates that theories would not be rejected by basic statements of any kind but could only be replaced by other theories. Finally, in the conclusion, I complete my argument that the replacement of theories is carried out by the negative role of empirical stimulus which is itself immediately incorporated into a theory.

1 Quine: from empirical stimulus to empirical content

Strictly speaking, when Quine talks about the aim of science, he fully demonstrates his stand of what I think is better called 'pluralism'. The tag is set because "I see it as defining a particular language game, in Wittgenstein's phrase: the game of science, in contrast to other good language games such as fiction and poetry"[11, p. 20]. Science, as a form of knowledge, does not repudiate other

---

[1]The author is with Department of Philosophy, National Taiwan University, Taipei, Taiwan.

forms on the basis of its own criteria of truth. This is a pluralistic position because science is one game among many. However, having said so, science does not cease to define itself, even though "It is idle to bulwark definitions against implausible contingencies"[11, p. 21]. Implausible contingencies might be very hard to believe, but science, insofar as it is an enterprise of fallibility and corrigibility, is ready to face them, as long as they hinge on checkpoints in sensory prediction. Science does not preclude any form of knowledge, even non-empirical phenomena such as telepathy and clairvoyance.

Science after such a convulsion would still be science, the same old language game, hinging still on checkpoints in sensory prediction. The collapse of empiricism would admit extra input by telepathy or revelation, but the test of the resulting science would still be predicted sensation[11, p. 21] The point boils down to this question–what does a 'checkpoint in sensory prediction' consist of? To Quine, the answer to this question has a great deal to do with his 'stimulus meaning' which constitutes the 'evidence' as the checkpoint to theory. Quine gives it a more explicit and neutral status by explaining: "A stimulation σ belongs to the affirmative stimulus meaning of a sentence S for a given speaker if and only if there is a stimulation σ' such that if the speaker were given σ', then were asked S, then were given σ, and then were asked S again, he would dissent the first time and assent the second"[8, p. 32].

We, as receptors of stimulation, reflect "in an evolving set of dispositions to be prompted by stimulations to assent or dissent from sentences. These dispositions may be conceded to be impure in the sense of including worldly knowledge, but they contain it in a solution in which there is no precipitating"[8, p. 39]. However, "stimulus meaning, by whatever name, may be properly looked upon still as the objective reality that the linguist has to probe when he undertakes radical translation continues Quine [8, p. 39]. The 'objective reality' Quine refers to here serves as an essential part of his idea of 'evidence' in his philosophy of science. The 'evidence' becomes clear in his exposition of 'observation sentences'.

Due to their direct and firm association with our stimulations, observation sentences serve as initial links in connecting sentences within the frame of scientific theories [11, p. 3]. Because observation sentences can serve as a means "of verbalizing the prediction that checks a theory" [11, p. 5], we can say hence that to Quine, a scientific community is practically also a linguistic community whose members assent or dissent unanimously with regard to a sentence given in a stimulatory situation. This corresponds with what Quine says: "that observation sentences serve in both ways - as vehicles of scientific evidence and as entering wedges into language - is no cause for wonder" [11, p. 5]. However, this does not mean that science equals language. Science, being one game among many, urges Quine to adhere to its normative nature by ruling out two things [?, p. 97]:

1. stored information which exceeds that which is involved in knowing how to speak the language;
2. the possibility of verdicts to observation sentences which turns on special knowledge possessed by some but not all speakers.

Quine holds firmly this position of maintaining empirical evidence in the realm of science, and therefore considers those who belittle the role of evidence and accentuate cultural relativism as "epistemological nihilism". Observation sentences play their role as evidence in science through theoretical hypotheses which, once formulated by observables, can be tested by experiments. The general character of theoretical hypotheses tested against the observational ground is called by Quine the "observation categorical" which is formulated in this form: 'Whenever this, that'. The function of the observation categorical in science links together theories projected to see what consists in the world and observations derived from the stimulatory situation. The link plays a double role in mediating the possibility of explaining the external world with reference to empirical evidence. To Quine, so long as the observational categorical can be formulated in terms of testable sentences, the empirical content of theoretical hypotheses is thus safeguarded; so is the empirical status of science.

2 Feyerabend: from observation sentences to theoretical proliferation

Being both heirs of the Vienna Circle, Quine and Feyerabend not so surprisingly share an essential idea concerning the causal nature of observationality' from the world we observe to the observation sentences we utter. With this causal relationship, both Quine and Feyerabend were able to avoid the traditional problems of 'sensory core' or 'the myth of given' by concentrating instead on a persons natural ability to link the external world with human reactions which are uttered by people as observation sentences in the stimulatory situation. Feyerabend describes it clearly: "In a word: observation statements are not just theory-laden but fully theoretical and the distinction between observation statements (the 'protocol statement' in the terminology of the Vienna Circle) and theoretical statements is a pragmatic distinction, not a semantic distinction. Quine, *whose philosophy shows close connections to the philosophy of the Vienna Circle, also used a criterion of observability that is rather similar to mine.*" ([4, pp. 228–229]; italics mine for emphasis). From this paragraph, we learn that what Feyerabend meant by 'theory' equals what Quine meant by 'ontology'. With this equivalence, we can see that the two 'heirs' of the Vienna Circle referred to something which dates back to the protocol sentences debates in which they had participated.

2.1 The Legacy of the Vienna Circle

The key figure of the protocol sentence debates is undoubtedly O. Neurath, who–against the traditional view of the Circle–held an anti-foundational view of observation sentence by stressing that, "Thus statements are always compared with statements, certainly not with some 'reality', nor with 'things', as the Vienna Circle also thought up till now". [6, p. 53] Though Neurath began with this view as a member of the minority in the Circle, he convinced Carnap later on with his anti-justifactory stand. This was manifested by Kraft, who offered an important clue concerning a hidden link among Carnap, Neurath, Popper and even himself. Kraft said that when Carnap held the absolute validity of protocol sentences, Neurath fought this view by holding strongly that "Protocol sentences are no more original than other statements; like others they can be corrected; they, too, are only hypothetical" [5, p. 193]. Kraft continued that Carnap later accepted this view which was rather obvious for Popper as he held that "it was a necessary consequence that the basic sentences are only hypothetical, for they are statements about objective facts which always include theories" [5, p. 193]. Therefore, a protocol sentence contains no indubitable truth, but only hypothetical validity. All this is further confirmed by Carnap in his paper "On Protocol Sentences" in which he categorized both Neurath and Popper in the camp of anti-absolutism while granting that Popper's position was even stronger. The outcome of this development is called by Feyerabend 'the pragmatic theory of observation': "The new theory of observability which results from the described procedure (and *which was formulated very clearly in the early thirties by Popper, Carnap and Neurath [my italics]*) may be called the pragmatic theory of observation. The choice between the pragmatic theory and the semantic theory is of course purely a matter of convention. However, if it is our intention not to except any part of our knowledge from revision, then we shall have to choose the pragmatic theory"[3, p. 125].

From this citation, we see in the Vienna Circle the 'naturalistic side' which boils down to a pragmatic theory of observation. In fact, Feyerabend's pragmatic theory of observation is, as a matter of fact, similar to the naturalism of Quine. However, while Quine designated stimulus meaning as the essential nature of sensory prediction in science, Feyerabend stressed instead that observation sentences assume "a special position not by their meanings, but by the circumstances of their production"[2, p. 212].

2.2 Feyerabend's Naturalistic Methodology of Scientific Knowledge

Feyerabend begins his analysis of the pragmatic theory with an exposition of the production of an observational sentence from a specific observer. The physical and psychological processes from which observational sentences are derived would NOT 'speak', [for not] "being able to give meaning to sentences that have not yet received any interpretation" [2, pp. 212 – 213]. These processes exert observational sentences as appropriate responses to the 'cause' of their production. These responses are 'actual interpretations' adhering to the causal formation of the observational sentences. They have a great deal to do with Feyerabend's 'idiosyncratic interpretation' of the meaning of theory which is theory-laden.

The theory has an inbuilt syntactical machinery that imitates (but des not describe) certain features of our experience. This is the only way in which experience judges a general cosmological point of view. Such a point of view is not removed because its observation statement says that there must be certain experiences that then do not occur ... It is removed if it produces an observation sentence when observers produce the negation of these sentences[2, p. 215].

Note two points from this citation: 1) that experiences would not be able to present themselves unless they are incorporated into a theoretical framework; and 2) that a theory would stop us from seeing things beyond its range unless the supra-theoretical part of sensory experience hints at its negation. Hence, pre-theoretical experiences, though theory-independent, are not arbitrators of the fate of a theory, but signs moving towards the removal of the theory. The removal would not even begin unless the old theory is in the first place in competition with a new one or with more new theories. Hence, we can see why Feyerabend proposes his proliferation of theories idea.

What can we do if we are limited within the tenacity of a specific theory without being aware of the limits? This question triggers doubts in Feyerabend's relativism which implies that all theories are on a par with one another. However, this is not what Feyerabend meant by the human experience which emerged from observation. What would be the function of human experiences here? They function as something either beyond or beside the tenacity of the theory and cause the 'observer' to utter certain sentences which do not match the predictions made on the basis of the 'theory'. This means that the observer, though uttering 'meaningful sentences' on the basis of a 'theory', is still able to utter 'other' sentences even without knowing 'another theory'. What would be the function of these 'other sentences'? They reveal the possible discrepancy between the 'physical order' of the theory working as a cosmological point of view and the 'natural order of sensations'[2, p. 215].

Pluralistic methodology is direct proof, confirming that experience cannot be fully exhausted by the structure of the 'theory'. The way to detect this fact is through the presence of experience which urges the observer to utter a sentence negating what the theory would predict in its tenacity. However, this 'negation' does not hence imply whether the prediction sentences are true or false; all it refers to is the disagreement between the 'physical order' of a theory and the 'natural order of sensations'. So, we should ask at this stage, what would be the function of the 'disagreement' here? Nothing, according to Feyerabend, if it is not accompanied by the guidance of a theory. What kind of guidance? Feyerabend responds: In order to be able to expand our field of action, the theory must guide us into new domains. It must also make us *critical* of our actions so that we may find out which actions are based on strong causal antecedents and which are not. *Only the latter ones will be valuable indicators of external events*" (the italics are mine; [2, p. 215]). Here Feyerabend coherently argues his support for a pluralistic methodology which plays the role of expanding our empirical knowledge through the constant removal of current theories by their alternatives.

3 Conclusion

With this view of expanding our empirical knowledge, we certainly receive a normative view of science from Feyerabend's exposition. This view corresponds to both the practice as well as the ideal of science. Indeed, science should be an enterprise in which the growth of knowledge characterizes its nature and aim. Though Feyerabend frequently later claimed to be an epistemological anarchist, his objective of making provocative concepts was by no means urging people to do nothing but safeguard pluralistic methodology. With this objective in mind, we can see that the results of the protocol sentence debates once widespread in the Vienna Circle evolved into a positive methodology under Feyerabend's elaboration. While Quine thought that 'assent' and 'dissent' regarding stimulation held equal weight in deciding the nature of a global stimulus, Feyerabend stressed the negative side of stimulation. Feyerabend was able to elaborate his methodology mainly because of his emphasis on the negative role played by the empirical stimulus.

Bibliography

[1] Bechtel, Stiffler. Observationality: Quine and the epistemological nihilists. *PSA*, 1:93 – 108, 1978.

[2] Feyerabend, P. Problem of empiricism. In Colodny, R. G. (Ed.), *Beyond the Edge of Certainty. Essays Contemporary Science and Philosophy*. Englewood Cliffs, NJ: Prentice-Hall, 1965.

[3] Feyerabend, P. *An Attempt at a Realistic Interpretation of experience* Philosophical Papers Vol. 1, vol. 1. Cambridge: Cambridge University Press, 1981.

[4] Feyerabend, P. *Against Method*. London: Verso, 2 edn., 1988.

[5] Kraft, V. Popper and the vienna circle. In Schilpp, P. A. (Ed.), *The Philosophy of Karl Popper*. La Salle, Il.: Open court, 1974.

[6] Neurath, O. *Philosophical Papers 1913-1946*. Dordrech: d. Reidel, 1983.

[7] Quine, W. V. On what there is. In *From a Logical Point of View*. Cambridge, Mass.: Harvard University Press, 1953.

[8] Quine, W. V. *Word and Object*. Cambridge, Mass: MIT, 1960.

[9] Quine, W. V. Epistemology naturalized. In *Ontological Relativity and other Essays*. New York: Columbia University press, 1969.

[10] Quine, W. V. *From Stimulus to Science*. Cambridge: Mass: Harvard University Press, 1995.

[11] Quine, W. V. *Pursuit of Truth*. Cambridge: Harvard University Press, 1996.

Sociology of Science and its Limitation to the Rational Justification for Science[1]

Zheng Huizi

Henan University

zhenghzi@163.com

The purpose of this paper is to explore in what sense the sociology of science can be a subject which provides a kind of justification for the rationality of the existence of science. The sociology of science, as a label, can be used in at least two different meanings: in one sense, it deals with things relating to the scientific community; in another, it is discussed within a range of social systems. As a matter of fact, the emergence of the sociology of science in the second sense is one of the results of the violent challenges which science has encountered after it became a kind of social institution. I will discuss this point mainly. What I am really interested in is the problem of whether the sociology of science can justify the rationality of science when it is regarded as a kind of social institution. This issue, I think, is not only an important one, but one that needs to be seriously clarified by sociologists of science.

We all know that science, as a social institution, began to influence the development of society more extensively and deeply than ever since its origins in its modern form, especially since the Industrial Revolution in the 19th Century. This influence, however, goes far beyond human expectations in two aspects. On the one hand, we have enlarged our welfare by using the knowledge which was produced by science; on the other hand, we have encountered many unprecedented social problems while we search for science's social values. As a result, after its its socialization, science has been called into question and has been opposed by and from every aspect of society. As Susan E. Cozzens argues, "science and technology are in society, and that they do not sit comfortably there. ...science and technology have become elements in most of the critical issues facing humanity". [4]

Thus, to this extent, the sociology of science, after it came into being as a subject, assumes a rather important theoretical and practical responsibility to verify or justify that science is a kind of fundamental human activity. In fact, as early as the 1930s, in *The Social Function of Science*, J. D. Bernal talked about the problem of the necessity and pressure for the justification for science's right to exist. He said, "Now that science appears in a destructive as well as a constructive role, its social function must be examined because its very right to exist is being challenged. The scientists, and with them a number of progressively minded people, may feel that there is no case to answer and that it is only through an abuse of science that the world is in its present state. But this defence can

---
[1] The author is with Center for Studies of the Yellow River Civilization and Sustainable Development, Department of Philosophy, Henan University, Kaifeng 475001 China.

no longer be considered to be self-evident; science must submit to examination before it can clear itself of these accusations."[2, p.1]

Generally, there are two kinds of justifications people often make for science. The first one essentially treats the cognitive activity of science as an independent system. A remarkable characteristic of this kind of justification is that it excludes the exploration of the interactive relationship between the social factors of science and social goals. The other one, radically different from the first, puts science in the social system inherently. And thus, we harvest quite different significance and results.

The first kind of justification treats science as a sort of pure cognitive activity seeking for truth. In fact, this is the modern continuity of the tradition of science in the Aristotelian sense. This viewpoint of science maintains that the activity of scientific cognition itself is the purpose, and that science is only a kind of free exploratory activity which aims to discover the world and seek for truth without any other pragmatic purpose in mind. It is true that this point of view has been pervasive among scientists ever since Aristotle. However, the image of science has changed a lot since science became a social institution and after it became professionalized. Methodologically, because this kind of justification is carried out from the point of cognitive activity, it implies that this exploration is accomplished by focusing on fairly simple items such as scientists, research objectives, scientific theories, and used languages, and on the fundamental relationships between them. And just because it does not take into account the interaction and the results between science and society, this kind of justification seems invalid. Even for philosophers of science themselves, they begin not to take it for granted, especially after they encounter violent criticism from scientific historicalism.[5, 6]

The second justification for science regards science as a social activity. The best and the most detailed justification, as far as I know, is made by Bernard Barber, even though it was made fifty years ago. Here, what interests me most is why science brings about unavoidable social responsibility, on what condition do scientists have to assume responsibility for all these social problems. Further, this is to ask if science should assume responsibility for all the unavoidable social problems it arouses.

Barber maintains that, if we want to clarify science's social responsibility, we should first of all find out what characteristics science has when it is regarded as a social activity. Generally, the social consequences of science are unavoidable. These social consequences are caused through the interaction between science and society. These social consequences, especially in the long term, are unforeseeable. Consequently, the social consequences of science are a social problem concerned with social arrangements and social values. This problem cannot be dealt with by science itself. And it is these social problems that give rise to the social responsibility of science.[1, pp. 297 – 298] Furthermore, Barber points out the nature of science's social responsibility. He says that, "All this will, perhaps, make it clear that neither scientists taken as a whole group nor any individual scientist alone can be considered responsible, in any sensibly direct fashion, for the social consequences of their activities. The very specialization and interdependence of the parts of our society implicate every one of us in these social consequences. ......Science can be given no exclusive responsibility, that is to say, for the social and political problems for which all members of the society must take some measure of responsibility. The social consequences of science, so-called misleadingly, we have now seen, are social and political problems that can only be managed by the social and political process, to the extent that they can be managed at all. Even if they wished to do so, scientists could not be allowed to pre-empt the social and political function in society."[1, p. 298]

Thus, it is clear that "it is of the very nature of our society that social responsibility is largely a matter of moral obligation voluntarily assumed, and this holds for all of us, scientists and non-scientists alike". Accordingly, in exploring the issue of this unavoidable social problem, Barber finds that "the scientist has no peculiar or exclusive social responsibility".[1, p. 300] For Barber, the statement that scientists should assume all the responsibilities, and that scientists should focus their eyes only on pure science and not have to care about the social outcomes of their scientific results, must be rejected. At the same time, Bernal maintains that scientists should not disclaim or evade the moral obligation of this social problem, even though this social application cannot be controlled wholly by scientists themselves. [3, pp. 3–4]

From the above mentioned, we can conclude that the second rational justification for science is more persuasive than the first. There is not an inherent relationship, as is shown by the explorations of the sociology of science, between scientific activities and their social consequences. It is science's institutionalization that brings about these problems. And this institutionalization is determined by a given social arrangement and social objectives. Further, the scientists' work and their scientific achievements are a kind of "special social responsibility" which science as an institution has to or should assume in order to satisfy its society.[7] Needless to say, this society is the one in which these scientists are arranged and the one by which these scientists are influenced. As a matter of fact, science cannot become a kind of social institution if science does not provide its society with these special products.

Seen from the challenge to the living right that science has encountered, the sociological analyses of science have not rightly responded to this problem. Clearly, this is caused by the inert limitations of the sociology of science as a subject. Thus, this verification or justification which the sociology of science provides is neither sufficient nor complete. The problem lies in that the sociological research for science is essentially a spatial analysis for science. In other words, this verification or justification is achieved by carrying out the spatial analysis of the interaction between science and other institutions such as politics, economics, the military and religion, etc. And that is what the sociology of science as a subject can do.

The substantive characteristic which the sociology of science discovered only indicates the spatially existing state of science as a kind of institution. And this spatial state reflects some interactive relations between science and other institutions. To sum it up, the verification or the justification which science makes only interprets the utility of science to other institutions. Needless to say, this utility of science can not only explain the social value of science, but it can also illustrate primarily that the aim of scientific activity is subject to that of our society as well. And it is this latter point that justifies the rationality of science.

What the sociology of science can not do, however, is a fundamental problem that is why science as an institution is an indispensable part of our society. That is to say, research in the sociology of science can not explain the problem "why our society must have such a special activity like science", a rather radical question for the sociology of science itself. Further, the sociologists of science even can not raise this question because it is beyond the range of their research. And this is an insurmountable barrier for the sociology of science to cover when it wants to justify science. Thus, we should scrutinize science against an extensive background of cultural anthropology if we want to provide science with an ultimate rational verification or justification. From the point of view of cultural anthropology which is based on evolutionary thought, we can see clearly how man as a species, contrasting with nonhuman life, can evolve by depending on his culture. Science, needless to say, is a fundamentally dominant part of that culture. Here it determines the rationality of the existence of science as a kind of most elementary human activity.

Bibliography

[1] Barber, B. *Science and the Social Order.* New York: The Free Press, 1952.

[2] Bernal, J. D. *The Social Function of Science.* London: George Routledge & Sons Ltd, 1944.

[3] Bernal, J. D. *Science in History.* London: Watts, 1954.

[4] Cozzens, S. E. The disappearing disciplines of sts. *Bull. Sci. Tech. Soc.*, 10:1 – 5, 1990.

[5] Kuhn, T. S. *The Structure of Scientific Revolution.* Chicago: The University of Chicago Press, 3nd edn., 1996.

[6] Laudan, L. *Progress and Its Problem.* Berkeley & Los Angeles, London: University of California Press, 1977.

[7] Zheng, H.-z. Social responsibilities of scientists. *J. of Wuhan Uni. of Sci. & Tech*, 3(4):111–114, 2001.

A Scientific and Philosophical Study of Fengshui Theory and Practice[1]

Li Jingjing

China University of Petroleum

lijingjing@yahoo.cn

ABSTRACT. This paper probes fengshui practice and theory – local knowledge – from a scientific and philosophical perspective, which offers a beneficial case for the relevant study of local knowledge. Should local knowledge or ancient China's non-standardized knowledge be classified as a certain kind of science or partly science? Attempting to deal with these problems, this paper examines fengshui in a scientific and philosophic way, especially in light of the new developing theory – philosophy of scientific practice in recent years.

1 Introduction: Fengshui practice and theory and changes in contemporary scientific perspective

Fengshui has always been a controversial issue. In recent years, studies in the contemporary philosophy of science, sociology, anthropology, and cultures in different areas and nations, and especially hermeneutics, a branch of continental European philosophy, have revealed the following characteristics of science:

First, contemporary natural science is not the only modality of science. Second, scientific knowledge is not the only way to understand truth, because when people search for truth, they need to consider the whole of nature, and those knowers and actors interact with it. All of these factors reflect the truth. Third, the contemporary philosophy of science is in a process of evolution. Different genres of philosophy develop and perfect their own theories along with their animadversions; they come up contradictory standards for delimiting science and non-science. The subsequent perspective on the philosophy of science tends to consider "science" as a process, a kind of social practice and a way for communication between humans and nature. The newly developed philosophy of scientific practices clearly defines science as a kind of practice skills rather than solely rational belief. For example, Rouse, one of the presenters of philosophy of scientific practices, points out that, "I advocate that we should try to understand the concept of science in the area of scientific practices. ..." We find that some local and existing knowledges are based on the understandable and possible practices of using equipment, technologies, social roles, etc. These practices are "not for the purpose of applications, but they reflect that practical skills and

---

[1] The author is with Department of Social Science and Humanities, China University of Petroleum, Beijing 102249 China.

manipulations would achieve a decisive and expected goal because of their won characteristics."[7, pp. xii-xiii, x, xi]

More importantly, in the philosophy of scientific practices, some new perspectives which are totally different from the traditional philosophy of science are brought forward. First, science is classified in the area of practice activities; second, scientific practices are definitely local and social practices. Therefore, scientific knowledges based on local practices are not universal knowledges at the very beginning. They reserve their own important local characteristics. The universal characteristics of scientific knowledges serve only as appearances; they are not the result of localization but standardization.

After the emergence of these philosophies of science, we get excited and doubt at the same time. What makes us excited is that some of the so-called "non-scientific knowledges" which are excluded by the traditional philosophy of science seem to have the practical characteristics of communicating with nature. They can be treated and studied as "scientific" knowledges. What puzzles us is that whether the traditional Chinese knowledges like traditional Chinese medicine and even Fengshui practice and theory can be treated as research objects in the system of the philosophy of science, is still out of our grasp. Fortunately, science is classified in the area of practical activities and a kind of local knowledge. Then we can question further: is Fengshui a way of communication between Chinese people and nature? The answer is "yes". Of course, logically the statement "science is classified in the area of practical activities and a kind of local knowledge" doesn't indicate that all the practical activities and local knowledges are science. However, since Fengshui is a special and local way in which Chinese communicate with nature, it at least indicates that Fengshui has some sort of parallel and similar qualities with contemporary science which enable us to compare science with Fengshui practice and theory in our research.

The newly developed philosophy of scientific practices does not discuss the delimitation issue. However, it would definitely and logically lead to issues which are materially concerned with delimitation. Then what can be used to judge which kind of study or research is a scientific one? The answer is practice – rational, effective and standard practice. Then we have this question–is Chinese special Fengshui practice and theory a rational, effective and standard practice?

All the reasons above offer us a rational base for further analysis and study of Fengshui practice and theory in a scientific philosophical view. Meanwhile, the study of Fengshui practice and theory at least offers us a beneficial case for the relevant studies of local knowledges.

2 Practice areas, local knowledges and Fengshui practice and knowledge

2.1 The practice characteristics of scientific knowledges as practice areas and Fengshui practice

Scientific knowledges come from the process of interactive practice between humans and nature. That's a truth beyond reproach. However, the traditional philosophy of science considers the importance of scientific knowledges not based on their origin but on their abstract essential as universal knowledges above practices. Practices are the origin of scientific knowledge. They only work at their origin. Subsequently, scientific knowledges remove their appearances of practices and become a series of abstract and universal statements and concepts. This scientific philosophical view is what we have now; and all of our perspectives and opinions about scientific knowledges are based on this traditional and standard philosophy of science.

However, the newly developed philosophy of scientific practices has a different point of view. Rouse, for example, points out his opinion in the prologue of KNOWLEDGE AND POWER (Chinese Version) that "knowledge is not only an superficial idea (such as a text, a way of thinking or a chart), but an existing interactive mode. This kind of mode contains phenomena, presentational objects, and corresponding scenarios. Only in these scenarios can superficial ideas be understood and make sense when they are connected with practices".[8, p. 2] Therefore, in the view of philosophy of scientific practices, scientific knowledges have practice characteristics. It doesn't mean that scientific knowledges haven't been abstracted from practical operations, but it indicates that all the abstracted practical operations and scenarios do not lack of the scientific elements; they're always connected with science. Upon that, according to the illumination of philosophy of scientific practices, the scientific knowledges of today based on modern science still have fundamental connection with practices. Why not the Fengshui practice and theory as one of the traditional survival practices for Chinese to communicate with the nature?

When we are using practical knowledge to study Fengshui practice and theory, our attention should focus on how Fengshui practice and theory form the initial criterions via many construction siting practices, and how these criterions are constructed as a series of major modes which direct people to carry on Fengshui practice in return.

Fengshui practice and theory has a huge and complicated content. After a continuous practice over thousands of years, Fengshui practice has constructed its own criteria, that is, when choosing an optimal site according to Fengshui theory, generally people need to consider four aspects: dragon finding, sand monitoring, water inspecting and pinpointing. Looking at the siting process above, Fengshui practice is based on observations of intuitive and a sentimental experience. It's a factual survey and study process from outside to inside, from the whole picture to the detail. The most important part of this process is the local mountains and the water there. The next is people's feelings and emotions (the intuitional feelings of the relationship between humans and nature, aesthetic feelings and enjoyment, etc).

2.2 "Local knowledges" and Fengshui practice and theory

When talking about "local knowledges", "it doesn't mean any specific, characteristic knowledges with local qualities, but a new type of knowledge concept," and " ' local' or 'native' is not a concept based only on specific areas; it also comes down to the specific context where knowledges are created and advocated. This context contains values and ideals formed in specific historical conditions of culture groups and subculture groups, positions and perspectives determined by specific interest groups. The meaning of 'local knowledge' is that: because knowledges are always created and advocated in a specific context, we'd better emphasize the detailed context when we review knowledges and their universal principles."[9] Therefore, Joseph Rouse thinks that: basically speaking, scientific knowledges are local knowledges, they're reflected in practices, and those practices cannot be abstracted completely as theory or formula separated from specific context for the purpose of application. Based on this concept, Fengshui theory and other traditional knowledges, as a kind of local knowledges, are gradually accepted by people.

Certainly Fengshui practice and theory are a kind of local knowledge. However, this affirmation is misapprehended in some ways. This misconstruction is initially from the attitude that Fengshui practice merely has Chinese characteristics. In other words, we probably treat Fengshui practice merely as Chinese characteristic local knowledge. Of course, that is not right. The localization of Fengshui practice and theory means that we need to track back to the root of this knowledge and in what specific context it was created, applied and advocated. The values and ideals formed in a specific historical context of culture groups and subculture groups also need to be considered and explained.

What's the relationship between Fengshui practice and Chinese in the relevant period? Fengshui practice and theory is actually a way for Chinese to prudently inspect surroundings, acclimatize to nature, make use of and change the nature properly, and to create a harmonious living environment in which the climate, geography, and human beings are perfectly syncretic.[3] Looking at Fengshui practice, we know that Chinese always put their greatest emphasis on their surroundings. This kind of practice and activities actually reflects a great or maybe the first concern about "micro-environment, micro-topography". What's more, Fengshui practice is not merely about nature, but the relationship between human and the nature. In the Fengshui practices and activities, people are able to make full use of their subjective initiative to create a comfortable and harmonious living environment. Heidegger's "poetic habitation" image displays its practicality fully in the Chinese Fengshui practices and activities. For example, the influence of micro-topography on microclimate and its environment can be explained and recapitulated in the Fengshui practice and theory, in which the micro-topography is compared to an "aperture". This aperture is often surrounded by hills on three or four sides, with a north high south low topography. It can be a restrained-type basin or mesa facing the sun and seated in the shade, or even a manual topography. That is the typical pattern of "aperture", which is a place "gathering winds" and the optimal and ecological Fengshui structure. Some scientists make a comparative analysis of this topography in contemporary scientific perspective and have discovered that the explanation and cognition of relations among micro-topography, microclimate, ecology, and natural landscape are totally scientific.[4, pp. 26–32]

What's the relationship between Fengshui practice and natural views and cognition in relevant period? First, it involves Chinese culture. In Chinese culture, human life should maintain a harmony with nature is one of the main features of the natural view in ancient China. There has always been a fundamental distinctness in natural views between ancient China and western countries. The former one emphasizes the relationship between humans and nature; humans are included in any physical nature mentioned. The latter emphasizes nature itself; humans are excluded in any physical nature mentioned. Secondly, this question also involves the standardization of Fengshui practice under the control of Chinese culture. Looking at the Chinese traditional way of thinking, we find that the former practices carried out by typical characters are handed down and theorized by later generations. Even if later generations no longer or hardly practice the art, their practices are carried out in the shadow of theorized former practices. The five elements theory, vitality theory and Yin-Yang theory are also the important scenario factors in ancient Chinese culture, thus Fengshui practice has been more or less influenced by them. Moreover, Fengshui has a close relationship with "gas"; "gathering the wind and gas to gain water" become the standard for choosing an optimal living place. These theories also follow the concept of "human life being in a high harmony with nature". However, when it comes to the root, we have to consider Fengshui practice as a kind of practice from people's daily lives. "The understanding of our surroundings is neither a series of beliefs or object-oriented regulations, nor the traditional 'cognition'; it includes all the skills and abilities inside us (that is, the practical knowledges). Only by assembling us, the practical objects and the specific surroundings as a whole can we fully understand our world and the real meaning of science. And then our interpretative behaviors can also be understood."[6]

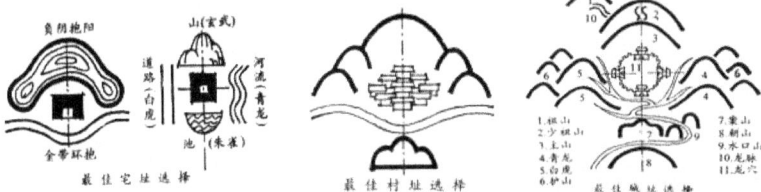

Figure 1: Optimal layout and sites of house, village and city

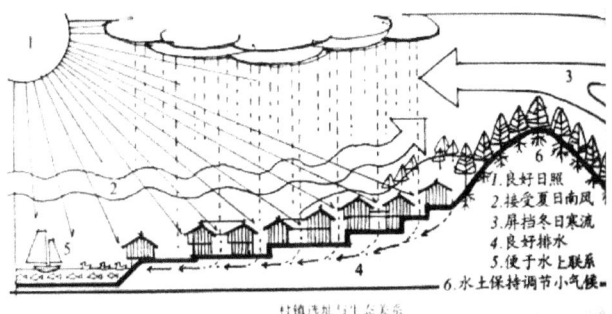

Figure 2: The scientific explanation of Fengshui layout

Why are there so many mysteries in Fengshui practice and theory? Is Fengshui practice and theory a kind of scientific local knowledge or a sort of witchery? In the traditional classification of Chinese knowledge, especially the orthodox view of history, Fengshui goes by the name of "Kan-Yu", which is a kind of so-called

Fang technology recorded in the book FANG TECHNOLOGY MEMOIR. However, Fang technologies and mathematics don't distinguish between rational and nonrational methods. They even confuse Confucians and wizards in some ways. Here "wizard" doesn't mean something nonrational. Therefore, the real meaning of "wizard" today is different from the meaning it supported in ancient China. In addition, Fang technology tends to be something mystical. So we cannot consider Fengshui practice as nonrational and a mysterious practice merely based on its mysterious parlance. However, Fengshui practice and theory certainly are a mixture of both scientiifc knowledge and wizardry.

We have only one world, and this world is divided into different places. Each knowledge system in different places flashes a side of this knowledge diamond. Natural philosophy based on five element theory, vitality theory and Yin-Yang theory form the scientific background of ancient China. We can't ignore the natural parts of Fengshui practice and theory just because of the blend of humanization, nor can we deny its strong practical characteristics merely because of the theorization of its philosophy.

Bibliography

[1] Cai, D. *Fengshui Practices in History*. Shanghai: Shaihai Science and Education Press, 1994.

[2] Chen, J. *Science Delimitation – Distinction between Science and Nonscience*. Beijing: East Press, 1997.

[3] Feng, J., Wang, Q. Study and research on fengshui theory. In *Studies on Fengshui Theory*, 3–4. Tianjin: Tianjin University Press, 1992.

[4] Guo. The chinese fengshui layout, entironment and landscape. In *Studies on Fengshui Theory*, 26–32. Tianjin: Tianjin University Press, 1992.

[5] Liu, P. *Fengshui – Chinese Environment Concept*. Shanghai: Shanghai Sanlian Bookshop, 2001.

[6] Qiu, H. Practical scientific concept. *Studies on Dialectics of Nature*, 2:19 – 22, 2002.

[7] Rouse, J. *Knowledge and Power, Toward a Political Philosophy of Science*. Ithaca and London: Cornell University Press, 1987.

[8] Rouse, J. *Knowledge and Power*. Beijing: Peking University Press, chinese edn., 2004. Translated by Xiaoming Sheng.

[9] Sheng, X. The constitution of local knowledge. *Philosophy Study*, 12:34–36, 2000.

New Epistemological Ideas Formed by Sino-Western Academic Interaction in Late Ming Times[1]

Shang Zhicong
The Graduate University of CAS
shangzc@gucas.ac.cn

ABSTRACT. The Principles of *Gewuqiongli* were important epistemological ideas utilized by scholars in the Sino-Western interaction of knowledge and epistemological ideas in Late Ming times. Along with Sino-Western knowledge exchange, intense interaction in epistemological ideas between China and the West was conducted during the Late Ming. As a result, the Principles of *Gewuqiongli* were formed. Being general epistemological ideas, the Principles of *Gewuqiongli* adopted the ideas of *Jiwuqiongli* and cognition by instinct from China. The idea of a clear concept and proposition and cognition by deduction from the West as well were expressed in two parts. The first principle emphasized cognition by instinct but provided no formal rules and was in the end hardly used. The second emphasized deduction using formal rules and was widely disseminated and used.

In Late Ming times, Chinese scholars and Western missionaries carried on an intense Sino-Western academic interaction regarding knowledge and epistemology. These discussions helped to form the *Gewuqionglizhixue* (learning about investigating the principles of things, 格物穷理之学) and the *Xiushenshitianzhixue* (learning about cultivating one's morality and serving God, 修身事天之学) as new forms of knowledge. In fact, the Chinese and the missionaries simultaneously carried on an intense and deep level academic interaction. The interaction in epistemological ideas led to the development of new ideas and concepts and the emergence of Principles of *Gewuqiongli* (格物穷理). The new epistemological ideas were used to support interactions regarding knowledge. I will discuss the new epistemological ideas in this paper.

1 The Background for the Interaction of the Epistemological Ideas

Some scholars, such as Liang Qichao(梁启超), Xu Zongze(徐宗泽), He Zhaowu (何兆武) and Fan Hongye (樊洪业) once discussed how academic interaction made Chinese academic methods change, with the result that such concepts as deduction and some specified mathematical techniques were introduced into Chi-

---
[1] The author is with The Graduate University of CAS, Beijing 100049 China.

nese research about mathematics, astronomy and other academic fields[2]. Unfortunately, they never epistemologically explained the occurrence of the new methods.

In fact, the Chinese and missionaries adopted a Confucian concept, *Gewuqiongli*, and used it basically as an epistemological concept. The concept was derived from *Gewu Zhizhi* (格物致知), which occurred first in the chapter *Liji* (礼记) of the *Daxue* (大学), with the original meaning of 'discovering the essence of things from phenomenon, and confirming the order of things'[11]. Well, later Confucians gave various explanations for the concept in the Han, Song and Ming dynasties, among which, the two representative theories were respectively given by Zhu Xi (朱熹) and Wang Yangming (王阳明). Zhuxi divided the mind-thing and the knowledge-principle and emphasized their identity. The principle is the essence of the entity and disperses inside things as the principles of things. Knowledge is inside the human mind. According to Zhuxi, investigating principles inside things is the process of obtaining knowledge from the mind. So, Zhuxi explained Gewu Zhizhi as Jiwu-Qiongli-Zhizhi (即物-穷理-致知). Jiwu means accessing things. Qiongli means investigating the principles of things. Zhizhi means obtaining knowledge.[13]

In contrast to Zhuxi's theory, Wang Yangming insisted on the identity of mind-thing and the knowledge-principle from the view point of idealism. According to his theory, *Zhizhi* is the process of finding good knowledge, and Gewu is the process for obtaining the principles of things. Gewu follows *Zhizhi* logically. In Wang Yangming's theory, good knowledge is transcendental morality, and Gewuzhizhi is the process by which a man recognizes transcendental morality and practices it in everything he does. So Wang Yangming' s Gewuzhizhi is basically an ethical practice and has less of an epistemological function.

Wang Yangming's theory was criticized by the Donglin School (东林学派) and other scholars for its less than epistemological function in the late Ming. Gao Panlong (高攀龙), leader of the Donglin School, and others suggested that Zhuxi's theory should be carried forward and investigations made into the principles for everything. He said, 'even a straw or a tree has its principle, and we should investigate it'.[2] Against this social and academic background, Xu Guangqi (徐光启) and other Ming scholars tried to develop Zhuxi's theory using Western learning and developed new epistemological ideas as a result.

2 The New Epistemological Ideas Formed in the Academic Interaction

In the *Introduction to the Translation of Elements*, Xu Guangqi and Matteo Ricci suggested the new epistemological ideas as the following two principles:

1. To begin obtaining knowledge by investigating the principles of things;
2. To obtain new knowledge from the known knowledge by deduction.

---

[2]Liang Qichao suggested, 'since Matteo Ricci and others introduced China the so-called Western learning in late Ming, there were something changed in Chinese academic methods. Some new methods were used in mathematics and astronomy firstly, then in other academic fields.' See reference[4]. Xu Zongze, He Zhaowu and Fan Hongye have the same opinions as Liang Qicao. See references[1, 3, 10].

He said, 'The purpose of Confucianism is to obtain knowledge; to obtain knowledge is by investigating the principles of things. As the principles of things are inside and human minds are crass, without deduction from the known knowledge, how can we obtain new knowledge' (夫儒者之学，亟致其知；致其知，当由明达物理耳。物理渺隐，人才玩昏，不因既明，累推其未明，吾知奚至哉)[9]

It is obvious that the first principle was from Zhuxi's theory of 'Jiwu-Qiongli-Zhizhi', which emphasized that knowledge was about the objects and obtained by practice and intuition. The second principle was introduced into China by the Jesuits in the late Ming and totally different from the theories of Zhuxi and Wang Yangming. Deduction is a conceptual reasoning method that was developed in ancient Greek philosophy and used widely in medieval theology. As they were combined and utilized for all kinds of knowledge by Xu Guangqi etc., the two principles were treated as general epistemological ideas. Jesuits in China such as Matteo Ricci took the epistemological model of 'Gewu-Qingli-Zhitian' (格物-穷理-知天). 'Zhitian' means serving the God[5]. So, the Jesuits agreed with Xu Guangqi about the phases of 'Gewu-Qingli' and took the two principles for granted. Ming Chinese scholars took the two principles, especially the second one, as very important epistemological ideas for developing new knowledge.

Xu Guangqi explained how to use the two principles in *Introduction to the Translation of Elements*. He emphasized that the first one should by utilized before the second, because knowledge should be obtained from the real principles, which were inside the things. He also explained the formal rules of deduction. Unfortunately, he could not explain the formal rules for using the first principle, just emphasized the tuition. So, it was hard for Xu Guangqi and other Ming scholars to use the first principle. However, they formed new epistemological ideas and developed the learning about natural things by it. We could call the ideas as 'principles of *Gewu Qiongli*'.

3 The Epistemological Functions of the Principles of Gewu Qiongli

Ming Chinese scholars such as Xu Guangqi knew well about and insisted on the first one of the new epistemological principles in their learning. They preferred the second principle to the first one for its remarkable epistemological functions. Xu Guangqi explained its functions in his *Comments on the Elements* (几何原本杂义) as follows: First, deduction could be conducted by formal rules, which was a benefit for reasoning and essential to learning. Second, deduction could be used in any kind of knowledge, and anyone who wanted to develop useful knowledge (knowledge of Jingshizhiyong 经世致用, i.e. the knowledge for severing the society) had to learn the epistemological principles and the formal rules for their use. Third, one could find real principles of things and learn true knowledge by deduction. In return, one could realize the mistakes in his learning, such as the imagined things and their principles. Especially, one could know that his knowledge is very limited, and there was much that human beings didn't know.

Xu Guangqi was very proud of his discovery of epistemological ideas. He suggested that every scholar should learn deduction from the Elements. After he explained the epistemological functions of the new ideas, he concluded, 'This book is useful to knowledge. I translated and published it with my colleagues for its circulation. Matteo Ricci prefaced it to help the current scholars to learn it. Although there are few people to learn now, I think everyone will be eager to learn the book a century later.' (此书为用至广，在此时尤所急须，余译竟，随偕同好者梓传之。利先生作叙，亦最喜其亟传也，意皆欲公诸人人，令当世亟习焉。而习者盖寡，窃意百年之后必人人习之，即又以为习之晚也。)[7] His prediction became true. Since the late 19th century, most Chinese scholars learn the Elements and Deduction. Zhu Kezhen (竺可桢) appraised Xu Guangqi: 'He began his scientific study from geometry as it is deductive'[12]. Xi Zezong (席泽宗) appraised Xu Guangqi: 'He developed a new thinking method—deduction in China. This method and induction which was introduced later by Yanfu (严复) were the main scientific methods in China before Marxist dialectic was introduced.'[6]

The new epistemological principles were not only used by Xu Guangqi and other Late Ming Scholars to develop *Gewuqionglizhixue*, which included natural philosophy and metaphysics, but also used to develop *Xiushengshitianzhixue*, which included ethics, economics, politics and theology etc.. Xu Guangqi learned medieval Catholicism from the Jesuits, being moved by the clear explanation about the canons. He said in his *Bianxue Zhangshu* (statements about true knowledge, 辨学章疏), the Catholic canons were explained so clearly by syllogism that everybody should trust them as they were true knowledge about the real principles of things[8]. He took clear knowledge as true and tried to foster the authority of the Ming feudal empire with it.

The Sino-western academic interaction in late Ming was carried out on two levels. On the surface level, Chinese scholars and the missionaries introduced Western knowledge into China and combined it with Chinese traditional scientific and other knowledge. On a deeper level, they introduced deduction and combined it with *Zhuxi's Jiwuqiongli* theory to develop new epistemological ideas. These became the two principles of Gewuqiongli. They tried to coin the formal rules for the utility of the two principles to support the interaction of knowledge and succeeded to some extent.

Bibliography

[1] Fan, H. Since 'Gezhi' to 'Science'(从"格致"到"科学"). *Journal of Natural Dialectics (自然辩证法通讯)*, 10(3):42.

[2] Gao, P. Answer to Mr. Gu Tingyang about Gewu(答顾泾阳先生论格物). In *Gaozi Posthumous Writings (高子遗书)*, vol. 8, 3a – 3b. Shanghai Classics Press. Reference to Siku Mingren Wenji Congkan (四库明人文集丛刊) copied by Shanghai Classics Press in 1993, Vol.1292, p.466.

[3] He, Z. On Xu Guangqi's philosophy (论徐光启的哲学思想). *Journal of Tsinghua University (清华学报)(humanities and social science edition)*, (1):1–10, 1987.

[4] Liang, Q. Outline of the academic success in the Qing dynasty (清代学术概论). In Zhu, W. (Ed.), *Liang Qichao's Two Books on the History of Academic Activities in the Qing Dynasty (梁启超论清学史两种)*, 23. Shanghai: Fudan University Press, 1985.

[5] Standaerd, N. Gewu quongli: Discussion between the western Jesuits and Chinese scholars in the 17th century ("格物穷理":17世纪西方耶稣会士与中国学者间的讨

论). In Witek, J. (Ed.), *Ferdinand Verbiest (1623-1688): A Missionary, Scientist, Engineer, and Diplomat* (传教士·科学家·工程师·外交家:南怀仁(1623-1688)), 454 – 479. Beijing: Social Science Classics Press, 2001.

[6] Xi, Z., Wu, D. (Eds.). *Collection of Papers on Xu Guangqi* (徐光启研究论文集). Shanghai: Xuelin Press, 1986.

[7] Xu, G. Comments on elements (几何原本杂义). In Wang, C. (Ed.), *Xu Guangqi's Collection* (徐光启集), 76 – 78. Shanghai: Shanghai People's Press, 1981.

[8] Xu, G. Statement about true knowledge (辨学章疏). In *Xu Guangqi's Collection* (徐光启集), 432 – 433. Shanghai: Shanghai People Press, 1981.

[9] Xu, G., Matteo, R. Introduction for the translation of elements (译几何言本引). In Xu, Z. (Ed.), *Abstracts of the Jesuits' Translations and Writings in Ming and Qing* (明清间耶稣会士译著提要), 259. Beijing: Zhonghua Bookstore, 1989.

[10] Xu, Z. The Chinese academic reform in Ming and Qing dynasties (明清之际中国整个学术思想之革新). *Saint Religion Journal* (圣教杂志), 26(10):579–588. 27(4):170-179, 27(5):326-334.

[11] Zhang, D. Outline of the categories in Chinese classical philosophy (中国古典哲学概念范畴要论). In *Zhang Dainian's Collection* (张岱年全集), vol. 4, 702. Shijiazhuang: Hebei People's Press, 1996.

[12] Zhu, K. Preface. In Institute of History of Natural Sciences of Chinese Academy of Sciences (Ed.), *Proceedings on Memory of Xu Guangqi* (徐光启纪念论文集), 5. Beijing: Zhonghua Bookstore, 1963.

[13] Zhu, X. Daxuezhangju/Bugewuzhuan (大学章句·补格物传). In *Sishu Zhangju Jizhu* (四书章句集注), 5b – 6a. Beijing: Zhonghua Bookstore, 1983.

Research on Front Issues in the Cultural Philosophy of Science[1]

Hong Xiaonan

Dalian University of Techonology

hxnharvard@yahoo.com.cn

1 What is the Cultural Philosophy of Science?

An important front issue in the modern philosophy of science is how to reveal profoundly the essence of science through philosophical reflections while facing science as a complex social-cultural feature.

1.1 "Philosophy of science culture" and "cultural philosophy of science".

In Chinese, the "philosophy of science culture" and the "cultural philosophy of science" can be translated as "kexue wenhua zhexue". For instance, people derive the first one from Ernst Cassirer's book *An Essay on Man: An Introduction to a Philosophy of Human Culture*. While in the US and the UK people accept the last one when "cultural studies of science" is translated as "*kexue wenhua yanjiu*" or "*kexue de wenhua yanjiu*". If we distinguish carefully, we can see that Cassirer's *An Essay on Man* is about the philosophy of human culture. It is a kind of "philosophy of culture" which synthesizes human culture into an organic whole. In contrast, the "cultural study of science" is a special mode of cultural studies. Through analysis, we can see that "*kexue wenhua zhexue*" includes two main parts: one is rethinking "science culture" from a philosophical perspective (science is a kind of cultural activity). There are no essential differences between the philosophy of science culture and the philosophy of science, but it stresses that science is a more important cultural pattern or force than culture. That is a kind of "Philosophy of Science Culture" (PSC). The other means rethinking science from the angle of cultural philosophy and forming a kind of "Cultural Philosophy of Science" (CPS). As I use it in this paper, "kexue wenhua zhexue" includes both parts, but I think the "Cultural Philosophy of Science" (CPS) is the more important part. We need to conform the Philosophy of Science (PS), the Sociology of Scientific Knowledge (SSK), the History of Science (HS), the Cultural Study of Science (CSS) and so on–all of which comprise science studies.

---

[1]The author is with School of Humanities & Social Sciences, Department of Philosophy, Dalian University of Techonology, Dalian 116024, China. Supported by the Program for New Century Excellent Talents in University.

1.2 The relationship between the "cultural philosophy of science" and the "philosophy of science".

In orthodox (traditional) studies of the philosophy of science, there are essentially some cultural philosophy factors. For instance, P. Frank regards the philosophy of science as connecting science and philosophy and science and the humanities in his book *Scientific Philosophy*. He mentions the understanding of science and the strategic systems method are the main contents in the philosophy of science. So Frank regards the philosophy of science as a means to connect philosophy and science which depends on philosophy and the humanities toward science (nature science). Therefore, Frank basically insists on Scientism. In his book *Conceptual Foundations of Scientific Thought – an Introduction to the Philosophy of Science*, M. Wartofsky regards the philosophy of science as "the missing link" or "bridge" between science (nature science) and the humanities. The substance of his argument is that concepts and models of scientific ideology should be expounded as object texts of humanistic study; the critique of logic, corrective analytic instruments and an integrative generalization of philosophy should be adopted into scientific history and contemporary scientific ideology. The Philosophy of Science offers access to two kinds of culture in a coherent way. And it would be unnecessary if we don't apply ourselves to seeking coherence and integrate in all fields of knowledge. From the most beautiful and profound sense of philosophy, the humanistic understanding of science is a philosophical understanding. That is to say, Wartofsky makes the study of the Philosophy of Science an object of study from a humanistic perspective.

2 "Philosophy of Science" and "philosophy of sciences" (the cultural philosophy of science).

We study science as a special cultural form or activity in cultural philosophy of science, not only as an epistemology or scientific logic. So it is different from the traditional Philosophy of Science and general cultural philosophy. The study of the cultural philosophy of science is still philosophy and closer to the traditional Philosophy of Science than such meta-science as the history of science, the sociology of scientific knowledge, etc. If we regard traditional Philosophy of Science as a narrow Philosophy of Science, the cultural philosophy of science can be regarded as a general one. That is because narrow Philosophy of Science discusses what science is from the perspective of epistemology and academic logical configuration. The history of science limits the basic contents of science to its historical development. The sociology of scientific knowledge describes the macroscopical image of science as reflected in social institutions and the influences of science on the behavior standards of scientific activities. As a kind of general Philosophy of Science, the cultural philosophy of science includes most of the content of both the traditional Philosophy of Science and the generalizations and summarizations of historical, social, cultural and political research on science.

While researching the Philosophy of Science, someone mentioned the sociology of knowledge, the sociology of science and the sociology of scientific knowledge. These are forms of research about external processes of science instead of research into the theme of the Philosophy of Science. The Philosophy of Science is, however, research into the international processes of science. That means we are in the first step thinking about the emergence and development of scientific cognition, the content and structure of scientific theory, the logic and models of scientific explanation, the nature of the entity known as scientific theory, the progress and objects of scientific inquiry, etc. Obviously, this point of view still belongs to the traditional Philosophy of Science, which makes the philosophy of science equal to logic and the methodology of science. In fact, the development of the modern philosophy of science has broken the limits of traditional Philosophy of Science. The Philosophy of Science has now become a Philosophy of Sciences. So we have achieved a new perspective which can be regarded as a cultural philosophy of science while investigating science through such prisms as logic, society, psychology, culture and politics, etc. The cultural philosophy of science in a spirit of openness and tolerance attempts to integrate continental philosophy (phenomenology and hermeneutics) and analytical philosophy (AP) in Britain and US. For instance, Edmund Husserl analyzed the crisis of European science and that of human nature; Martin Heidegger revealed the technical essence of science; Hans Poser grafted the Philosophy of Science in Britain and the US on to continental hermeneutics; the Frankfurt School criticized science and culture and implement rationality; the feminists deconstructed the so-called "androcentric ideology" in science; the post-colonial theory of science unpacked the so-called racist elements of Euro-centric science. All of these movements and developments have brought us new energy and hope that we may understand science better.

We comprehend the question "what is the Cultural Philosophy of Science?" by pectination. Then what is the significance of research in the cultural philosophy of science? In my opinion, through research, we can know more about the status and trends in the philosophy of science; we can deepen the urgency and the importance of the theoretical structure in the philosophy of science; we can integrate the historical, social, cultural and philosophical researches of science from a humanistic perspective; and then, we can research, rethink and comprehend science including its origins, motivation, purpose, significance, values and relationships between science and other forms of culture, science and gender, science and race, etc; we can transcend narrow positivism and humanism, and combine scientific culture and humanistic culture so as to form "the third culture" (C.P.Snow).

3 The Front Issues in the Cultural Philosophy of Science

A significant trend and popular domain in the development of modern philosophy is the philosophy of culture. As the philosophy of science changes its direction toward culture, it sets a completely new course of study for us – research into the cultural philosophy of science, a systematic examination into the front issues in the cultural philosophy of science. The latest achievements in the philosophy of culture and the philosophy of science plus the results of those already studied as well as the author's understanding of this research. In order to examine and arrange the fruits of research in the philosophy of science, the sociology of science (scientific knowledge), the history of science and so on, I will now turn to an

Research on Front Issues in the Cultural Philosophy of Science 143

analysis of the dimensions of the cultural philosophy of science which lie within the history of science: the sociology of scientific knowledge (SSK), the cultural study of science and the philosophy of science.

Some scholars believe that post-modern science has three forms: one is the post-modern science of organic theory, which has made real progress in changing our world view. The second is the right theory of scientific knowledge, which has two distinct features - relativism and the ideology of scientific knowledge. The direct source of the later doctrine is Jean-Francois Lyotard's post-modern "grand narrative" turn and Michel Foucault's theory of knowledge and power, reflected in the scientific concepts of the "strong program" SSK, post-colonialism and feminism. In the current "science wars", the scientific defenders mainly criticize this thought. The last is the deconstruction of scientific texts, the purpose of which is to undermine any certainty of understanding of scientific texts and their scientific basis. In my opinion, the first involves post-modern science. The latter two don't actually talk about post-modern science, but the thoughts or theories relevant to post-modern scientific thought – the post-modern scientific concept. Therefore, we should distinguish between "post-modern science" and "post-modern scientific concept." The so-called post-modern science has been established as a kind of new artistic conception of science opposed to the concept, activity, object and form of modern science on the basis of criticizing and reflecting the characteristics of modern science and modernity. Post-modern scientific thought is "anti-science". Here the word "anti-" has two meanings: one is the reasonable "reflection" on modern science, which is of positive significance and help us better and more completely understand reflections on science; the other is "against" modern science, which is the most radical reflection on modern science which is opposed to the foundation and essence of modern science and consistent with post-modern thought.

In this context, I believe that Sokal's hoax in his paper for the post-modern scientific description of certain aspects is not entirely "obvious nonsense" as he has said. For example, he said the first feature of post-modern science is non-linearity and non-continuity, and I think he has grasped the fundamental problem. Post-modern science can be called a revolution in science which does not totally abandon modern science but picks up its nonlinearity and complexity. As for post-modern science, as Sokal's hoax argues, we need further analysis: post-modern science not only puts forward a powerful rebuttal rooted in the traditional scientific authoritism and elitism, but also provides an empirical basis for the democratization of scientific works. The content and methodology of post-modern science provides a powerful ideological support for the progressive political program. It transcends boundaries, breaks barriers and achieves complete democratization of all-round social, economic, political cultural life. Through discussing the issues of post-modern, post-colonial and feminist science, it tells us how to treat the ideological criticism of science. For example, feminist and post-colonial scholars revealed and strongly criticized the ideological criticism of science.

Feminist scholars think the core issue and methodological problem of modern science is the fact that modern science was created by patriarchs – dead, white, European males–to deal with feminine problems. Therefore, the feminist scholar Evelyn Fox Keller figures feminist science criticism is not the subject of women or women in science, but the creation of men and women impacting on science. The project leads to the convergence of two fields of knowledge developing independently on the surface: feminist theory and the social research of science. These are the thoughts of the feminist critique of science. Of course, this so-called "scientific" critique is what we generally acknowledge as the "recognized viewpoint" – the neutral, rational, objective, universal science. The recognized viewpoint–syllogistic, analytic, atomic, non-sense and quantitative cognitive style–is "male", while the intuitive, integrated, holistic, sensible and qualitative cognitive style is "female". In science, women are "executors", "edgers", while men are the "mandators" and the "centers" – that is to say science is essentially an "androcentric ideology". This typical dichotomy infiltrates all social sectors, not only science.

Through the brief analysis outlined above, we can see that one of the front issues in the cultural philosophy of science includes several types of problems:

The first problem is the new methods used to solve old problems. Some problems are traditional Philosophy of Science problems. Even now, there are still front issues in the cultural philosophy of science because they involve how to understand science as a meta-problem. For example, the problem of the demarcation of science, scientific rationality, scientific objectivity, and issues concerning the relationship between science and technology, including the relationship between science and values, etc., all fall into this category.

The second problem is that the cultural philosophy of science is caused by changes in the philosophy of thought. The critique of science modernity, for example, of implementation rationality, of science ideology, and the relationships between science and enlightenment all fall into this category.

The third is the new hot-point problems and puzzlings about the development of science. For example, how do we treat the social results of science (such as the social results of Nano-ST and clone techniques)? How do we view the foreseeable and unforeseeable influence to the environment, society, and mind caused by science activities? And finally, what are the moral duties of the scientist? These are all concerns too!

Bibliography

[1] Cai, Z. What is postmodern science? *Science Technology and Dialectics*, 19(5), 2002.

[2] Cassirer, E. *An Essay on Man: An Introduction to a Philosophy of Human Culture*. New Haven: Yale Uni. Press, 1944.

[3] Frank, P. *Philosophy of Science: The Link between Science and Philosophy*. Shanghai: Shanghai People's Publishing House, 1985.

[4] Hong, X. The cultural turn in 20th century western philosophy of science. *Journal of Seeking Truth*, (6), 1999.

[5] Hong, X. From philosophy of science to philosophy of science culture. *Journal of Dialectics of Nature*, (1), 1999.

[6] Hong, X. *Towards a Philosophy of Science Culture*. Shanghai: Shanghai Literature and Art Publishing House, 2004.

[7] Li, S. Where is the developing perspective of philosophy of science of China. *Social Science Forum*, (1):2, 2006.

[8] Lin, X. Science is the cultural patterns and forces. *Democracy and Science*, (3):11 – 14, 2005.
[9] Meng, J. Turn from philosophy of science to philosophy of science and culture. *Studies In Dialectics of Nature*, (6), 2003.
[10] Sokal, Cai, Z., Xing, D. *"Sokal Event" and "the Science Wars"*. Nanjing: Nanjing University Publishing House, 2002.
[11] Wang, S., Wan, D. *From Scientific Philosophy to Cultural Philosophy: On Postmodern Turn of T.S.Kuhn's and P.K.Feyerabend's Thought*. Beijing: Social Sciences Documentation Publishing House, 2006.
[12] Wartofsky, M. W. *Conceptual Foundations of Scientific Thought- An Introduction to the Philosophy Of Science*. Beijing: Qiushi Press, 1989.
[13] Yan, B. Studies on the conception of modern science and postmodern science. *Journal of Dialectics of Nature*, (2), 2006.
[14] Zhang, Z. *Theory of Philosophy of Science*. Beijing: People's Publishing House, 2004.

Walk out from Imer Lakatos' Dilemma[1]

Hu Mingyan

Tsinghua University

humy06@mails.tsinghua.edu.cn

For a long time, science undoubtedly has been taken as a progressive career. Since logical positivism, numerous philosophers have depicted the history of science as progress. Nevertheless, a review of the facts tells us that the history of science has not been all smooth-sailing. A number of problems have arisen. The dispute between Karl Popper and Thomas Kuhn is just one example. Imre Lakatos was on Popper's side but also tried to compromise with Kuhn in an effort to save the concept of scientific rationality which he believed Kuhn's theory endangered. But in fact he failed and slipped into a kind of dilemma. In this paper, I analyze the work done by Lakatos to try to discover why he failed. In the paper I indicate the possibility for walking out of his dilemma.

1 Debate between Karl Popper and Thomas Kuhn

Karl Popper seriously criticized the absolute demarcation between observation and theory made by logical positivism. He emphasized that observation was theory-laden, and science was a process of continuous conjectures and refutations. In his opinion, problems are the starting-point and end-point of the growth of science and knowledge.[3] Although taking scientific development as a procedure of continuous revolutions, Popper held that objective "scientific knowledge" could be regarded as having no subjects. Our aim is to increase the true content of our theories.[2, p. 71] Like the logical positivists, Popper was still a logicist, believing that a set of basic formalized standards and methods exists to appraise scientific theories, and that behind the ever-lasting revolutions in scientific development there is a set of logic of scientific discovery independent of human will which assures the rational and continuous evolution of scientific knowledge.

As the founder of the Historical School, Thomas Kuhn held a different idea from Popper on how scientific development keeps on. On the one hand, Kuhn pointed out that, Sir Karl described the whole scientific cause in a way that fit only occasional revolutionary characteristics. Therefore, Kuhn stressed that the "normal science", a problem-solving activity defined by a "paradigm", was the normality in science, while revolutions were just sporadic. It is "normal science" rather than abnormal science or revolutions that best differentiate science from other causes. On the other, Kuhn was not content with Popper's strong (logical) rationalism. As for Kuhn, it is ridiculous to suggest that scientists can preset every case imaginable to verify or falsify certain theories. If we want to explain how science could be progressive or how scientific rationality is possible, the

---

[1]The author is with the Center for STS, Tsinghua University, Beijing 100084 China.

answer "in the final analysis, is psychological or social" [2, p. 26]. It seems that Kuhn regarded the value judgements made by the scientific community as the source of scientific rationality and took Popper's logical rationality as only one, not even that important part of rationality. In his book *The Structure of Scientific Revolutions*, Kuhn compared scientific revolution to political revolution–full of rhetoric! He compared the paradigm change for a scientific community to a conversion in religious faith and stressed the incommensurability between a new paradigm and an old one. This interpretation greatly highlights the irrational and illogical factors in scientific development.

In regard to Kuhn's idea mentioned above, Popper not only reprehended the incommensurability between two paradigms as "framework myth" with obvious logic and philosophical errors, but also claimed, "To me, it is surprising and disappointing to resort to sociology or psychology to explore the aims of science and its possible progress ... Sociology and psychology are filled with fashions and unrestricted dogmas. It is obviously wrong to think that we can find anything like 'objective and pure description' here." [2, p. 72]

So far, we can see that when the philosophy of science doesn't stick to static analysis of the logical structure of scientific knowledge, and bring time into it, a huge gap appears when it comes to explaining how the evolution of scientific theories maintains itself. Karl Popper took the strong logicism or rational position. He insisted on the logic of discovery and claimed that the internal (logic) rationality of the objective world of knowledge assures the evolution of scientific knowledge. Thomas Kuhn took a more social stand. He advocated the psychology of research and emphasized the influences of value tropism of a certain scientific community on the choices between different scientific theories. Who is right? Let's see what Imre Lakatos' has to say on this conflict.

2 Lakatos' Mediation

At the beginning of *The Methodology of Scientific Research Programmes*, Lakatos indicated that the conflict between Popper and Kuhn was not only a special problem in epistemology, but also related to a very significant knowledge value question. As a student influenced profoundly by Popper, Lakatos held fast to the idea that science can never be a mess where "anything goes"; contrarily, we should follow Popper's logic of discovery and depict scientific revolution as rationally progressive rather than a process of conversion to a new religion.

On the one hand, Lakatos disagreed with Kuhn for bringing a social viewpoint into the philosophy of science. To him, Kuhn made the issue of scientific revolution a problem in mob psychology which degraded the philosophy of science into a psychology of science and excluded any possibility for reconstructing scientific growth rationally. Rather, Lakatos agreed with Popper's opinion on objective knowledge, that is, "the psychology of science is not automatic: because rationally reconstructed scientific growth essentially takes place in the idea world, in 'the third world' of Plato and Popper, in the clearly expressed knowledge world. And this world is independent of epistemic subject. The aim for Kuhn's research programme seems to describe (no matter individual or group) psychological changes in ('normal') science. But the third world in the mind of an individual scientist or even in the minds of 'normal' scientists is always a caricature of the original one. If we want to depict this caricature without relating it to the original third world, it will probably produce a caricature of a caricature."[2, pp. 127 – 128] On the other hand, Lakatos appreciated Kuhn for his emphasis on the continuity of scientific growth and the tenacity of scientific theories. He himself also realized that we cannot exclude all the irrational factors such as social and psychological influences when making choices between scientific theories during the process of scientific growth which takes on a scene of everlasting theory subrogation. And it is here that Popper too easily denied reasonable points advanced by Kuhn. In this way, historically speaking, Lakatos was the first to formulate this problem – "whether there are any general objective reasons to justify the acceptance or rejection of certain theories as rational" on the map of the philosophy of science. He started his journey to mediate the positions of Popper and Kuhn, hoping that he could realize the saying, "The philosophy of science without the history of science is empty; the history of science without the philosophy of science is blind".

Therefore, Lakatos first criticized the naive falsification and methodological falsification which had creeped intothe debate between Popper and Kuhn. He maintained that both ideas regard a testament as a war between a theory and a trail, and the only meaningful result of this conflict is final falsification. With regard to the actual history of science, Lakatos refuted this viewpoint. He believed that there were three participants in the testament in practice, that is, the two conflicting theories and the experiment; and at least some experiments didn't seem to result in falsification.

After indicating Popper's failure to construct a theory on the rationality of scientific growth in accordance with the history of science, Lakatos absorbed Kuhn's conception of "paradigm" and called a series of theories with obvious continuity a "scientific research programme", the basic unit of appraisal. Based on this, he came up with his own "methodology of scientific research programmes" – including a conventionally accepted "hard core", a "protecting belt" comprised of auxiliary hypotheses and the initial conditions of theories, a "positive heuristic" which defines problems. He outlines the construction of a belt of auxiliary hypotheses, foresees anomalies and turns them victoriously into examples, and also sees a "negative heuristic" which changes the protecting belt to make anomalies consistent with the hard core or overthrows anomalies totally and forbids anomalies from attacking the hard core. Between different programmes, as long as the theorietical growth of a research programme keeps predicting novel facts with some success, we say it is "progressing"; it is stagnating if its theoretical growth lags behind its empirical growth. While within a programme, a theory is eliminated by a better theory which has more excess empirical content. Most importantly, there is no instant rationality, no crucial experiments, no defeat of a research programme with just one blow. "Nature may shout no, but human ingenuity may always be able to shout louder."[2, pp. 127 – 128]

With the help of this new standard of appraisal, on the level of historiography, Lakatos believed that historians could look to rival research programmes for progressive and degenerating problem shifts. Thus it can explain the high degree of autonomy of theoretical science, turning external for other historiographies into internal ones. For instance, when scientists take no notice of anomalies and still try to construct auxiliary hypotheses and insist on a certain scientific theory at all costs, we think this reasonable rather than irrational since it is a long and complex procedure to abide by the internal logic of scientific development. As long as a research programme anticipates novel facts and follows progressive problem shifts, then, even if it encounters many anomalies, it is still rational.

3 Lakatos' Dilemma

As we have seen above, Lakatos remained loyal to Popper's idea that there is a permanent logic rule in the development of scientific theories; on the other hand, he grasped Kuhn's thought and not accepted a set of theories or research programmes as the basic unit undergoing continuous change, but also pointed out that it is a long and zigzag procedure to follow the internal logic of science: the owl of Mineva doesn't fly until dawn. In this way, Lakatos brought time into the logic of the rational development of science itself. According to him, now we don't need to resort to sociology or psychology to explain the seemingly irrational behaviors of scientists to ignore anomalies when they advance a certain theory. Thus, we guarantee the rational explanation of the history of science. Nevertheless, the incommensurability between different paradigms in Kuhn's mode is changed into progressive and stagnating problem shifts in Lakatos' rationality theory of science, which seems to avoid the relative tendency in Kuhn.

Is it really so? Wait. Please, let's think twice!

First, it is the viewpoint of "the termination of instant rationality" that sustains Lakatos' new idea. According to Lakatos, to appraise a theory or a research programme, we don't follow the static logic between it and some observation statements while in a long historical development period. The subject of appraisal is moving all the time. One's opponent, even if lagging badly behind, may still stage a comeback. There is nothing inevitable about the triumph or defeat of a programme. One can only be wise after the event. Then, the hindsight effect makes this demarcation and definition of a progressive scientific research programme a form of lip service. When on Earth can we say a research programme is degenerative? No time limit? Second, if different research programmes coexist for a long time, doesn't this imply a kind of relativism or irrationalism? No crucial experiment, no instant falsification, no crisis because of the accumulation of anomalies. As long as it makes theoretical and experimental progress, it can fail countless times. How can we then say that this doesn't possess any meaning? In a word, Lakatos seems to solve the riddle left by his predecessors, but actually fails. He slips into a kind of dilemma.

4 Walk Out from the Dilemma

Lakatos tried very hard to mediate his predecessors' positions. Seemingly, he succeeded, but on second thought, his success pays only "lip service" to the importance of the issue. How come?

First, let's look at the "rationality" Lakatos wanted to defend. Brown argues that, the classic mode of the concept of "rationality" possesses three features: universality, necessity and regularity. Universality means if all rational thinking starts from the same information, it should lead to the same conclusion; Necessity means the final results can be induced necessarily by the given information; Regularity requires that every step of theoretical induction should abide by the rules.[4] From logical positivism to Popper and to Lakatos, they actually followed this kind of "rationality", regarding "logicality" as "rationality" and believing that objective and eternal logic assured the universality, continuity and regularity of choices in scientific theory. Logical positivism used inductive logic to cover all the actual movements, taking scientific development as a one-way accumulation of truths; Popper stressed the critical character of science, but he only opened a free land at the entrance to scientific development; Kuhn meanwhile opened free space at both the entrance and exit to scientific development, breaking the unilateralism in terms of the logic of scientific development. Thus he endangered the classic "rationality", although he himself denied he was an irrationalist. Lakatos fully recognized the complexity revealed by Kuhn about the actual history of scientific development, but he couldn't tolerate the loss of unilateralism, so he mediated between Popper and Kuhn and expanded the range of traditional "rationality", making seemingly nonlogical scences into "rational" ones. However, in nature, traditional "rationality" excludes time. It is doomed to fail if you want to introduce the historical dimension into a scheme based on static logic.

As for me, "rationality" itself is a concept which changes as time goes by. In connotation, it means regularity, the recognition of regularity and the criteria and rules for appraisal. In denotation, it can be divided as follows: epistemology /ontology, concept/logic, instrument/value, and social/ practical.[1] Any singular rationality cannot cover all the meanings of "rationality" because of its imperfection. If we don't confine ourselves to one kind of rationality, but understand this concept with a whole view, then we don't need to rest on classical "logicality" and repel other kinds of rationality, making a clear demarcation between so-called "rational" and "irrational" – this is always arbitrary. Since history is made up of all kinds of factors, why not be more tolerant and accept different rationalities? If so, Kuhn's incommensurability doesn't really discard the rationality of scientific development. Instead, he discards the traditional understanding of "rationality".

Secondly, we should update our definition of science. Indeed, what Lakatos said about the complexity of making choices in scientific theory is in accordance with the facts. But still, the development of science is objective and independent of human will. How should we explain this? If we regard science not only as a set of theoretical systems but also a colorful activity or a kind of practice which requires intervention and operation, then we may find that the dependence on material factors during practice and the theories formed from this dependence come back to reconstruct our life world, making science a compulsory force and explaining the objectivity of science.

In conclusion, as a turning point from critical rationalism to historicalism, we may understand science itself better by thinking over Lakatos' work and avoiding his dilemma.

Bibliography

[1] Chen, Q. On scientific rationality and scientific progress. *Research on Natural Dailetics*, (2), 2002.

[2] Lakatos, I., Musgrave, A. (Eds.). *Criticism and The Growth of Knowledge*. Huaxia Publishing House, 1986. Translated by Zhou Jizhong.

[3] Popper, K. *Conjecturess and refutations – the Growth of Scientific Knowledge*. Shanghai: Shanghai Translation Publishing House, 1986. Translated by Fu Jizhong.
[4] Wang, W. Philosophical analysis on scientific rationality. *Journals of Natural Dialectics*, (2), 2004.

A Project against Traditional Scientific Realism and Anti-realism – An Appraisal of Rouse's Practical Realism of Science[1]

He Huaqin

Tsinghua University

hehq06@mails.tsinghua.edu.cn

The controversy between scientific realism and anti-scientific realism has been momentous in the history of the philosophy of science in the 20th Century continues into the 21st Century. W. Sellars, the American philosopher of science, claimed that "science is the measure of all things, of what is that it is, and of what is not that it is not" in the 1960s. He believed the external world really exists and that scientific truth really describes and reflects this world; scientific outcomes prove or will prove the objective reality of the material world. Since then, the "scientific realism camp" has been established. After being developed by philosophers like Hillary Putnam, Mario Bunge, and Dudley Shapere, scientific realism has become the mainstream in the philosophy of science.

Human thought, however, never seems to sustain a high level for a prolonged period of time. Just because of internal conflicts in scientific realism, scientific anti-realism was born and argued by L.Laudan, Van Fraassen as well as assorted pragmatists and positivists.

Scientific realism is typically construed as the view that "scientific theories are true or false depending on whether the objects they describe (including unobservable objects like electrons and quarks) actually exist and have the characteristics the theories ascribe to them "[6, p. 127]. The main points of scientific realism include the belief that the object, state and process of scientific research really exist, even when it is a microcosmic object which cannot be observed; cognition is a reflection of the external world; the theoretical object is real as long as it is effective in cognition; scientific theory is gradually and approximately convergent towards the truth. Scientific anti-realism refutes realism mainly from an empirical and constructivist (or instrumentalist) point of view. According to anti-realism, scientific theories cannot capture the nature of things. They can only describe the observable surface. Scientific concepts are defined by manipulation, not by the properties of the things themselves. Scientific theory is simply a kind of instrument which can be applied to the solution of puzzles. It does not necessarily correspond to the ways things actually are in the external world. As Van Fraassen claims, science aims at providing the empirically adequate theories. Scientific activity is a kind of construct or image or analogy, and scientific laws are fictions. His constructive empiricism can be stated, "the aim of science is to

---

[1]The author is with Institute of Science, Technology and Society, Tsinghua University, Beijing 100084 China.

provide empirically adequate theories for us, and the acceptance of a theory only connects with the belief that the theory is empirically adequate."[7, p. 12]

It is clear that the contrasting foci of the two sides in the realism-anti-realism controversy are as follows: whether scientific theory truly describes an objective reality; and the reality of unobservable theoretical objects (including such concepts as particles, fields, structures, statements, and theoretical objects of social science). How can we come to an understanding of scientific realism? How can we find a way to walk out of this dilemma? Joseph Rouse, a modern American philosopher of scientific practice, shows us a constructive way to approach this problem. Rouse believes the controversy between empiricism and constructivism has become a huge mess. Each side in the controversy seems to misunderstood its opponent and seems unable to clarify the major points of contention. As an alternative, Rouse advocates a kind of practical view of science which avoids the the radical core theories on which realism and anti-realism are founded. He advocates a common sense approach to science which avoids the extreme pitfalls of the two extremes. Rouse's approach attempts to transcend this debate in three stages.

1 Criticism of Convergent Realism

The direct argument for realism is that only realism could be applied to explain the general success of science. That is how Putnam justifies realism. It is called an "ultimate argument" or "no-miracle argument". In his article "What is mathematical truth?" Putnam claims, if we don't believe our theories are true (i.e. their references are unobservable objects), then we must admit that the success of our theories is simply a miracle.[5] For this "argument", Van Fraassen uses the illustration of the Darwinian concept of natural selection. Laudan argues, however, lacking a coherent theory in confirmation, it is impossible to precisely explain the success of the theory of natural selection. On the contrary, he argues, science is largely "unsuccessful" because it cannot demonstrate a high rate of confirmation.

According to Rouse, the methodology of abduction used by realists to establish successful predictions and posited as generally reliable is not generally successful. He holds that successful prediction in science is as much a matter of practical manipulative skill in the laboratory as of the prior possession of the correct theory. While we attempt to attach the theory to the world through experiment, the theory itself also evolves and changes to confirm the observable results of experimentation. When explaining the pragmatic success of science, convergent realism always fails, because "the realist cannot explainwhy the abductive appeal to real entities causally related to our use of certain theoretical terms succeeds in underwriting the success of our theories by appealing to the theories themselves."[6, p. 136] The pragmatic success of scientific theory always appeals to the theory itself. This means it does the same thing at the metatheoretical level. For example, the efficiency of high-pressure steam engines is explained by the second law of thermodynamics; the light is on or off can be explained in terms of electromagnetic theory and the fact that the switch closes and opens the electric circuit.

Furthermore, Rouse points out, if we want to explain the pragmatic success of science, the optimal strategy would be similar to the argument of natural selection as propounded by Van Fraassen. Realists would object to two aspects of the argument:

1. An argument based on a natural selection model doesn't tell us why some theories are successful and others are not; it merely explains why scientists usually distinguish them from each other. Why did the Bohr model of the hydrogen atom fail, for example, and the quantum mechanical models succeed?
2. The selection argument does not explain the differential success of various sciences, i.e., why the success of science arises at a differential rate in particular times.

As for the first objection, the realist explanation inclines toward the metaphysical conclusion which is usually reserved for teleological arguments and cosmological arguments for the existence of God. As for the second objection, the realist claim that the theorists hold the appropriate theoretical terms, and these terms can refer to reality precisely. In the opinion of Rouse, these answers cannot demonstrate how we close in on the truth and how we measure how close we are to a presumed truth. In 20th Century genetics, for instance, the science is rightly based on new cytological techniques, the detailed study of phenotypic/genotypic correlations in a single organism, and experimental tools (X-rays to detect mutations). Applying these tools and techniques, mapping the chromosomes of *Drosophila melanogaster* was eventually successful. Rouse draws the conclusion that science usually is led by technical developments since they open new opportunities for experimentation and theorizing, not only by new theoretical insights. If we realize how important the technology and progress of practice are in their effect on science, we would not need the help of realism which appeals to theoretical terms. We could use the correct terms and theories to explain the success of science.

2 Reject the Instrumentalist Anti-realism

Is it necessary to draw a line between observable objects and unobservable objects? Rouse's answer is negative. He argues the differentiation between them is practical and there is no ontological meaning. His view of science is also different from the traditional view, because he suggests science is a kind of outcome of practice rather than a system of representation. The observable world, furthermore, is not independent of our representation. As he says: "we are already engaged with the world in practical activity, and the world simply is what we are involved with"[6, p. 143], so the traditional view that we must appeal to observation and access to the world is not correct anymore. He agrees with Fraassen's concept of empirical adequacy in the sense that it accounts for the success of science, but his understanding of empirical adequacy is more than observation as Van Fraassen claims. Rouse's empirical adequacy of science includes "the reliable ability to manipulate and control phenomena in the laboratory, whether or not the entities we manipulate are themselves observable". And "our practical grasp of the equipment and the theories that are integrally involved in its design and operation are quite relevant to any judgment about the empirical adequacy of the scientific practices in question"[6, p. 144].

Besides, Van Fraassen insists that science does not care about how to describe what exists correctly. The only concern of science is for observable empirical adequacy, leaving the ontology of science to be dealt with by the realists. Rouse refutes such a misunderstanding and claims science is really concerned about what exists, but the existence of that being is not separated from what we manifest in scientific practice. Hacking had a similar answer to this question. As he said: "if you can spray electron beams successfully in an experiment, then they are real."[3, p. 22]

According to the instrumentalist view, the ontological status of theoretical entities of science is less than the ordinary objects. The instrumentalists assert: "statements of a scientific theory that deal with unobservable entities should be neither believed nor disbelieved. Their value is purely instrumental."[1] Rouse replies that there is no substantial difference between theoretical entities of scientific research and objects in ordinary life. The theoretical entities of science today become tomorrow's objects of daily life. Instrumentalists fail to demonstrate why the fictional or instrumental object they identify turns real when it enters the world of daily life.

3 Rouse's Practical Realism

For the problem of ontology of science, scientific realists basically hold materialistic point of view. They believe the reality which scientific theory describes is independent of our understanding and interpretation, whereas the anti-realists believe it is not independent. How to grasp the reality? Is it independent of our understanding and interpretation? Rouse attempts to develop a more systematic idea of scientific ontology and makes a more reasonable explnation of the reality which he refers to as scientific theory.

In Rouse's opinion, interpretation in scientific research (including theoretical and practical interpretation) and what we interpret are interdependent. The epistemic picture about the world of realism is that, the world itself and our understanding (i.e., the understanding of how we interpret the world) are all determined. Generally, we make an interpretation of the world at first and then act according this interpretation. As our actions by trial and error fail, we recognize that theremust be something wrong with the interpretation. But if our actions do not fail, that does not necessarily mean that our interpretation accords with the reality it purports to interpret. If our actions, however, meet with universal success, then the best explanation of this success should be that there is an approximate accordance between the interpretation and the reality. That is the argument realists use to prove their point. Rouse questions: "where do we acquire our understanding of what our various interpretations do say about the world and of what would count as success in our action?"[6, p. 154] In his view, sentences and practices do not have ready-made meanings and their meaning is not conventional. Only in their performance or use, i.e., in the particular context and in a process which is laden with purpose, understanding and practice, this interpretation constitutes meaning and could be understood. Rouse adopts the view of Heidegger's practical hermeneutics to imply that we can not interpret without a holistic understanding of the context of objects, conditions, equipment, etc.

If Rouse is right, the scientific realism hypothesis that our interpretation is independent of what we interpret is wrong in the process of specifying the meaning of interpretations. Only in the process of practice, the interpretation of the world can be understood and make sense. Practice dominates a prior status in the accordance between interpretation and what we interpret. In this sense, Rouse advances his understanding of reality as "the real is what we manipulate, what resists us, what we notice, and what we take account of without ever taking explicit notice."[6, p. 156] The reality of existence is implied in our dealing with it.

Rouse's understanding of "reality" is explicated in the field of practice. He suggests that interpretation is not independent of what we interpret. It is in the interactive practice which is also in a concrete context and laden with purpose. In the process of coming to understand it, the "reality" of things can be shaped. Rouse's view could be called "practical realism" in contrast to Hacking's "experimental realism".

4 Comments on Rouse's Practical Realism

Maybe it is meaningful to compare Rouse's practical realism and Hacking's experimental realism. Both criticize the traditional view which sees science as a set of representation systems and emphasize the importance of practice (or experiment). They maintain that practice can usually remain independent of the representation system of theory. Furthermore, Hacking claims that "experiment has its own life". Besides, they all pay attention to the significance of technology in scientific activity. Rouse claims that science usually is led by technical developments since they open new opportunities for experimentation and theorizing, not only by the new theoretical insights. And Hacking emphasizes the importance of physical apparatus in the laboratory. He asserts the debate about realism cannot be resolved at the level of theory. "Only at the level of experimental practice is scientific realism unavoidable"[4]. Hacking's experimental realism is based on the manipulation of reality in or through a physical event rather than the reality of theoretical structure.

There are some differences, however, between their views on realism. First, Rouse's practical realism uses Heidegger's hermeneutics to explicate reality. He suggests that science is a way of acting on the world rather than a way of observing and describing it. He denies differentiating the observable and unobservable object. Rouse does not strictly distinguish reality of theories and reality of entities, whereas Hacking differentiates the "tiny yet observable" that "in principle can not be observed".[4] Second, Rouse doesn't put theory against practice. He claims theory as a kind of research activity is also a practice. Nevertheless, Hacking goes farther; he emphasizes excessively the importance of manipulation and even claims that scientific practice should be reduced to engineering.

Rouse criticizes both scientific realism and anti-realism from his practical, common sense view of science. The practical hermeneutics which Rouse advocates and uses has very deep roots in the soil of Heidegger's metaphysics. Rouse believes scientific research is an engaging practice. Through it we interpret the world when we engage in it; the world is what we engage in, and there is no question of "getting access to the world." He refuses to make a metaphysical interpretation of reality like scientific realists. He tries his best to avoid the disposition of scientific anti-realism which is historicism or relativism. In sum, the most brilliant aspect of Rouse's practical view of science is that, to a certain extent, it deconstructs the ontological problem of observable and unobservable entities argued by scientific realism and anti-realism and provides a new perception to the epistemic problem of traditional scientific realism concerning the relation between theory and practice.

Bibliography

[1] Alan, M. Van fraassen's instrumentalism. *The British Journal for the Philosophy of Science*, 36(3):258, 1985.

[2] Cornman, J. W. Sellars on scientific realism and perceiving. In *Proceedings of the Biennial Meeting of the Philosophy of Science Association*, vol. 2. 1976. Symposia and Invited Papers.

[3] Hacking, I. *Representing and Intervening*. Cambridge, England: Cambridge University Press, 1983.

[4] Hacking, I. Experimentation and scientific realism. In Gasper, R. B. P., Trout, J. D. (Eds.), *The Philosophy of Science*, 247. Cambridge, Mass: MIT Press, 1991.

[5] Putnam, H. *Mathematics, Matter and Method*, vol. 1. Cambridge: Cambridge University Press, 1975.

[6] Rouse, J. *Knowledge and Power*. Ithaca and London: Cornell University Press, 1987.

[7] Van Fraassen, B. *The Scientific Image*. Oxford: Clarendon Press, 1986.

The Significance of Leibniz's View of Science in the Academy and for Public Happiness[1]

Yang, Jing

Dalian University of Technology

yjing407@163.com

ABSTRACT. Gottfriend Wilhelm von Leibniz (1646-1716) was a talented natural scientist as well as a great philosopher during the 17th and 18th Centuries. In his career, Leibniz sought a universal method for scientific research. He also had a passion for science and reason. Leibniz emphasized the combination of theory and practice. He stressed that the most important aspect of technology is to apply it in practice. In Leibniz's eyes, science could be advanced by exchange between different cultures, so he advocated the institution of all kinds of international committees, the compilation of encyclopediae, the building of large-scale libraries and museums, and the establishing of academies for the sciences in European countries. Another important characteristic of Leibniz's view of science is that he argued that the final aim of science is to promote as much as possible happiness and a better life for human beings.

1 The combination of theory and practice

Throughout his life, Leibniz sought a universal method for the acquisition of knowledge and invention, which is systemic integration. There are two methods: one is theoretical science, which can perfect the mind; the other is practical science, which can protect the tools of mind – body. Leibniz, who was passionately devoted to innovation, emphasized the combination of theory and practice. In his opinion, technological inventions are as important as scientific investigations, and scientists need to pay attention not only to pure scientific research, but to the application of scientific knowledge in order to enhance the pubic welfare.

It is unique that Leibniz's invention is not isolated, but in the form of a system, with a self-regulating function. Leibniz's inventions are not traditional crafts, but on the based on theories drawn from mathematics, mechanics, chemistry, hydrodynamics, etc. In Leibniz's opinion, the most important aspect of technology is to apply it in practice. And there must be regulation in the process of such application which means everything has its own position. In one of his manuscripts, Leibniz stated two ways for solving the problem: checking the correctness of knowledge and accumulating our knowledge through invention.

---

[1] The author is with School of Humanities and Social Sciences, No.2 Building, Dalian University of Technology.

The former corresponds to traditional proof technique and the latter to systemic invention. In 1666, Leibniz conducted research about the relationship between the art of combinatorics and invention. He assumed that a machine is made up of a certain quantity of basic elements. If we can find a way to regulate those elements, it would be possible to produce an entire machine. Of course, Leibniz's use of the word "machine" in this context is just a figure of speech. Leibniz regarded the universe as a machine and he also regarded organisms as types of machines. Compared to the typical artificial machine, there are some differences in Leibniz's concept of machine: (1) it is an actual entity; (2) everything is a machine, which means that organisms or natural machines can be broken down into smaller pieces without end; (3) it has self-regulating function[8].

If we look at Leibniz's requests to people who discover and produce machines, we find that in Leibniz's opinion, human beings imitate God's creative activity, although the creation of humans is defective. Human beings, like God, obey optimal rules, that is to say, human beings should try their best to choose the optimal solution to specific problems. However, the difficulty is in the difference between our intelligence as compared to Gods. We need to seek, explore and test continuously. Human beings must be responsible for those practical purposes which require a scientific understanding and foundation. This idea was inherited by Dessauer, a famous German philosopher in the 1920s.

2 Leibniz's projects for the academy and universal science

Leibniz made more than 60 suggestions to the academies in Germany, Austria, Russia and Holland in his lifetime. Different from contemporary ideas about academies in England and France, Leibniz's assumed that the academy included the humanities, literature, law and theology besides natural science,. It had been an important innovation which included both sciences in the narrow sense and engineering on the one side and humanities on the other. Leibniz mentioned the disciplines in a memorandum, namely mathematics, physics (including astronomy and geography), the mechanical arts (including architecture, military and nautical machines and so on), chemistry (including mining), biology (including anatomy and agriculture). Furthermore, Leibniz included two further tasks: (1) the possibility of a Protestant mission in China; (2) the encouragement of the German language, since language is "a clear mirror of the mind". Only if language is rich and precise, can thinking be subtle and exact. Therefore, Leibniz proposed to collect specialized dictionaries, a task for the academy and libraries. In contrast to very old-fashioned universities in Europe at that time, who did not really recognize the new problems and possibilities, Leibniz had in mind that there had been no empirical research and absolutely pure technology. The task of the university consisted in educating students about theology, law, and medicine. The academy, meanwhile, should not only enlarge knowledge but function as a culture exchange center for knowledge.

Leibniz developed his discipline schemes in a systematic order, beginning with logics and combinatorics, and adding step by step the special conditions of each field of knowledge. In one of his papers, Leibniz wrote, "the beginning and first step of the universal science is to install and enrich sciences, to promote the mind and the things by inventions for the public felicity"[9]. Leibniz gave a list under the title of "What to do", which indicated that he was not thinking theoretically but about something practical to be realized. The notes range from introductory comments on the reasons of the author, namely to enlarge "human felicity", by "rational language and grammar" to mathematics, physics, mechanics, astronomy and geology to medicine, law, politics, social structures and finally to "natural theology", ending up with Christianity. How to connect the academy and the universal science is as following: The list ends with "On the society of friends of God". His society clearly means an academy, representing universal science, as Leibniz was convinced that the combination of science and natural theology opens the way to both felicity and Christian faith. This is the reason that he always supported the Jesuit's activity in China, based on the mediation of faith through the sciences. This shows that the academy and universal science were just like two sides of the same medal.

The art of demonstrating and inventing describe what to do in the academy. The former had the task of demonstration, which is the theoretic part of scientific research, whereas the latter was concerned with the way from theory to practice. He explained why to establish a theory is much more significant than to have only a practice. Essentially speaking, practice depends on special cases, so the result can be meaningful only in such concrete cases. Therefore, "even a blind theory will be incomparably better than a blind practice without theory"[9], because one would have no ability to solve a new trouble, even it's just a little dissimilar with before, if he or she only follows his practical experience. This corresponds to the arguments which Leibniz used in order to describe the "merry wedding" between theory and practice in his academy project.

3 To seek for happiness as the final aim of science

For Leibniz, the development of knowledge and technology has as its purpose to enhance the common welfare of the people. Leibniz repeated this many times. Science will warrant a better life for all the inhabitants of a country. For Leibniz, all these responsibilities, ranging from theory to practice, from collecting and enlarging knowledge up to organizing new institutes, aim ultimately at promoting the greatest happiness for human beings. His idea presupposes that all forms of knowledge are connected, like the parts of a complex, well oiled machine in such a way that the machine can work appropriately only if all the parts are working harmoniously together.

Human beings need science and technology. They are vital to people's lives. Due to Leibniz, the practical use of theory is important. He distinguished between the systematic knowledge and pure knowledge collection. Systematic knowledge is prior to pure collection. Only by acquiring systematic knowledge can human beings recognize the rules of the universeas they reflect the harmony of God's wisdom. The value of science is not just to preserve people's lives and ensure their possibilities. More importantly, science as theory declares the significance of the world. It points out the direction of human development and shows how the world is harmonious. Therefore, people's behavior should be harmonious as well.

The collection and exchange of knowledge is significant to the economy and politics in Bacon's philosophy, a concept which was borrowed by Leibniz. Leibniz was convinced that, "from China, great things can be learned in an exchange of sciences"[7]. Because of this belief, he always emphasized the enormous amount of empirical data which Chinese scientists had brought together through the centuries, whereas the theoretical capacity is better in Europe; thinking of theory and practice, to bring both sides together would cause such positive developments as admirable progress on both sides. Leibniz' concept of universal harmony includes the human obligation to expand the harmony in the world not only by means of better life conditions but also and in the first place by increasing tolerance and harmony among human beings in spreading out this kind of reasonable Christianity which, due to Leibniz, is based on science and reason alone, so that it can be accepted e.g. in China.

There is an important reason for Leibniz's thought on happiness – the mercantilism of Leibniz's times. Leibniz was way ahead of his times. He pursued a much more universal happiness by means of science. Furthermore, the same as Socrates, Leibniz was convinced that to act morally is a question of knowledge. His argument runs as follows. Moral necessity "demands to follow the rules of the perfect wisdom." An "obligation is what is necessary to do for a good human", as "Each prudent is a good human." Therefore, this is "an obligation of reason, which always has its effect on the wise", because these are always "good reasons", as Leibniz says, namely depending on the Principle of the Best. Now, if I know that an intended act does not really increase happiness for me or anyone else, I would immediately change my intention which means that an enlargement of knowledge will immediately increase morality in society. Consequently, the task of science is to collect, enrich and distribute the knowledge. In one article written to Czar Peter, Leibniz said, "The true end of all studies is human felicity, which means a constant happiness, so that people do not live in idleness ... but act in accordance with the common welfare"[1]. There is nothing that can make a greater contribution than the perfection of knowledge to universal happiness. He advocated cultural exchange between Europe and China, which partially reflected his final aspiration to achieve happiness for the whole world through science and knowledge.

Bibliography

[1] *Concept einer Denkschrift für den Czaaren Peter.* Foucher, 1708.
[2] 张祖贵. 莱布尼茨及其科学观. 科学学研究, 7(1):78 – 86, 1989.
[3] 李少斌. 莱布尼茨的科学观. 湖北大学学报(哲学社会科学版), 27(5):26 – 29, 2000.
[4] 段德智. 论莱布尼茨的自主的和神恩的和谐学说及其现时代意义. 世界宗教研究, (1):99 – 108, 2000.
[5] 莱布尼茨. 从中国回来的耶稣会神父闵明我就所提问题对我的回答. 中国科技史料, (3), 2002.
[6] 莱布尼茨. 中国近事. 郑州: 大象出版社, 2005. 中文本序.
[7] 孙小礼. 莱布尼茨与中国文化. 北京: 首都师范大学出版社, 2006.
[8] 波塞尔, H. 莱布尼兹与技术. 世界哲学, (4):92–96, 2005. Translated by 李理.
[9] Poser, H. *Nihil sine ratione.* Hannover, 2002.

Research on Modal Terms in the Mohist Canon[1]

Wang, Xiaojing

Yanshan University

wangxiaojing-@163.com

ABSTRACT. Known as the culmination of logic in ancient China, the Mohist Canon involves words like *possibility, necessity, must, past, now, present* and *future*, etc. These words belong to generalized modal logic, which has standard modal terms, deontic terms, tense terms and so on. In this paper I will study and analyze each of these modal terms. First, I will analyze the modal term *bi (必)*, which has two meanings. Then I will turn to other modal terms such as the tense terms *qie(且), jiang(将), yi (已), fang (方), chang(尝), zheng (正)* and other modal terms such as *bu bi(不必), shi(使), wei(谓), gu(故), yi (疑suspect), yi(宜)* and so on.

Mo-tse was a great thinker, an educationalist and a great statesman from the pre-Qin period. The School of Mo made Mohist (Mojia 墨家) logic comparable to that of the Hetuvidya (因明India logic) and Aristotle's logic. Together, they are known as the three major origins of ancient logic. Mohist schools also explored noun (concept), speech (judgment), speaking (reasoning), and thinking problems of the law, thus enriching and developing Chinese logic. The Mohist Canon (Mojing墨经) is part of the logic text Mo Zi (墨子). The Mohist Canon contains six parts: Jing Shang 经上(Canon Vol.I), Jing Xia 经下(Canon Vol.II), Jingshuo Shang经说上(Expositions of Canon. Vol.I)Jingshuo Xia 经说下(Expositions of the Canon. Vol.II), Da Qu 大取(Major Illustrations), Xiao Qu小取(Minor Illustrations).

1 The Main Modal Term "Bi" in the Mohist Canon

A modal term is used to indicate a modal proposition or modal concept. It is the constant of a proposition. Modal Logic is a new branch of modern logic and has even made great achievements. There are a lot of modal terms in the Mohist Canon. First let's look at "bi" (必).

In the *Logic Dictionary*, "bi" is defined as "*Dialectical logic language is equivalent to apodictic proposition. Mo Zi · Jingshang (墨子·经上) 'Bi, bu yi ye'. Here 'yi' means 'change' (from Gao Heng).*"[1] For example: youzhi bi ran, wuzhi bi buran 有之必然, 无知必不然. Here it is used to underscore the necessary link in the proposition. Dr. Zhongyuan Sun (孙中原) expressed the formula as follows:

---

[1]The author is with College of Humanities and Law, Qinhuangdao 066004 HeBei Province.

Research on Modal Terms in the Mohist Canon 163

Bi (necessity) = universal (all) + full-time (now + past + future)
= completely + consistent

In the *Mohist Canon*, "Bi" is used up to as many as 67 times. And [jing xia, [jingshuo xia] have each of "Bi" (必) mistakenly written "xin" (心heart). In [jing xia], the original source "wu shuo er ju, shuozai fu xin (心)" should be corrected to "wu shuo er ju, shuozai fu bi (必)"; In [jingshuo xia], the original source was "*bu xin (心) suo che zhi zhiyu shi ye*", should be corrected to " *bu bi(必) suo che zhi zhi yu shi ye* ". Add these two " bi " which together makes a total of 69 "bi". Then in [jing shuo], the text called "Gong bu (不) dai shi ruo yi qiu", should be corrected as "Gong: bi (必) dai shi, ruo yi qiu". Here "bu" (不) is a "bi" (必) wrongly used.[5] Together with this one, then in the whole Mohist Canon there are up to 70 uses of "bi". But the text is written "bi" having a surplus, such as "bi re shuo zai dun", which is how it originally appeared in the Canon, so it should be corrected from "bi" to "huo" (火fire). The text means: Fire is hot; the fire heat is pure, totally hot. So I do not attribute this "bi" to the 70 occurences. "Wei li bu bi li bu bi li" in the original text were two "bu bi li", but it should only have one injection and remove the "bu bi", the latter "li" connected to the back characters, that is: wei li bu bi li, li yu bao ye. Moreover, in "Er bi ren zhi ke jin bu ke jin yi wei ke zhi, er bi ren zhi ke jin ai ye." the two " bi" were both in the original text, but in Mr.Yunzhi Zhou' s opinion, only the latter "bi" should be retained. So it appears that after the collations the Mohist Canon uses 68 "bi". So many "bi", they are not only for modal logic, but also reflect other meanings. Broadly speaking, I think its meanings can be divided into two categories: one marks the hypothetical proposition as necessity; the other category is as a deontic term meaning "must" .

"Necessity" belongs to standard modal logic, with "possibility" to correspond. Logic speaks the rule, if "you have A, it must have B", means you have A, then you may necessarily get B. In the Mohist Canon, "bi" (necessity) may also necessarily draw some conclusions from the known premise and "bi" performs the role of bridge. Having the meaning of "bi", the standard modal propositions express things necessarily, certainly so. In the Canon, "bi" with the meaning of "necessity" appears altogether 52 times, for example:

"Necessity" belongs to standard modal logic, with "possibility" to correspond. Logic speaks the rule, if "you have A, it must have B", means you have A, then you may necessarily get B. In the Mohist Canon, "bi" (necessity) may also necessarily draw some conclusions from the known premise and "bi" performs the role of bridge. Having the meaning of "bi", the standard modal propositions express things necessarily, certainly so. In the Canon, "bi" with the meaning of "necessity" appears altogether 52 times, for example:

[jing]sheng,ying zhi sheng,shang buke bi ye.生，盈之生，商不可必也。

Translated: Physical presence has the perception that while there is life, human life does not necessarily last forever, and thus life is so impermanent.

In 1951, Mr. Wright, following standard modal logic, built the deontic logic. He compared the relationship between "possible" and "necessary", and then established the definition of relationship between "obligatory" and "permitted". In Chinese, "bi" also means "obligatory". As "obligatory", "bi" is a deontic term, expressing the order and obligations. In the Mohist Canon, this meaning of "bi" is used up to 16 times. Examples are as follows:

> [jing] wu zhi suoyi ran,yu suoyi zhi zhi,yu su yi shi ren zhi zhi,bu bi tong,shuozai bing.物之所以然，与所以知之，与所以使人知之，不必同，说在病。

i++¿

Translated: Things like this condition, nature, and the reason why I know and how my people know, is not the same. For example: I know that a person is sick through observation, blood pressure measurement, blood testing, but how can a person tell other people this disease, it is not necessary to use the same methods.

> [da qu] san wu bi ju, ranhou ci zuyi sheng.三物必具，然后辞足以生。

Translated: It must have three arguments (premises); then the subject can be established. Here "bi" means obligatory. "Speech" 辞has been established from the three prerequisites. In fact, the sentence expresses a logical inevitability; "zuyi" 足以shows an inevitable link which sufficient prerequisites will present the consequence of "speech" necessarily, that is: these conditions are in place, so that it will come to a certain conclusion.

2 The Other Modal Terms in the *Mohist Canon*

In addition to "bi" , the Canon has some other modal terms, such as the tense terms "yi" (已, past), "fang" (方, ,just), "chang" (尝, ever), "qie" (且, will or just), subjective and objective modal terms "shi" (使), "wei" (谓) and "gu" (故), negative modal terms "bu bi" 不必, "fei bi" 非必, "wu bi" 毋必(not must or not necessary) and "bu yi" (不疑not in doubt), problematic modal terms "yi" (疑, in doubt), "yi" (宜) and so on.

2.1 An Important Tense Term of "qie" (且)

"qie" is generally as "will", which belonged to the modal term of probability proposition.

> [Jing] qie, yan ran ye. 且，言然也。[jingshuo] qie, ziqian yue qie, zihou yue yi, fang ran yi qie. 且，自前曰且，自后曰已，方然亦且。

For example, [xiao qu] "qie ru jing", namely: To be into wells, "qie chu men," namely: to be out of the house, "qie yao" is about to die. That said, "qie" (且) means the future. Yim (已): Already, the time is past. [jing shuo xia] yi ran ze chang ran. (已然，则尝然). "yi ran" is already the case. It refers to the past, equivalent to an assertoric judgment.

"Fang ran yi qie"方然亦且in canon has given "qie"(且) an alternative definition that "fang ran" also known as "qie", and it said that it is just now, referring to the present time. Here the "fang" also expresses the tense that means now, present. But not all the "fang" has been used to express the tense.

2.2 Expressed Subjective and the Objective Model in the Mohist Canon

[jing78] shi,weigu.使，谓、故。"shi"(使) used in two ways, one refers to instructing a person to do something; another is a kind of objective reasons will inevitably produce a certain result.

[jingshuo]shi: ling,wei ye,bu bi cheng.shi,gu ye,bi dai suo wei zhi cheng ye.使：令，谓也，不必成。湿，故也，必待所为之成也。To order or make someone do something, that it is only instruction, a person would not necessarily complete this matter. But for the wetlands of this natural phenomenon, there is the objective reason, which must produce the outcome.

Here "shi" means command, expressing both subjective modal and objective modal. The meaning of "wei"谓(tell) is "said". In an imperative sentence it expresses the subjective modal. "Gu"故 (reason) would inevitably produce such results, and "ling"(令) is an order to them. Obviously, "gu" here is the expression of objective necessity, but ling (command), wei (say) are used in imperative sentences. I think they are deontic terms, for example: I order you to switch off the television. The order namely represented the meaning of "I hope you must do it", But "shi" represents two kinds of modals, not only the standard modal term, but also the deontic term, depending on "shi" in the linguistic environment of the sentence.

2.3 Expression of the Interrelationship between Necessity and Possibility

[jing84]he, zhengyibi.合，佱（止）、宜、必。

[jingshuo]he, shi zhi zhi zhong,zhi gong he,zheng ye.zang zhi wei,yi ye.fei bi bi bu you,bi ye.sheng zhe yong er wu bi,bi ye zhe ke wu yi.合：矢至厂矢中，志工（功）合，正也。臧之为，宜也。非彼必不有，必也。圣者用而勿必，必也者可勿疑。

Translation: Arrow shoots at the center of target, so its effect and will precisely coincide, and it is called zheng he正合 (real consistency) . Zang (a man) does something, but his effect and will may likely coincide, may not coincide, when he wanted to do and what he did coincide, we call it yi he (宜合). If we have not A then we necessarily have not B, we can call it bi he (必合). Saints are engaged in unnecessarily coinciding with the results, but if awareness coincides with the actual, it is possible without suspicion.

In this one, we know zheng he正合 (actually coincide) ,yi he宜合 (maybe coincide) ,and bi he必合 (necessary coincide, they also like $p, \Diamond p, \Box p$. So it also fits well with our logic opposition square, when necessary $P \rightarrow$ real p, real $p \rightarrow$ possible P, necessary $P \leftrightarrow$ impossible not P, possible $P \leftrightarrow$ unnecessary not P, real $p \leftrightarrow$ unreal not p, possible not $P \rightarrow$ real not p, real not $p \rightarrow$ necessary not P, which embody the modal logical relationships among "apodictic", "assertoric" and "problematic". Here "zheng" means fitness and exactness, and is the assertoric modal term. "yi" is the problematic modal term. "Bi" still means apodictic modal term.

2.4 Some Modal Terms Expressing Denial

For instance: bu bi (不必), fei bi(非必, and wu bi(毋必), "not inevitably P" to be equal to "possibly not P", before " bi"(必) adds the denial word(¬) expressing unnecessary, does not must, such as,

[jingshuo1]gu: xiao gu,you zhi bu bi ran,wu zhi bi bu ran.故：小故，有之不必然，无之必不然。

Xiao gu(小故), if there is a A, there is unnecessarily a B; but if there is not a A, there is necessarily not a B.

Here "bu bi" 不必 means "possibly not". This is a probable modal term. "Bu bi" also means "not must", according to the logical square of the deontic proposition, ¬ Op equal to (↔) $P\neg p$. So here "bu bi" is still a deontic term. "bu 不(not)" and "bi 必(necessarily)" are also logical constants expressed in the meta-language.[2]

Obviously, there are so many modal terms in the *Mohist Canon*, namely logical constants in the modal proposition. We can see that logic is regarded already to have understanding and application in ancient China as recorded in the religious texts. Formal logic is researching variables and constants in propositions to achieve the goal of cognition, so the study of modal terms as logical constants is important. To study our ancient books, in particular the Mohist Canon, it becomes possible to recognize that these modal terms really existed and can provide evidence that there was such a thing as a modal proposition in ancient China. At the same time, it demonstrates in part the truth that "there was logic in ancient China". Many kinds of modal terms can be found in the Mohist Canon and construed in order to reinforce the link between Chinese logic and Western modern logic. This study forms a part of the effort to show that East and West share a common ground on the origin and essence of logic.

Bibliography

[1] Peng, Y., Ma, Q. *Luojixue Dacidian*逻辑学大辞典, *Logic Dictionary*. Shanghai: Shanghai Cishu Press, 2004.

[2] Sun, Z. Meta-research in chinese logic. *Zhongguo Renmin Daxue Xuebao* 中国人民大学学报*(Journalof Renmin University of China)*, (2):56 – 62, 2005. Translated by Saihua Song(宋赛花).

[3] Sun, Zhongyuan 孙中原. *Zhongguo Luoji Yanjiu*中国逻辑研究*The Research on Chinese Logic*. Beijing: Shangwu Press, 2006.

[4] Zhang, Zhongyi 张忠义. *Zhongguo Luoji Dui "birandi dechu" de Yanjiu*中国逻辑对"必然地得出"的研究. Renmin Daily Press, 2006.

[5] Zhou, Yunzhi 周云之. *Mojingjiaozhu · jinyi · yanjiu – Mojing Luojixue* 墨经校注·今译·研究–墨经逻辑学*Mohist Canon note · translation · study – Mohist logic*. Gansu People Press, 1993.

C

Philosophy Issues of Particular Sciences

Does the Spirit of Universal Logic Coincide with Logical Pluralism?[1]

Tzu Keng Fu

University of Neuchâtel

tzukeng@alumni.ccu.edu.tw

ABSTRACT. Universal logic and logical pluralism start from different considerations. The foundations of universal logic are mathematical. The foundations of logical pluralism are philosophical and logical. As mathematicians, we see the mathematics of logic, and as logicians and philosophers, we see the philosophical insight of logic. As mathematicians, philosophers or logicians, we claim that the spirit of universal logic coincides with logical pluralism. In this paper, we discuss Tarski's notion of consequence operation, Łoś-Suszko's notion of the structural consequence operator in *abstract logic*, and Béziau's structural consequence operator in universal logic. All of these views are abstract views of logic. Then we argue that consequence itself is a common feature of the spirit of universal logic and logical pluralism, even a mother structure in the Bourbakian sense.

1

1.1 History.

The idea of universal logic comes from the Swiss logician Jean-Yves Béziau. Universal logic is a function or consideration of mathematical theory, that is "considered logic as one of the mathematical structures" in a Bourbakian sense. This consideration manifests itself in much the same way as universal algebra is a general theory of algebra structures. That is universal logic is a general theory of logical structures which "unifies a multiplicity of logics by developing general tools and concepts that can be applied to all logics." ([4]) As Béziau expresses the idea, universal logic is a general theory of logics which intends to discover the common features of logics. As such it is not a new logic but a theory which attempts to account for existing logics. Roughly speaking, the development of universal logic can be divided into three periods. The first period is dominated by many para-consistent logical systems. The second is captured by the idea of Polish logic, which is frequently referred to as the **abstract** logic period. And the third is the "universal logic" period. Although the interest of universal logic comes from para-consistent logic, abstract logic plays the most important role in the development of universal logic.

---

[1]The author is with Institute of Logic and Semiological Research Center, University of Neuchâtel, Neuchâtel, Switzerland.

1.2 Abstract Logic.

In order to realize a universal logic, it is customary to say that universal logic is "in an abstract point of view" to see what logic is.[2] "An abstract point of view" means that it is (1) on a general level; and (2) without axioms. For (1) it means we are working on a general level without specifying any language or logical operator. For (2) it defines logic to be in a consequence relation to a given undetermined set. Generally speaking, the starting point of universal logic is to escape from specific views in logics and try to see logic in general and even give up considering axioms of consequence relations. This kind of idea is due to the Bourbakian sense. [3]

1.3 Notion of Consequence Operator.

Historically speaking, the idea before universal logic is abstract logic. We must get involve into the discussion of the notion of consequence operator. We make a short comparison of this notion coming from Tarski, Suszko, and Béziau, respectively. The discussion here will be surrounded around the basic question, that is, '*Should the domain of a logic be an algebra?*' First of all, before answering the question above, we have a common premise, that is, '*consider logic as a structural consequence operator*'. If we accept this premise and define consequence operator on fixed structure, such as algebra structure, then we say we define consequence operator on algebra, apparently, it is in the realm of algebra. Naturally, taking logic as a part of algebra is asserted. Even more, we see the general theory of algebra (universal algebra) contains the general theory of logic (universal logic). Suszko appreciated this idea.

However, standard approach in Polish, such as Tarski's approach, is different from Suszko's approach. They did not define logic on fixed structures. If we follow this idea, we get different sets of structures which make consequence operator generate different properties. Even at the end of XX century, Tarski didn't specify any structure for consequence operator. However, we wonder if it is suitable just to take logic as a structural consequence operator. If the answer is negative, what should logic be considered? And then, '*which structures shall be considered on if we get reasons to consider logic as structural?*' Béziau said, "*My proposal was clearly distinct, because I was considering logical structures as different from the already known structures and because by so doing I was defining them in a very abstract way, in particular without stating any axioms for the consequence relation.*" ([3, p. 14])

Apparently, he keeps the idea of taking consequence operator on structures. He not only specified consequence operator on *non-fixed* structures, but he asserts

---

[2] It doesn't mean universal logic is renamed from abstract logic in the Polish sense. If it were, we would get involved in a situation where, as many Polish logicians have argued, universal logic is a part of universal algebra.

[3] Although a logic always called as a normal one if any only if properties: reflectivity, monotony, transitivity hold for consequence operator, we still set up this definition without any axiom on consequence relation within a structure.

to specify it on some *unknown* structures.[4] As a structuralist[5], Béziau's consideration will have some advantages. First of all, he does not need finding out any further philosophical reasons for specific structure where consequence operator defined on, because he didn't specific on any, instead only to be a structuralist and then follow the Bourbakian spirit in mathematics, that is to be a mathematician. Second, he can avoid seeing logic just as a set of truth and furthermore go on to capture one of the main tasks of logic[6], capture reasoning.

1.4 Mathematical Structures and Logical Structures.

We view some mathematical parts of logic, i.e. to see the mathematical structures of logic. Suszko takes mathematical structures of logic are algebraic structures; instead Béziau thinks not only algebraic ones but also others, such as topological structures. But, 'which kind of structures are logical structures?' Béziau argues logical structures have to be fundamental mother structure in the Bourbakian sense with other mathematical structures (topological, algebraic, and order). It means logical structures will not be any mathematical structure, but logical structures are also abstract structures together with mathematical structures, including algebraic, topological and order, i.e., considering mathematical structures with logical structures $<L;\vdash>$ where L may be one of three mother structures.

2

2.1 Logical Pluralism and Universal Logic

Logical pluralism is a pluralism of logical consequence.[7] It is to say that "there is more than one relation of logical consequence." ([1, p. 25]) Here we will not lose generality by taking a version raised by Beall and Restall. As we discuss

---

[4]It is crucial here, "specify on non-fixed structures" means "have to be specified on known structures but not fixed". Meanwhile, Béziau claims specifying on some unknown structures, which means specifying on structures but not only on known ones, such as algebraic, topological, and order structures, but also on unknown ones.

[5]Clearly, Béziau avoids arguing whether he considers logic as structural consequence operator, which is fine. Actually he agrees that logic may be not structural ([3, p. 14]), hence it is needless for him to argue where a consequence should be defined on. He strongly suggests he is talking mathematics, as we make a distinction in the abstract of this paper, to be a mathematician Even though he considers logic as structural consequence operators is fine, all Polish logicians, whether Tarski or others who hold the same position, also meet the problem, *'Which structure should be considered?'* But, if they are from the school of *structuralism*, it seems to be fine to consider logic as structural consequence operator.

[6]And how Béziau defines logic can be traced back to his paper.[2, 3]

[7]At first, from the term itself, universal logic, it is easy to misunderstand it as an assertion of logical monism, that is, to say 'Yes' for only one true logic. However, if we adopt logical monism, we totally get the wrong direction for universal logic. As we say before, universal logic is not a new logic; it is a general theory of logic. Maybe you say the abstract part of this hierarchy is properly true logic even if it is not a new logic because it is considered as the mother structure according to Béziau's words. Unfortunately, the basic settings of universal logic are not to find out this mother structure as the only one true logic. Besides it also needs to be with one of mathematical structures that can form a suitable one, that is, its universe will not be only one but many. And one of the tasks of universal logic is to find out the common features that can be applied to all specific logics. Actually, this task is familiar to the main idea of logical pluralism.

before, universal logic as treated by Béziau is a general theory of logic, where logical structure is an abstract structure $<L;\vdash>$ with mathematical structures. The idea of the consequence operator in universal logic is familiar with logical consequence in logical pluralism. How? First of all, we analyze $<L;\vdash>$ where L can be substituted by different structures. We want to know after substituting to be specific structures $<L;\vdash_1>$, $<L;\vdash_2>$ or else, whether $\vdash_1$ and $\vdash_2$ say the same $\vdash_i + +_i$ or not? Actually, our answer is both. At first, the setting of universal logic follows Tarski's approach partially, i.e. to conclude 'there is an abstract version of logical consequence' in this abstract setting, and is without any axiom. We all admit the intuitive notion of logical consequence is without any axiom. For logical pluralism, it is seen as GTT^8, which is also a general version of Tarski's idea where axioms are not considered. Secondly, in specific mathematical structures, $\vdash_1$ and $\vdash_2$ are two specific consequence operators respectively when we substitute L for algebraic or topological structures. $<L;\vdash>$ becomes $<L;\vdash_{algebra}>$ or $<L;\vdash_{topology}>$. We also have the same consideration on GTT for specific true instances. Mathematically speaking, for abstract consideration in universal logic, it is due to different mathematical structures, but for GTT, there are also many true instances of it, each of which bears a different consequence relation.

Universal logic, from a mathematician's point of view, sees logic as a given more abstract structure, and then specifies where a given abstract consequence operator can be defined. Furthermore it seeks the philosophical reasons to support them. We call it a *downward version of specification*. For logical pluralism, we also give an abstract version of consequence relation GTT. It is due to a different analysis of worlds, but we don't consider where GTT should be defined on because we are not looking at it from a mathematician's point of view. We don't see the structural senses of it. Instead, we consider which notions of cases are suitable to be substituted. Note here, both universal logic and logical pluralism don't claim (answer) directly what logic is. They only say how logic could be considered. Béziau claims that logic is a mother structure, with other mathematical structures, by a general consideration. It is the same idea in logical pluralism where GTT is the heart of logic with other cases, even if they go different ways. However, both are 'from general to specific, they just go differently because of different points of view. They all try to make logic free without setting any axioms on an abstract level.

2.2 One Logic or Many?

> "Graham Priest poses the question: "Logic: One or Many?" Our answer is "both". One: There is precisely one core notion of logical consequence, and that notion is captured in schema (V). Many: There are many true instances of (V), each of which specifies a different consequence relation governing our language. This one-many answer is what we call 'pluralism'."[1, p. 17][9]

Béziau holds the same position. He claims: *"I used the expression logic both as a generic term and also as a specific term. I defined an abstract logic to be*

---

[8]An argument is valid if and only if there is no case in which the premises are true and the conclusion is false.

[9](V) means GTT.

a consequence relation on a given undetermined set. I stated this definition with no axioms for the consequence relation, even if my work was concerned mainly with what I called normal logics in which the three basic properties (reflexivity, monotony, transitivity) hold."[4, p. 8]

3 Conclusion

We think considering logic in an abstract way is a good approach. Although many non-classical logics try to answer the basic question, 'What is logic?', whether we see logic in a mathematical or philosophical way, the consequence relation can be seen as the common feature of all logics. This unifying principle can apply to different known logics. Although Tarski's method is a general and abstract version of logic, axioms for logic are still there. We suggest that this notion that logic should 'be axiomatized' is bad and starts to develop at the beginning of the 20th Century. We conclude that we need to partially adopt Tarski's idea and go to a more abstract consideration whether we are talking mathematically or philosophically. Hopefully, this initiative to discover a universal logic can be achieved.

Bibliography

[1] Beall, J. C., Restall, G. *Logical Pluralism*. Oxford: Clarendon Press, 2006.

[2] Béziau, J.-Y. From paraconsistent to universal logic. *Sorites*, 12:5 – 32, 2001.

[3] Béziau, J.-Y. From consequence operator to universal logic: a survey of general abstract logic. In Béziau, J.-Y. (Ed.), *Logica Universalis*, 3 – 17. Basel: Birkhauser, 2005.

[4] Béziau, J.-Y., Costa-Leite, A. *Handbook of the First World Congress and School on Universal Logic*. UNILOG, 2005.

Thoughts on Sinicizing the Hetuvidya[1]

Zhang Xiaoxiang

Linyi University, Nankai University

zxx-0128@163.com

> The Sinicization of the Hetuvidya is an inheritance and innovations were made to it in China's Han and Zang areas. Its influence was not simply spread. Popularly speaking, Chinese scholars fundamentally adapted and changed the Hetuvidya in the course of its introduction and dissemination in China. Here, "Sinicization" is used in a different way from the Marxist concept of "localization" in China. It does not refer to the combination of the Hetuvidya and Chinese logic, but only to the various ways it was used in China. The Hetuvidya and Chinese logic are thus still two different logical systems. The Hetuvidya and western logic, which are both different from Chinese logic, did not originate in China. However, as an important component of the logic in China, the teachings of the Hetuvidya need to be developed. We can't be satisfied with simply absorbing and spreading the original theory. This article discusses some ideas for sinicizing the Hetuvidya and looks at the traces of sinicization in the historical development of the Hetuvidya, and discusses preliminary ideas for the transformation of the Hetuvidya's system based on its intrinsic characteristics.

1 Traces of Sinification in the Hetuvidya

Dignaga had further transformed the old Hetuvidya into the new Hetuvidya, and Dharmakirtti continued to develop the theory of the Hetuvidya. Compared to the Indian Hetuvidya, the Chinese version of the Hetuvidya both in the Han area and the Zang area changed more or less. The translation of the two books of the Hetuvidya was changed by Xuanzang. The Hetuvidya in the Zang area begins to generate its own forms of individual inference.

If we follow the spirit and innate character of Hetuvidya theory, we can cooperate with the Buddhists to create a theoretical system which has been adapted to local Chinese conditions. It enables the system of Chinese logic to draw on the experience and resources and development of western logic, to link up with the Chinese cultural heritage and Chinese characteristics, and to absorb and preserve the essence of traditional Chinese logic. Such an enterprise can avoid faults, exalt the advantages and enhance the usefulness of the Hetuvidya.

First, there are many changes in the books of Hetuvidya as translated by Xuanzang, including the most representative translation–three conditions for the

---

[1] The author is with Law College, Linyi University, the middle part of Shuangling Road of Lanshan District, Linyi city, Shandong Prov. 276005 China. Supported by the MOE Project of Key Research Institute of Humanities and Social Sciences at Universities, Project No.08JJD720044.

middle. "Condition of being an attribute of the subject" is the paraphrase of Sanskrit of "paksadharmatva". The literal translation is "property of subject", "being an attribute" is added by Xuanzang; the original text in Sanskrit of "positive instance possesses property inevitably" is "sapakse sattvam", "positive instance possesses property" is the literal translation. Xuanzang adds "inevitably"; the original text in Sanskrit of "negative instance does not possess property universally" is "vipakse asattvam", "the property of positive instance" is the literal translation, and Xuanzang adds "universally". The meaning of "universally" exhausts all possible meanings, "condition of being an attribute of the subject" means property-bearer of subject possesses ground property universally, the extension of ground must greater than property-bearer of subject, not smaller than it absolutely; additionally, the both can not equivalent, otherwise there would not have instance stand for it. This is different from formal logic. From the complete inductive Logic point of view, "does not possess universally" means that all is nothing, "negative instance does not possess property universally" means that the entire negative instance does not possess ground property. "certain" means inevitably, impossible to avoid or prevent, possesses "inevitably" means that it has part or all of the property; "positive instance possesses property inevitably" means that positive instance has to contain property of ground, and the amount is uncertain. Xuanzang's changes in translation embody the high quality of the logic.[2]

Second, Xuanzang founded Vijnapti-anumana and the theory of the three types of inference based on Dignaga's Hetuvidya. According to the Dictionary of Logic's explanation, *Vijnapti-anumana* is the paragon that Xuanzang applied to the demonstration theory of the Hetuvidya, the inference form in Kuiji's Da Shu of Hetuvidya as follows:

Thesis: "ZhenGu JiCheng"(phonetic transcription in Chinese), color can not depart from the visual sense

Ground: Admitted "ChuSanShe" (phonetic transcription in Chinese), because the eyes are not involved

Instance: Such as the sense of eyes

ZhenGu is the restriction on "Color cannot depart from the sense of the eyes" means an idea brought up by the sect of DaCheng, which is different from the sect of Xiao Cheng, in order to avoid violating common sense. JiCheng is the restriction on the *Dharmin* "color" of the Subject, means that the reference of "color" for both sides is recognized. The Dharma "sense of eyes" of Subject is recognized each other, so the subject is inference for each other. The significance of the logic is to say, "color" and "the sense of eyes" possess inseparable relation, it is relational proposition but categorical proposition. "Admitted" is the restriction on ground, because the foe approved that color is absorbed by the eyes, admitted "color is not absorbed by eyes", but the other do not agree. It clears and defines concept in order to avoid fallacies, so the ground is inference for one's own. According to the rules of the Hetuvidya, "the sense of eyes", for instance in the *Anvayi-Udaharana* and "the sense of eyes" in the ground should not indicate one thing, or else it will lead to circular demonstration. In the standard of three inferences,*Vijnapti-anumana* is inference for oneself, only for defense, and can not refute others.

Third, there is an individual inference form from the Hetuvidya in the Zang area. The inference form is dedicated to refute, not to advocate any subject.

It dates from Nagarjuna's *Madhyamikasaatra*. The students of Nagarjuna often use this inference form to refute the foe. The followers gradually evolved into a school. The works of Dignaga often use this inference form, but not as a standard inference form. Even Dharmakirtti argues against this form of inference. The inference form becomes even more influential in the long process of spreading and can be found widely applied.

The Hetuvidya's individual inference form in the Zang area is different from the traditional Tri-*avayava* of Indian logic and the syllogism of western logic. It consists of a subject and a ground, and it abridges instance.[1] The inference form is divided into two types, the true inference form that can confute the foe and a false inference form that cannot confute the foe.

In a word, the Hetuvidya has accreted many changes since it was first introduced into China. More changes need to be explored.

2 Thoughts on Transforming Hetuvidya Theory

The Hetuvidya contains forensics, epistemology, and especially logic. The theory system should obey the laws of logic. In order to inherit and enhance the wisdom that will be destroyed and make effective use of it, we should combine the Sanskrit with the Hetuvidya of the Zang area, elaborate and summarize the two works of the Hetuvidya in a short form, and popularize it. No science is an absolute truth. Every science must endure the test of time and continuous practice. This is true for the Hetuvidya in the Han area. It contains many unscientific thoughts which need to be enriched and improved constantly. Fallacies of thesis are the important part of the two Works of Hetuvidya which need to be collated, simplified and systematized.

The Hetuvidya's fallacies of thesis (Sankarasvamin's) disobey the logical rules and need to be examined from a new perspective. In the five violate fallacies of thesis: violate common sense, self-contradictory, violate perceptual cognition and violate inference intercrossed in extension. According to the definitions provided by the two Viruddhas and their status in the Hetuvidya, we can integrate violate common sense and self-contradiction into violate perceptual cognition and violate inference. Many causes should cancel the thesis that violates its own doctrine, such as the internal contradictions which emerge in the process of transforming the Hetuvidya system and the changes in the Hetuvidya's function. We can cancel the fallacy of XiangFuJiCheng because it is unnecessary as the factors of logic and epistemology increase. That leaves only five new Hetuvidya fallacies of thesis after collating and simplification. They not only conform to logical rules, but they are also more succinct and easier to understand. Sinicizing the Hetuvidya not only can enrich and develop the content of the Hetuvidya, but it can also lay a solid foundation for research into its practical application.

Bibliography

[1] Huang, M. *The defence law of an individual inference form of Hetuvidya in Zang area, Hetuvidya Research.* JiLin education Press, 1994.

[2] Zhang, X. Explore and analyze Lv Cai's contribution to hetuvidya, 2007.

On Sinologic and Sinomathematics in Ancient China[1]

He Xiangdong
Xinan University
he_xiangdong@sina.com

Liu Bangfang
Yanshan University
liubangfan@sina.com

1 What is Sinologic

Sinologic refers to logic in ancient China, or what we would call traditional Sinologic, which was at the time still free from the influence of Aristotelelian logic, which was still unknown in China. The mainstream Sinologic was represented by Mohist logic, which was based upon Mozi's polemical thought and method. However, Sinologic also includes the logical thought and methods of various other philosophical schools and thinkers from ancient China. According to such understanding, Sinologic is heterogeneous and can be divided into three categories: 1. Mohist logic*(Mo jia luo ji*墨家逻辑*)*; 2. The logic of proper naming*(zheng ming luo ji*正名逻辑*)*; 3. The logic of proving*(lun zheng luo ji*论证逻辑)*

As already mentioned, the most important thought on logic in China was the Mohist logic, which embodied the essence and basic content of Sinologic. Also important were the so-called theories of proper naming accreted by the representatives of the Nominalist*(Ming jia*名家*)*and Confucian*(Ru jia*儒家*)*schools. The proper naming movement developed theories and methods of inference. Later on, these theories were developed into a system of naming which formed the core of the new rationalism that emerged during the Song and Ming Dynasties(960∼1644). The logic of proving was mainly based upon ancient philosopher Wang Chong's 王充Demonstration Theory and included different methods of cognition. The main emphasis was placed on developing the proper methods and procedures for dialectical thought.

The basic concept within Mohist logic was the concept of type*(lei*类*)*and a certain way of inference, which was called tuilei推类and was similar to the process of reasoning by analogy.

The Mohist logic was an influential element within the development of Sinologic as well as in the history of Chinese culture in general. It had a great impact upon the various schools of thought from the pre-Qin era until the Jin Dynasty. In the same way as traditional Western logic was dominated by Aristotelian logic and Indian logic by the logic of Nyaya, the Mohist logic can be regarded as the highest achievement of Sinologic.

The basic elements of Sinologic were names or concepts *(ming*名*)*, dictions *(ci*辞*)* and arguments *(shuo*说*)*. As mentioned, its leading process of reasoning

---

[1]The author are with College of Politics-Law Xinan University, Chongqing City 400715 China, and College of Humanities – Law of Yanshan University, Qin huang dao City, Hebei Prov. 066004 China, respectively.

was similar to the process of analogy *(tuilei*推类*)*. In the Mohist theory, names or concepts represented the basic elements of naming, while dictions and arguments represented words and phrases used to express facts.

What are dictions? One of the main followers of Confucius, Xunzi荀子, defined them as follows:

"辞也者，兼异实名以论一意也。"

"Dictions (ci) are sentences that express thoughts or judgments by combining different kinds of names or concepts (ming).[10]"

Analogous arguments can be understood as explanations of facts through the introduction of different names or concepts *(ming)* as well as by different dictions.

Chinese analogy *(tuilei*推类*)* is based on the concept of type *(lei*类*)*. The first element of the Chinese analogy *(tuilei* 推类*)* idea was the notion of the cognitive process *(tui* 推*)*.

2 What is traditional Chinese mathematics?

Traditional Chinese mathematics mainly refers to that part of Chinese mathematics, which was established and developed before the end of the Ming Dynasty (1644), i.e. before Western mathematics was first introduced to China. It also includes the continuation of the end of the Ming Dynasty to the beginning of the 20th century. Its development can be divided into five distinct stages:

1. the stage of forming the theoretical system (from approx 2700 BC to 200 AD, before and during the Qin and Han Dynasty).
2. the stage of gradual development and popularization (from 200 to 581; during the Six Dynasties period)
3. the stage of substantiation and expansion (approx from 581 to 618, during the Sui and Tang Dynasties).
4. the recession period (the Song and Yuan period).
5. the transformation stage (during and after the Ming Dynasty).

At first, traditional Chinese mathematics mainly represented a system of counting and calculation. It was based upon a unique theory, which included specific methods and styles of calculation and research. In short, traditional Chinese mathematics was called Zhong suan 中算, i.e. Chinese calculation.[8, p. 18] Even in the 1950's, this name was being used by Yan Dunjie严敦杰in his mathematical textbooks for high schools as well as by Li Yan in his study On the History of Chinese Calculation中算史论丛. However, at present, this word also includes the application of traditional Chinese mathematical measures which deal with the newest foreign knowledge and theories. As a matter of fact, most of the mathematicians from the Ming and Qing Dynasties dealt with this kind of research work. Thus, the term Chinese calculation is not limited to traditional Chinese mathematics solely, and experts in Chinese calculation are not necessarily traditional Chinese mathematicians.

3 The method of 'matching types' and the *tuilei* analogy

Since the method of matching types *(lei*类*)* was the basis for traditional Chinese mathematical methodology, the tuilei analogies represented the main reduction type in traditional Chinese mathematics.

1. The *tuilei* analogy was initially established within the mathematics of the pre-Qin and Qin-Han periods. The pre-Qin and Qin-Han eras were the periods of the formation and rapid development of traditional Chinese mathematics. These two periods can be further divided into two stages: the first one was the pre-Qin period, in during which the earliest gems of traditional mathematical methodology were formulated. This stage ended with the Western Han Dynasty. The second stage was the period of the Western and Eastern Han Dynasty, which can be seen as the era of the initial completion and theoretical blooming of this system.

   In respect to logical thought and the methods of reduction, the traditional Chinese mathematics of the Qin-Han era was deeply influenced by the Mohist logic, which was based upon the method of tuilei analogy which was the most elementary mode of reduction. Such logical processes were commonly applied in early Chinese mathematics; thus, its initial logical path was the path of matching types *(lei)*.

2. The *tuilei* analogy developed strongly during the period of the Six Dynasties (220-581).

   Although this an era of constant instability, in which China was a divided state and went through periods of long wars, the science of mathematics flourished and experienced a number of splendid achievements. This was the time in which several famous mathematicians appeared, such as Zhao Shuang赵爽, Liu Hui 刘徽as well as a father and son named Zu祖氏父子. They wrote and edited a number of important algebraic works which were assembled in the collection *Ten Books of Classical Calculation* 算经十书. This collection included a number of important mathematical studies, including Zhao Shuang's Notes on *Zhoupi Arithmetic*周髀算经注, *Liu Hui's Comments on the Nine Chapters of Mathematical Procedures*九章术注, and *The Arithmetic of the Islands*海岛算经, Zhen Luan's 甄鸾*Sunzi's Arithmetic*孙子算经, Xia Houyang's *Arithmetic*夏侯阳算经, *Zhang Qiujian's Arithmetic*张丘建算经, *The Arithmetic of the Five Caos*五曹算经and *The Arithmetic of the Five Classics*五经算术, as well as *Zu Chongzhi's*祖冲之and *Zu Heng's*祖 most famous work *Zuishu* 缀术.

   During the of the Wei (220~265) and Jin Dynasty (265~420)periods, Chinese mathematics focused mainly on mathematical theory. The works of Zhao Shuang and Liu Hui are considered founding texts in ancient Chinese mathematical theory. Zhao Shuang was one of the earliest mathematicians who tried to prove the initial mathematical theorems and formulas and explained *The Zhoupi Arithmetic* in detail. Liu Hui interpreted, formulated and reduced methods, formulas and theorems from *The Nine Chapters on Mathematical Procedures* and enriched the process of explanation in a very inventive way. In his above mentioned study The Arithmetic of the Islands he applied 'the method of doubled difference' (重

差术) in order to solve measurement problems. One of his most important contributions was the invention of the so-called 'tactics for cutting circles'(割圆术), by which he laid the foundations for the invention of π (ratio of the circumference to the diameter in a circle).

The era of the Six Dynasties represented a second peak in the early development of traditional Chinese mathematics, both in advances in mathematical theory and in respect to mathematical methods that were promoted by Liu Hui's ideas of reduction.

Liu Hui's *Comments on the Nine Chapters of Mathematical Procedures* represented the highest level of traditional Chinese mathematics not only in respect to mathematical theories, but also in respect to mathematical methods. In this work the author also established the basic outline for his own methods of mathematical logic, including the basic elements of deduction such as the notions of name or concept *(ming*名*)*, word or sentence *(ci* 辞*)*, law or principle *(li* 理*)*, reason *(gu*故*)* and type, sort or class *(lei*类*)*. His main mode of deduction was the tuilei analogy. In his Comments, Liu Hui not only provided definitions for these notions, but he also supported it with a direct and profound analysis of the notions of inference *(tui*推*)* and type *(lei*类*)* in the sense logical procedures.

He used the notion of *tui* on a number of occasions; in fact he used it over twenty times. From his applications, it is clear that he understood its meaning in the same way as it had been understood by the Mohists. In his 'method of circular fields' (圆田术) in the chapter *On square fields* 方田章in his main work, Liu Hui demonstrated the meaning and importance of the notion of π in Chinese mathematics(ratio of the circumference to the diameter in a circle):

> "由此言之,其用博矣。谨按图验,更造密率。恐空设法，数昧而难譬。故置诸检括，谨其记注焉"
>
> "could serve a number of purposes. A strict abidance of his method could lead to a more precise calculation and proof of π. But I am afraid my descendants might misunderstand me and think that I have set a new π without a reason, so they would not be able to understand where these results come from. Thus, I have added a detailed explanation and records (on this method)."[5, p. 62]

As we can see, Liu Hui's explanation of the tuilei analogue type of inference coincided with that of Mozi (the founder of the Mohist school) as can be found in the Xiao qu 小取chapter of his main book.

According to my findings, the term type *(lei)* appeared no less than eighteen times in Liu Hui's Comments (including the Preface). An analysis of his writings shows that except for one single application, in which lei was used as a word expressing weight (斤数)[5, p. 56, paragraph 8], it was used with the meaning of type, sort, analogy, or similarity on all other occasions. While in Mohist logic, the term type *(lei)* was connected to notions of 'identity and difference', 'existence and non-existence'. in Liu Hui's work it represented a summary of the relations between things. On most occasions it stood for 'sorts' or 'analyses'. This shows that Liu Hui's application of this term corresponds with his understanding of the Mohist logic and shows that Liu Hui was familiar with the Mohist logical concept of type *(lei)*. Therefore, we can conclude that Liu Hui's Comments inherited and developed the thoughts and methods of traditional Sinologic connected to the term type *(lei)*. On this foundation, he discussed and rearranged mathematical terms and provided new classifications for them. Guided by mathematical classifying methods, Liu Hui differentiated between various mathematical terms. For example, he classified figures into integers and fractional ones, and he subdivided fractions according to different units. He wrote:

"数同类者无远，数异类者无近。"

"Identical numbers are without a distance, while different numbers have no closeness."

"其异名也，非其类者，犹无对也。"

"If two objects have different names, the do not belong to the same type, and between them there can be no relation of addition or subtraction (Positive and negative numbers)."

"凡物类形象，不圆则方。"

"Graphs can be divided into compositions of curved and straight lines".

"令出入相补、各从其类。"

"The reciprocality of exiting and entering depends on the type."

"朱青各以其类。"

"Red and green belong to different types."

"令颠倒相补，各以类合。"

"The reciprocality of upwards and downwards can be obtained by matching types." [5, pp. 62-132]

Similarly, he pointed out that congruent circles belong into one kind and discordant ones into another.

One of his most important achievements in mathematics was the establishment of classified mathematical methods. He first summarized the identities of mathematical methods. He regarded Jin you shu (今有术), which was a method of calculation based on comparison, as the leading technique in mathematical reasoning. He also pointed out the importance of the universal arithmetic method called du shu (都术), which included techniques such as division *(jin lˆü shu*经率术*)*, proportional distribution (shuai fen shu衰分术) , contra-proportional division (fan shuai shu返衰术) and a complex method of proportional divisions based upon affirmations and negations (jun shu shu 均输术). He considered du shu an important method with which one could find solutions to a number of problems.

Secondly, Liu Hui recognised the stratification of mathematic methods and analyzed them. For instance, he pointed out that Jin you shu was higher in rank than Shuai fen shu.[3, p. 272]

Third, Liu Hui analyzed the identities of solutions to many questions and classified them into several categories. For example, in Jun shu, questions 20 to 26 refer to different objects. He concluded that philosophical questions belonged in the same category.[3, p. 272]

Fourth, Liu Hui pointed out the identity of some specific calculations according to the inter-linking of mathematical methods and pointed out that different questions were applied in different spheres, which meant that in each of them we have to apply a different, specific method of counting. However, Liu Hui attributed them to the question of the right-angle triangle, because some of them needed to obtain gu (股) the small side of a right-angle triangle or xian (弦) (the large sides of a right-angle triangle), while the dispersion between gu and Gou (勾) (also a small side of a triangle) or xian were given.

In Liu Hui's *Comments*, the tuilei method represents the basic analyzing principles according to propositions (*xi li yi ci*析理以辞)and the foundation for interpreting systems according to illustrations (*jie ti yong tu* 解体用图).

Both of the above mentioned methods constructed the central principle that guided Liu Hui in writing his commentaries to the *Nine Chapters on Mathematical Procedures*. This principle was based upon high abstraction and represented the paradigmatic generalization of cultural bounded elements in the mathematical research of the time (and before). According to the classical work of ancient Chinese mathematics, the above mentioned *Zhoupi arithmetic*周髀算经, 'the method of analyzing principles' (*Xi li*析理) was founded upon 'the importance of formulation' (*gui yue* 贵约). Liu Hui and Ji Kang嵇康both presumed that analyzing principles was based upon the formulation, and that the central axiomatic method was the *tuilei* analogy.

Therefore, it is not surprising that Liu Hui's Commentaries followed the identification processes based upon the type *(lei)*. In this work, Liu Hui proved a number of mathematical propositions, formulas and conclusions. The crucial point of his proofs was the concept of type *(lei)*, which formed the basis of 'the matching types method' (*yi lei he lei* 以类合类).[2]

One of the starting and at the same time fundamental points of Liu Hui's illustration was the method for proving and applying rates (*l˜ü*缕). According to him, rates represented a crucial criterion for calculating programs. Thus, besides the method of matching types, he also applied the method of 'using rates to match types' (以缕合类). Both methods represented the basic calculation mode and main reduction type of Liu Hui's *Commentaries*.

Therefore, it becomes obvious that the development of abstractions, which reached a high level during this period, was not limited merely to mathematical theories, but also referred to mathematical methods and especially to the mathematical logic, the backbone and main subject of mathematical methods. During this period, the *tuilei* analogy became the central mode of calculating, while 'the method of matching types' became the major inference type of traditional Sinological thought. So we can say it was Liu Hui's Commentaries to the *Nine Chapters on Mathematical Procedures* that established the classical position of this work in traditional Chinese mathematics, not only in respect to mathematical theory, but also in relation to the formation and establishment of its logical thought.

All this shows that Liu Hui's logical thought was directly connected to the Mohist logic, although Liu Hui was superior to it in many aspects with his logical methods in arithmetic. While the Mohists were merely putting forward certain mathematical propositions without proving them, Liu Hui defined concepts, used logical methods and proved all the formulas found in the *Nine Chapters on Mathematical Procedures*[2, p. 143–144]. Of course, in other aspects Liu Hui was inferior to the Mohists, especially in the abstraction of mathematical theories.

---

[2]An example of the application of this method can be foundin Liu Hui's proof of the Xian chu (羡除).

Mohist logic was also more concise in its expression of mathematical propositions.

4 The tuilei analogy remained the basic method of mathematical inference during the Song (960~1279) and Yuan (1271~1368) Dynasties.

Traditional Chinese mathematics reached its last climax in the period of the Song and the Yuan Dynasty. This period was not only marked by breakthroughs in mathematical thought and theory, but also by numerous significant innovations in the fields of mathematical method and reasoning. These kinds of innovations can be best illustrated by the fact that while the *tuilei* analogy still represented the central methodological pattern of traditional Chinese mathematics, it gradually began to change into a more deductive programming or a more mechanical mode of thinking. This change can easily be detached from the works of *Qin Jiu shao*(秦九韶), Jia Xian(贾宪) and Zhu Shi jie(朱世杰). However, this tendency did not develop much further as time passed by. This fact could be explained only by the specific social and economic circumstances of the time. It was by no means a problem of mathematics itself. Besides, the level of mathematical logic in this period never surpassed the level it had reached in Liu Hui's *Commentaries*.

Following the *Nine Chapters on Mathematical Procedures* and Liu Hui's *Commentaries* of this classical work, the next epoch-marking work was the Mathematical Treatises in *Nine Chapters*. Its mathematical achievement was not merely of a theoretical, but also of a methodological nature. In respect to the mathematical methods, *Mathematical Treatises* in *Nine Chapters* successfully inherited China's traditional mathematical logic as developed since the Nine Chapters on *Mathematical Procedures* and included a number of innovations. We can find this inheritance in the fact that Qin Jiushao still took the *tuilei* analogy as the basic method of his mathematical reasoning, which can be seen from the following facts:

1. The extensive use of the concept type (lei类)

    The word type was used over one hundred times in the Mathematical Treaties in *Nine Chapters*. It was applied with three different meanings: firstl, it represented type or sort; second, classifying or sorting out; and finally, it represented the tuilei analogy. The abundant use of the word type *(lei)* indicates that the author Qin Jiushao regarded it as the basic logical method in mathematical analysis, both in deduction, as and induction. There were two reasons for his emphasis of this method. He was first of all deeply influenced by the earlier classics, especially Liu Hui's Commentaries. And second, he was influenced by traditional Sinologic, especially by the work of Yi Jia易家, whose logical thought was based upon the concept of type and *tuilei*.

2. The extensive use of the concept of inference *(tui推)*

    Another frequently used word in the Mathematical Treaties in Nine Chapters was the notion of inference (*tui*推), which appears no less than 30 times. The Mohist logic developed a very precise definition for this notion. The Mohists understood it as a process of inferring from a situation that had already taken place *(suoran*所然*)*, to a stage that was yet to occur *(weiran*未然*)*. Thus, in Qin Jiushaos work, the notion of inference *(tui)* was not only used in the sense of deduction but also implied the logical properties that were developed by the Mohist school.

3. The *tuilei* (推类) analogy as a basic programme.

Qin Jiushao's Mathematical Treaties in Nine Chapters represents a profound elaboration in the construction of the mathematical system. This work is based upon detailed research and a most carefully selected choice of methods and techniques of mathematical argumentation. Here, the *tuilei* analogy, which was (once more) forming the crucial method of Qin Jiushao's study, was based upon assorted premises. The author wrote:

"周教六艺；数实成之，学士大夫，所从未尚矣。其用本太虚生一，而周流无穷。大则可以通神明，顺性命；小则可以经世务，类万物，讵害以浅近窥哉？"

Mathematics is one of the Six Talents of the Zhou Education, thus scholars and officials always attached great importance to it. Mathematics came into being as a result of pursuit of regularity. Mathematics can be used in a number of various areas, from solving great problems like recognizing nature and understanding life, to very trifle things like dealing with daily affairs and sorting out facts. How can we thus state that mathematics is a superficial subject?[7, p. 469]

In his preface to the Mathematical Treaties in Nine Chapters, Qin Jiushao compared and analyzed over thirty schools of mathematics that existed in China at that time .He wrote:

"今数术之书，尚三十余家。天象历度、谓之缀术；太乙、壬、甲，谓之三式，皆曰内算，言其秘也。九章所载，即周官九数，系于方圆者为　术，皆曰外算，对内而言也。其用相通，不可歧二。独大衍法不载九章，未能推之者。历家演法颇用之，以为方程者，误也。"

"At present there are over 30 schools of mathematics. They can be divided into two groups, namely the schools of internal (内算) and external (外算) calculation. Internal calculation included the so-called Zhui shu(缀术) technique, which was used in astronomy and calendar calculations, as well as the methods of Taiyi, Liuren and Dun-jua(太乙、六壬、遁甲) that were used in divination and astrology. External calculation included the method of the so-called Nine treaties(九数) stated in the Mathematical Treaties in Nine Chapters and the Zhuan shu(　术), which was used to measure distance and specify positions and directions. There is also the technique of Da yan shu(大衍术), which was not included in the Nine Chapters on Mathematical Procedures, but was widely used by different schools, although nobody could deduct its algorithms. As a result, most people mistook it for an equation."[7, p. 469]

In the preface, Qin Jiushao also explained the basic structure of his book and the system of its content.

"所谓通神明，顺性命，固肤末于见；若其小者，窃尝设为问答，以拟于用。积多而惜其弃，因取八十一题，厘为九类，立术具草，间以图发之。"

"Although the understanding of the sacred spirits and following the laws of nature and destiny seems to be ignorant, it really helps us when dealing with affairs and sorting out things. Thus, I have stated many applied problems that lead to a better understanding. After many years, the number of such problems was on the increase. Thus, I have chosen 81 of them and divided them into 9 categories. I wrote down the answer and some of them were even equipped with a sketch to help people understand them."[7, p. 469]

5 Conclusion

It is a challenge to explore traditional Chinese mathematics from the viewpoint of Sinologic. All in all, Sinologic exerted a deep and positive influence on traditional Chinese mathematics; its thoughts and methods have defined the continuing development of traditional Chinese mathematics to a great extent. On the other hand, traditional Chinese mathematics also formed logical methods of its own. Its leading inference mode (until the Song and the Yuan Dynasties) was based upon the method of the so-called *tuilei* analogy.

Merely a glimpse into some of the most important ancient books and records confirms the statements of Qian Baocong (钱宝琮)and Du Shirang (杜石然), who maintained that there was plenty of material on the history of Sinologic in the field of traditional Chinese mathematics. Thus, it could be said, that traditional Chinese mathematics is logic in itself.[6]

During the period of the Qin and Han Dynasties (approx.200BC ~ 200AD), traditional Chinese mathematics accepted and applied most of the viewpoints and methods of Mohist logic. The ideas and methods of the specific tuilei – a form of analogy, which was developed by the representatives of the Mohist school – was mostly applied in the particular method of matching types, which was formed by the mathematicians of the period.

In the period of the Six Dynasties (220~581), traditional Chinese mathematics was developed to an unprecedented high degree not only in respect to theory, but also in respect to methodology. In particular, the logic path which represented the backbone and subject of mathematical methodology was elaborated to a very high level of abstraction. The tuilei analogy, which was mainly based upon the method of matching types, became the major mode of inference in traditional Chinese mathematics.

During the Song (960~1279) and Yuan Dynasties (1271~1368) Dynasties, traditional Chinese mathematics reached the last stage of its development. In this period, we still witness a number of important breakthroughs in mathematical systems and theories as well as certain significant innovations in methods and thought patterns. Simultaneously, the traditional method of the tuilei analogy was gradually transformed into a deductive mode based upon mechanical thought. All of this can be noted in the works of the most important mathematicians of the period, namely Qin Jiushao, Jia Xian and Zhu Shijie.

In fact, the deductive form of the *tuilei* analogy had already reached a very high level in Liu Hui's times, while the mathematical thinking and reasoning in the period of the Song and Yuan Dynasties (judged from Qin Jiushao's work) had already gained some of the characteristics of axiomatic thought. Following the beginning of the Ming (1368-1644) and during the Qing (1644-1911) dynasties, Chinese mathematics lagged the West., This was not due, however, to problems or shortcomings in mathematical theory itself, but rather it depended more on such other social factors as a poor and backward economy, social turbulence, and the domination of the dogmatic, ruling Confucian examination system by means of which bureaucratic policy directly obstructed the development of traditional Chinese mathematics. The supposition that the development of traditional Chinese mathematics in the Ming Dynasty was refrained because of its thought patterns is highly doubtful.

The methodological system of traditional Chinese mathematics, which was based upon the tuilei analogy and its method of matching types, took the lead in the linea development of thought throughout the entire world for a long time and provided a number of theoretical spaces for new innovations and explorations. Therefore, we can conclude that Sinologic was by no means the element that hindered the further development of traditional Chinese mathematics. In other words, the logical paths of traditional Chinese mathematics were not the reason for the later stagnation of traditional Chinese mathematics. There were other, more significant factors.

Bibliography

[1] Cui Qing Tian崔清田. 名学与辩学 *(The nomenalist School and the School of Dialectics)*. Taiyuan: Shanxi jiaoyu chuban she, 1997.

[2] Dai Qin代钦. 儒家思想与中国传统数学 *(Confucianism and traditional Chinese Mathematics)*. Beijing: Beijing Shangwu yinshu guan, 2003.

[3] Guo Shuchun郭书春. 古代世界数学泰斗刘徽 *(The Great Ancient Mathematician Liu Hui)*. Jinan: Shandong keji chuban she, 1992.

[4] Li Yan李俨. 中算史论丛 *(On the History of Chinese Calculation)*. Beijing: Zhongguo kexueyuan, 1954.

[5] Liu Hui 刘徽. 九章算术注，中国历代算学集成 *(The Commentaries on the Nine Chapters on Mathematical Procedures and the Integration of Mathematical Books in Ancient China, ed. by Jing Yushu)*. Jinan: Shandong Renmin chuban she, 1994.

[6] Qian Baocong钱宝琮，Du Shirang 杜石然. 试论中国古代数学中的逻辑思想 *(On the Logical Thought in Ancient Chinese Mathematics)*. Beijing, 1961. Guangming Ribao, 29.05.1961.

[7] Qin Jiushao秦九韶. 中国历代算学集成 *(Mathematical Treatises in Nine chapters, the Integration of Mathematical Books in Ancient China)*. Jinan: Shandong renmin chuban she, 1994.

[8] Shen Kangshen沈康身. 中算导论 *(An Introduction to Chinese Calculation)*. Shangh: Shanghai jiaoyu chuban she, 1996.

[9] Wu Feibai伍非百. 墨子 *(Mozi)* - *Zhongguo gu mingjia yan*. Beijing: Zhongguo shehui kexue chuban she, 1983.

[10] Wu Feibai伍非百. 荀子正名(xunzi's theory of names). In *Zhongguo gu mingjia yan*, 745. Beijing: Zhongguo shehui kexue chuban she, 1983.

Kripke on Necessity and the Essence of Individual Objects[1]

Liu Yetao
Yanshan University
pkulyt@yahoo.com.cn

Feng Lirong
Nanjing University
lirongfeng1984@sohu.com

Saul A. Kripke, a distinguished contemporary logician and philosopher, helped to change the face of Analytical Philosophy in the last half of late 20th Century. He has been recognized as one of the most creative and influential philosophers. Kripke was initially a defender and interpreter of modern modal logic and later gave three lectures on necessity and language at Princeton which played a very important role in the transformation of contemporary philosophical research.

In "A completeness theorem in modal logic", which he wrote at an early age, and in several papers written later, Kripke developed the outline of a semantic theory for modern modal logic. For his first explicit use of the term "possible worlds", this theory has been now widely accepted as "Kripke's semantics". It is with the aid of this theory that modal logic developed into one mature branch of modern logic. Later, Kripke turned his attention to the philosophy of logic. In 1970, Kripke delivered three lectures at Princeton University which were published under the title Naming and Necessity. In those lectures, Kripke explicitly and systematically put forward a causal-historical theory of names. He argued that proper names and natural names were rigid designators which have only reference but no sense. This theory helped to overturn the predominance of Descriptivism in the domain of the theory of meaning. In the meantime, Kripke developed a new theory of essence. He argued that the essence of one thing is the necessary attributes belonging to it in all possible worlds in which it exists. The essence of an individual object is its origin, and the essence of a natural kind is its inner structure. He also applied his theory of names and essences to the philosophy of mind and proposed many surprising and influential theories. With these theories, he effectively rejected the early analytical philosophers' ideas of "rejecting metaphysics" while clearing a new path for "analytical metaphysics".

There is a close connection between Kripke's theory of names and his essentialism which can be shown from the title Naming and Necessity. The common basis of these two theories is Kripke's philosophical intuitions about possible world semantics, especially his moderate realism on possible worlds. Modal logic introduces new philosophical operators on the basis of classical logic. These operators are basic modal categories. So the philosophy of modal logic can be reduced to a philosophy of these categories. It is evident that the key for grasping Kripke's thoughts on the philosophy of logic is to grasping Kripke's readings of

---

[1] The authors are with College of Humanities and Laws, Yanshan University, Qinhuangdao 066004 China, and Department of Philosophy, Nanjing University, Nanjing 210093 China, respectively.

the concept of "necessity".

In Kripke's account, possible worlds are "ways the world might have been or states or histories of the entire world".[3, p. 18] He explicitly argued against conceiving possible worlds as something like distant planets or something existing in a different dimension; he rejected discussing such meaningless problems as the problem of "transworld identity". In order to avoid misunderstanding, Kripke proposed substituting "It is possible that ..." for the word "possible worlds". He even vividly asserted that possible worlds are "stipulated", not discovered through a telescope. According to moderate realism, a possible world is just a counterfactual situation of the actual world. The key for understanding Kripke's theory of possible worlds is that our discussions about possible worlds begin with objects existing in the actual world. We get the notion of possible worlds by way of describing or stipulating counterfactual situations about these objects. Possible worlds have no independent existence and can't be discovered independently of the actual world. Kripke said, we can refer to the object and ask what might have happened to it. So, we do not begin with worlds and then ask about the criteria of transworld identification. On the contrary, we begin with the objects we have and identify them in the actual world. Then we ask whether certain things might have been true of these objects. [3, p. 53]

What Kripke means is that we must think about possible worlds from the perspective of the actual world. Possible worlds differ from the actual world in either their components or the attributes of those components when they have the same components. For instance, we can imagine a possible world inhabiting two persons whom someone in the actual world considers to be his parents, but in this world these two persons have no chance to meet, not to mention marry or have offspring. In this possible world, the offspring in question do not exist, though the parents identified in the actual world are sure to be there. In another case, a particular individual belongs to a possible world, but he does not have certain attributes which belong to him in the actual world. For example, we can imagine a possible world in which Shakespeare never gets the idea to write *Hamlet*. We can see that adding or reducing of attributes (properties and relations) leads to possible worlds. On Kripke's account, possible worlds are seen as abstract objects constructed by human thought. They are similar to abstract entities. In other words, possible worlds are dependent entities which depending on actual individuals. They are the results of our imagination about those actual attributes of actual individuals.

From the point of logical technique, in order to determine the truth-value of sentences containing modal operators such as "necessity" and "possibility", we interpret them by means of "possible worlds" so that every modal proposition can be assigned a unique truth-value. For example, we say a proposition is necessary in the actual world if and only if it is true in all possible worlds. Whatever worlds which satisfy this condition, space area, time point, or other eventuality are all so-called "possible worlds". So it is evident that possible worlds semantics is a good tool by which different levels implicit in the notion of necessity can be distinguished. Then this intractable notion becomes an intelligible and analyzable one.

In order not to misunderstand Kripke's explanation of "necessity", the distinction of various levels implicit in this notion is necessary to grasp correctly Kripke's thought. From the logical point of view, the most important thing is

Kripke on Necessity and the Essence of Individual Objects 189

to distinguish actual (broadly physical) necessity and logical necessity. By "actual necessity", we mean the necessity in the actual world. It exhausts all those possibilities of an actual object. Logical necessity refers to the necessity across possible worlds. It exhausts all those possibilities of an object in all possible worlds in which this object exists. The key for grasping this distinction is to correctly understand the logical relation between the actual world and other possible worlds. From the point of Kripke's moderate realism, the actual world contains many possible situations. Every situation is a possible world. What this means is that the actual world is only one of many possible worlds, the realized one. All possible worlds are logically equal, and the actual world has no special place in the whole system of possible worlds. The idea that the actual world has a certain special place does harm to our understanding about the notion of logical necessity.

There is a certain criterion for distinguishing the above levels. We can base the distinction on the "accessibility relation R". Just like our formal treatment for the notion of possible worlds, the accessibility relation is also an abstract logical tool which is introduced for determining the truth-value of a modal proposition. As a semantic concept, we can put aside all kinds of meaning of accessible relation and only regard it as a primitive notion. If necessary, it can be obtained from the point of truth-value of modal propositions, that is to say, let w1 and w2 are any two possible worlds, if they satisfy the following condition C,

> C: For an arbitrary proposition α, If α is true in w_2, $\Diamond \alpha$ is true in w_1, that is, true propositions in w_2 are possible in w_1, then whatever the relation (if any) between w_1 and w_2 is, we can conclude that w_1 is acccosible to w_2.

Nevertheless, intuitively, accessibility relations are manifold, such as changes in the state of affairs, transitions in time and space, and so forth.

According to R, we can explore not only the logical necessity but also the actual necessities on any level. First, a world involving no contradiction is a logically possible world. We can see that this definition involves no R at all; secondly, for actual (broadly physical) necessity, R is intuitively understood as an "imaginable relation". When exhausting all possibilities determined by it, we will obtain those laws in corresponding levels. In a narrowly physical sense, for example, we can't imagine a situation in which the speed of light becomes changeable because it is a narrowly physical law that the speed of light remains unchangeable. In other words, our imagination is confined within the sphere of physical laws.

It should be noted that all possible worlds are of equal right is only in the logical sense. We are also interested in those actual necessities determined by actual possibilities. Determined by actual possibilities, actual necessities can be divided into many levels, such as narrowly physical necessity, narrowly biological necessity, etc. For example, it is necessary that p is physically true if and only if p is true in all imaginable worlds that follow physical laws. So biologically, Aristotle can't become a superman and a car can not physically run faster than light because we can not imagine a biological situation in which Aristotle is a superman, and similarly, the speed of light is a limit in the narrowly physical sense. If a proposition is true in all possible physical worlds or biological worlds, then it is necessarily true at the corresponding level. For instance, the proposition "a car can not run faster than light" is true in all physically possible worlds. However, logically possible worlds include those impossible worlds in the actual sense. From the logical point of view, Aristotle may be a superman, and a car might run faster than light, for such an imagination does not lead to any contradiction. Logically, any world involving no contradiction is a possible world, and any world involving contradiction is an impossible world. To repeat, we must not give the actual world any special place.

In some possible world, such as in the actual world, the essence of an individual is not absolute. On the contrary, an individual's actual necessary attributes are determined by his/her actual possibilities. Everything is changing, and we keep obtaining more and more attributes of things as a result of our practice. Among these attributes, which is (are) necessary to an individual(s)? We need to appeal to our social practice. We pick them out according to purposes of our social practice. For a doctor, for example, 'the inventor of the syllogism' is not Aristotle's essential attribute, because it is certainly possible the doctor is not interested in any attributes other than Aristotle's health condition.

But when we focus on the transworld essence of one individual object, the "necessity" is not the actual necessity but the logical necessity. Kripke's so-called 'essential attribute' means those attributes that must be held by one thing across all possible worlds in which it exists. For any individual, it has an origin after all. Originating from this origin, it enters into different changing chains, each of which is a possible world. The actual Aristotle, for example, had his origin in the fertilized egg formed from his father's sperm and his mother's egg, developing later into a philosopher who invented the syllogism. But we can certainly imagine that this fertilized egg evolves into another chain, that is, it is possible that Aristotle did not become a philosopher, invent the syllogism, etc. But in any case, we certainly can't imagine such a situation in which Aristotle had not his origin. So this origin is necessary to Aristotle. It seems that Aristotle's classical essentialism could not deal with logical necessity in this sense. Only when discussing "accident" did Aristotle speak about individual objects. This is consistent with the fact that his syllogism only presupposes the existence of individuals and focuses on genus and species. It is one of the most important contributions Kripke makes to contemporary analytical philosophy that he clears a new path for inquiring into the essence of individual objects.

The above distinction and the results derived from it can be deduced from Kripke's own theory. The reason why Kripke considers "transworld identity" etc. as pseudo-problems is that those possible worlds we speak of are counterfactual situations of actual individual objects. Those objects that we speak of remain to be those actual individual objects. Take Nixon as an example. We need not try hard to ask the question: in the possible world in question, how can we tell which one, because of such sufficient and necessary conditions of Nixon, he is. This is because we have assumed that: 1) that possible world contains this individual; 2) that the individual loses the presidential election in the USA in 1968. For the actual Nixon, what Kripke is interested in is the self-identity standard of Nixon in the logical sense. Alternatively, for this individual, what can endure the test of logical necessity? Kripke gives origin as the answer because whatever has the same origin is the same individual.

The theory of origin as the essence of the individual materially involves two important theoretical problems; 1) the problem of the individuals' existence;

2) the problem of the self-identity of individuals. Origin is the sufficient and necessary condition for the existence and the self – identity of individuals. From the point of existence, Kripke's so-called essence only stipulates the existence of individuals, but it does not stipulate properties. Origin only commits us to the existence of individuals. So Kripke's essentialism is in the ontological sense. From this, we can understand Kripke's restrictions on his notion of rigid designators: when talking about rigid designators, it is not essential that the objects in question exist in all possible worlds. We only need to show that in any possible worlds that this object indeed exists or will exist. We use this designator to refer to this object. If that object does not exist, that designator refers to nothing.

On the other hand, origin can guarantee the self-identity of an individual object. Starting from the origin, an individual object contains an infinite number of possibilities, but whichever possibilities it realizes does not change its self-identity. Otherwise, we will have to conclude every possibility can produce an individual, and then an individual has many kinds of self-identity. Still take Aristotle again as an example, Whether he is the inventor of the syllogism or once the teacher of Alexandria, he is necessarily Aristotle. Otherwise, we leap into falsehood. Even if there is a person who has exactly the same characteristics as Aristotle, they are two different persons. For an individual, only its origin can withstand the test of transworld logical necessity.

Since an individual's origin can stipulate its existence and determine his self-identity–which has nothing to do with any attribute–so from the point of origin, an individual whose existence is stipulated and who has nothing to do with any attribute, that person only obtains a name by initial baptism. This name is only a referring mean or tool which has no sense associated with a certain definite description. So, from the point of Kripke's essentialism, proper names themselves are rigid designators. And from the point of Kripke's semantics, both his essentialism and his theory of proper names are justifiable.

Bibliography

[1] Chen, B. *The Philosophy of Logic.* Peking University Press, 2005.
[2] Fitch, G. W. *Saul Kripke.* Acumen Publishing Limite, 2004.
[3] Kripke, S. *Naming and Necessity.* Basil Blackwell Publisher, 1980.
[4] Liu, Y. *Researches on Kripke's Theory of Names.* 2005. Dissertation for Peking University.
[5] Soames, S. *Beyond Rigidity.* Oxford: Oxford University Press, 2002.
[6] Zhang, J. *An Introduction to Researches on Logical Paradox.* Nanjing University Press, 2002.

A Study of the "Deontic Paradox"[1]

Xia Sumin

Chinese Academy of Social Science

xiasumin@gmail.com

ABSTRACT. Based on existing achievements in field of deontic logic, this paper conducts a holistic research into the deontic paradoxes and tries to clarify the clues derived from numerous interpretations about the concept and essential nature of deontic paradoxes, with some useful tools and instruments available , including the pragmatic interpretation of "logical paradoxes", "paradoxical degree", the "soluble degree" of paradoxes and the "RZH" (Russell-Zermelo-Haack) criterion for paradox solution.

1 Motivation and purpose

From the perspective of philosophy, paradoxes are interesting and significant. The Deontic Paradox, the problem under investigation, is a kind of special paradox which consists of several "paradoxes". Then how do we understand the conception of the Deontic Paradox? There are numerous interpretations of different deontic "paradoxes". Are there connections among those viewpoints? What new ideas can be gained? Based on achievements in this field, this paper conducts a holistic research into the subject of the deontic paradoxes and tries to clarify the clues derived from numerous interpretations about the concept and essential nature of the deontic paradox.

For the above purposes, there are some useful tools and instruments available which we can use, including the pragmatic interpretation of "logical paradoxes", "paradoxical degree", the "soluble degree" of paradoxes and the "RZH" (Russell-Zermelo-Haack) criterion for paradox solution. We will apply these tools in our discussion.

2 Analyses of some deontic paradoxes

Let's introduce those available instruments detail by detail. First, R. M Sainsbury notes, "an apparently unacceptable conclusion derived by apparently acceptable reasoning from apparently acceptable premises."[13, p. 1, introduction] While Jianjun Zhang considers that "paradox is such kind of theoretic fact or situation that under certain accepted background knowledge contradiction can be obtained logically."[2, p. 21]

The two ideas share three basic characteristics: common acceptable knowledge as premises, strictly logical reasoning and a contradiction as the conclusion.

---

[1] The author is with Institute of Philosophy, Chinese Academy of Social Science, No.5 Jianguomennei Street, Beijing 100732 China.

A Study of the "Deontic Paradox"

What is laid out above is apparently a pragmatic understanding, especially the first point of "common acceptable knowledge as premises". This determines the pragmatic interpretation of the "logical paradoxes"; it also determines that a logical paradox is radical, relative and eliminable.

Then we can reach a new definition of the deontic paradox, i.e. the "deontic paradox is such kind of theoretic fact or situation that under certain accepted deontic background knowledge, deontic contradiction can be obtained logically". But the problem is that there are several different deontic paradoxes. Are they all logical paradoxes?

The second available instrument is the "RZH" (Russell-Zermelo-Haack) criterion for paradox solution. To state it more explicitly, the new system for removing paradoxes must be:

1. Adequately narrow or limited and the new system has not any paradox;
2. Broad and comprehensive enough, the new system still can deal with significant problems well;
3. Not ad-hoc.

The last available instruments are "paradoxical degree" and "soluble degree" of paradoxes, which are all relative concepts.

"Paradoxical degree" denotes the different degree of different paradoxes, which is in direct proportion to "acknowledged degree" of background knowledge from which paradoxes derived. In brief, higher the "acknowledged degree" is; the "paradoxical degree" is higher. According to "acknowledged degree" and "RZH" criterion, solutions to paradoxes is different in "acceptable degree", therefore different "soluble degree" of paradoxes can be defined.

"Ross's paradox", paradox of commitment, the Good Samaritan paradox and paradox of contrary-to-duty imperative (CTD) are four main "Deontic paradoxes". Here will take the first and the fourth as examples to show how to use the instruments forenamed.

"Ross's paradox" (1930) involves a theorem of standard deontic logic (SDL):

$$O\Phi \to O(\Phi \vee \Psi)$$

This paradox is named after Alf Ross whose example was "If I ought to mail a letter, I ought to mail or burn it." Intuitively, the conclusion can't be accepted.
$p \to (p \vee q)$ PC
$Op \to O(p \vee q)$ derivable rule in SDL
Here, derivable rule[2]

Here the reasoning is strict but the conclusion is unacceptable. In fact, what is unacceptable is "Ought to burn it", but it can not be derived immediately unless we use:

$$(p \vee q) \to (Op \wedge Oq)$$

The trouble is not $p \to (p \vee q)$ and other rules, but $(p \vee q) \to (Op \wedge Oq)$ which is only intuitive, not logical. The background knowledge results from which this

---

[2] 1' ⊢ $A \to B$
2' ⊢ $O(A \to B)$
3' ⊢ $(A \to B) \to (OA \to OB)$
4' ⊢ $OA \to OB$

paradox occurs is only a "misconception" of the logical connective "∧", "∧" and the relation between them.

So the "Ross's paradox" can also be called the imitation of a paradox, with a very low paradoxical degree. The logic-revised project accordingly hasn't been commonly accepted. Otherwise it would require too much effort and lose many valuable conclusions. In fact, this "paradox" could be removed readily only by correcting the wrong intuitive judgment.

Next, the paradox of contrary-to-duty (CTD) imperative is as follows:

1. it ought to be that a;
2. it ought to be that if a then b;
3. if not-a, then it ought to be that not-b;
4. not-a.[3]

Formally, these premises and the process of reasoning are as following:
(CTD 1') Op
(CTD 2') $O(p \to q)$
(CTD 3') $\neg p \to O \neg q$
(CTD 4') $\neg p$
(CTD 5') $O(p \to q) \to (Op \to Oq)$
(CTD 6') $Op \to Oq$ CTD 2',CTD 5',MP
(CTD 7') Oq CTD 5', CTD 6', MP
(CTD 8') $O \neg q$ CTD 3', CTD 4', MP
(CTD 9') $Oq \land O \neg q$ CTD 7',CTD 8',∧+
(CTD 10') $\neg(Oq \land O \neg q)$ deontic consistent principle
(CTD 11') $\neg(Oq \land O \neg q) \land (Oq \land O \neg q)$ contradiction

Here reasoning is strictly but conclusion is unacceptable. To be clearer, the premises can be writing as following:

$$Op \land O(p \to q) \land (\neg p \to O \neg q) \land \neg p$$

Or change into

$$(Op \land \neg p) \land (O(p \to q) \land (\neg p \to O \neg q))$$

In details, $(Op \land \neg p)$ means "ought to p, but in fact not-p", while $(O(p \to q) \land (\neg p \to O \neg q))$ means "ought to $(p \to q)$, and if not-p, then ought to not-q". They themselves aren't theorems in SDL, but why not they be removed finally.

In fact, they all can be regarded as background knowledge.

As we all known, our actual world is not a perfect deontic one, so the obligations are actually violable, Op and $\neg p$ can coexist. If a certain commitment and relevant obligations have been violated, then the primary obligation and the following new obligation (called by a general name "conditional obligation") would be satisfied at the same time. Then $O(p \to q)$ and $(\neg p \to O \neg q)$ can coexist.

From all these universal common premises, the unacceptable contradiction is derived by elementary deontic logical rules bringing the paradox into being.

Because the background knowledge the paradox relies on is with fairly high "common degree", the "paradoxical degree" of this paradox is the rather higher.

---

[3]Intuitively, we can take "a" as "I go to help my neighbor" and "b" as "I tell her I will help her"

Moreover that knowledge could be formalized strictly in deontic system, so the paradox of CTD imperative has become a special logical paradox from philosophical one in fact.

Through the reviews of the paradox of the CTD imperative, a possible route from the philosophical paradox to the special logical paradox is presented.

3 Summary and future work

The Deontic Paradox is the title of a paradox sequence. Different deontic paradoxes have a different "paradoxical degree". The degree of the "Ross's paradox" is the lowest, and that of the CTD is the highest. In fact, the paradox of CTD is the most serious hit against deontic logic. The analyses of every deontic paradox are complicated and precise, but here we give only a brief outline. Many parts need to be enriched in the future[4].

According to our investigation on the paradox of commitment and the solutions to it, we found that the paradox of commitment results from the incompatibility between the deontic theorems and the intuitive counterexamples. Its paradoxical degree is far higher than that of the "Ross's paradox", for the corresponding background knowledge is universal intuition of the deontic concept of "commitment", instead of "misconception". However, this intuition cannot be strictly formalized in the deontic system, so the paradox of commitment is still not a special logical paradox. The good Samaritan paradox which we investigated has a far higher paradoxical degree than that of the "Ross's paradox". It is still not a special logical one, similar to the paradox of commitment. The different rests with their corresponding background knowledge. Those of the good Samaritan paradox are the relationship between "obligation and obligation, prohibition and prohibition", so its paradoxical degree is also higher than that of the paradox of commitment.

Aiming at different background knowledge, different solutions to a paradox come forth, the analyses about every solution also need to be continued and finally find feasible solutions. Each deontic paradox's relative orientation in the sequence of paradoxical degree is specified distinctly as further research on their relations is conducted, through all above investigations. The paradox of contrary-to-duty imperative located the highest pole of the sequence of paradoxical degree; the solution to it would be undoubtedly of great importance to clarify and solve other deontic paradoxes. A thorough study of the pragmatic property of paradoxes would be conducive to mastering the essential characters of the deontic paradoxes. Because the paradox of the contrary-to-duty imperative could be reworded as a pragmatic paradox, one kind of logical paradox, all the achievements of pragmatic paradoxes could facilitate the progress of deontic paradox research. As one part of philosophical paradox, the powerful methodological function of deontic paradox research could lay a great foundation on which we could develop further general paradoxes research, and advance deontic logical studies.

Bibliography

[1] 张建军. 科学的难题–悖论. 杭州: 浙江科学技术出版社, 1990.

---

[4]This is the task of my dissertation.

[2] 张建军. 逻辑悖论研究引论. 南京: 南京大学出版社, 2002.
[3] 余俊伟. 道义逻辑研究. 北京: 中国社会科学出版社, 2005.
[4] Gabbay, D. M., Guenthner, F. (Eds.). *Handbook of philosophical Logic*, vol. 8. Dordrech: D Reidel Publish Company, 2002.
[5] Hansson, S. O. Situationist deontic logic. *Journal of Philosophical Logic*, (6):423 – 448, 1997.
[6] Hilpinen, R. (Ed.). *Deontic logic: Introductory and System Readings*. Dordrech: D Reidel Publish Company, 1971.
[7] Hilpinen, R. (Ed.). *New Studies in Deontic Logic: Norms, Actions and the Foundation of Ethics*. Dordrech: D Reidel Publish Company, 1981.
[8] Hintikka, J. Some main problems of deontic logic. In Hilpinen, R. (Ed.), *Deontic logic: Introductory and System Readings*, 59 – 104. Dordrech: D Reidel Publish Company, 1971.
[9] Norwell-Smith, P. H., Lemmon, E. J., Escapism. the logical basis of ethics. *Mind*, 275(7):289–300, 1960.
[10] Nute, D. (Ed.). *Defeasible Deontic Logic*. Dordrecht: Kluwer Academic Publishers, 1997.
[11] Prior, A. N. *Formal Logic*. Oxford University Press, 2nd edn., 1962.
[12] Rescher, N. *Paradoxes: Their, Roots, Range, and Resolution*. Carus Publishing Company, 2001.
[13] Sainsbury, R. M. *Paradoxes*. Cambridge University Press, 2nd edn., 1995.
[14] von Wright, G. H. Deontic logic. *Mind*, 60:1–15, 1951. New Series.
[15] von Wright, G. H. A note on deontic logic and derived obligation. *Mind*, 65:507 – 509, 1956. New Series.
[16] von Wright, G. H. *A Correction to a New System of Deontic Logic*, vol. 2. Danish Yearbook of Philosophy, 1965.

The Cushing Thesis and Underdetermination[1]

Bai Tongdong

Xavier University

bai@xavier.edu

ABSTRACT. Applying a weak Forman thesis to the rise of the orthodox quantum mechanics interpretation and the rejection of hidden variable theories, especially Bohmian mechanics, Cushing argues that they are historically contingent and have to do with external factors. But this argument is based upon a problematic form of social determinism. He exaggerates the role of external factors in theory choice. He also fails to realize that those who reject the orthodox interpretation may be more motivated by these factors than the advocates of it. Moreover, his assumption that factors that are extra-logical and extra-empirical are necessarily external, and the use of them to eliminate underdetermination lacks validity, is mistaken. He is correct to say that our worldview could be drastically different if some contingencies had not happened. Accepting this weak version of underdetermination, we are nevertheless justified, based upon good sense and the internal mechanism of modern science, to neglect a promising alternative and to eliminate underdetermination between theories.

1. The thesis in James Cushing's 1994 book Quantum Mechanics: Historical Contingency and Copenhagen Hegemony is that the rise and hegemony of the orthodox or the Copenhagen interpretation of QM and the marginalization of hidden variable theories (HVT), especially Bohmian mechanics (BM), are historically contingent. First, through technical analysis, he shows that BM is a viable and even a superior alternative to quantum mechanics (QM). Second, he maintains that the rise of QM in the late 1920's and early 1930's and the rejection or neglect of BM in the 1950's is due to contingent (social, personal, ideological, and psychological) factors. Third, he argues that it is an open choice whether we adopt indeterminate QM or determinate BM. All this serves as a concrete example of a weak version of the under-determination thesis which he argues for in the book.

   In this paper, I challenge all these claims. Generally, I show that there are non-logical and non-empirical but scientific and reasonable considerations that eliminate under-determination and determine theory choice. The general idea that there can be such considerations that resolve under-determination or resolve uncertainty in general has been introduced by Pierre Duhem [7], Ludwig Wittgenstein [26], Rudolf Carnap [5], W. V. O.

[1] The author is with Department of Philosophy, Xavier University, 5 University Drive, Cincinnati, OH 45207, U.S.A.

Quine [21], and Ian Hacking [10]. In this paper, I focus on developing only Duhem's idea of "good sense".

2. To show the conceptual superiority and empirical adequacy of BM, first, Cushing argues that there are problems QM can't handle–for example, the measurement problem and the problem of Schroedinger's cat [20, pp. 34 – 41]. But he admits that, for example, the measurement problem can be avoided if we take the state description in QM not literally but symbolically. If we "abandon any pretense that QM applies to individual systems and retreat to the statistical, or ensemble, level" [20, pp. 36 – 37] then the problem can be avoided. In fact, this kind of symbolic interpretation was adopted by Niels Bohr and Wolfgang Pauli, two important members of the so-called Copenhagen School and advocates of "the orthodox interpretation"[3, p. 101][16, p. 308][20, p. 133][1, Section 5]. So, the measurement problem can be resolved by using a version of the orthodox interpretation.

Cushing then shows that not only can BM handle all these problems, but it also applies to individual systems. But as Cushing admits, "At this level of the causal interpretation or 'theory,' we still have no understanding of the physical origin of the highly nonlocal quantum potential U" in the basic equations of BM[20, p. 47]. So, BM only appears to offer us an understanding of the individual system. Moreover, to deal with the problems of the reduction of wave packets and the Heisenberg relations, BM has to appeal to the correlation between the states of the system and of the apparatus and the complexity of the apparatus, which, according to Cushing, "reflects Bohr's concept of the wholeness of quantum phenomena and the spirit of his principle of complementarity. How a microsystem behaves depends upon its environment – an observed value is contextual" [20, pp. 26 – 47]. Thus, what BM actually does is to write Bohr's holistic interpretation into its basic equations.

Moreover, BM has not made any successful predictions that are observationally different from those made by QM. More importantly, the question of how to make BM relativistic can only be answered on an observational level [20, pp. 191, 206, 270, n.33], while the orthodox quantum theory, with the success of relativistic quantum field theory, is able to solve this problem in a conceptually consistent manner. Conceptual consistency between different fundamental theories is crucial for physics, and it has to do with the adequacy of a theory in a broad sense. To sum up, the conceptual superiority of BM to QM is more apparent than real, and BM is not (yet) as adequate as QM.

3. With the status of BM clarified, let's now examine how HVT are received in history. The first important episode is from the late 1920's to the early 1930's. On the acceptance and propagation of QM and the rejection of HVT, instead of what he calls a strong Forman thesis that "would claim a major causal role of the cultural milieu in determining the very form and content of a scientific theory," Cushing proposes a weak Forman thesis that "would see the cultural milieu as something playing an important role in the acceptance and propagation of an already-formulated scientific theory" [20, p. 100],[9].

However, using this thesis to explain the rise and dominance of QM and the rejection of HVT presupposes that the advocates of QM lived in the same cultural milieu that produced the same influence on them. But Albert Einstein and Werner Heisenberg, both of German descent, and Erwin Schroedinger and Pauli, both of Austrian descent from Vienna, took rather different attitudes toward QM. Among the people of the so-called "Copenhagen School" who supported the orthodox interpretation, scholars have shown that they were of a diverse social, psychological, and ideological group, which means that the reasons for their uniform acceptance of QM have to be different from the external ones so often cited [2, 12, 25].

In fact, Pauli and Heisenberg, who had methodological differences ([18, pp. 147–148], letter on 02/21/24; and [12, pp. 42–43]), and de Broglie, who did not like the new QM at first, were all convinced by mounting theoretical and experimental challenges to give up the attempt to save classical laws and to adopt QM and its orthodox interpretation instead([6, p. 157] [23], [12, pp. 30–50], Darrigol 1992, Chapter 8; and Cushing 1994, 109-111, 120, and 126). This historical fact suggests that, faced with mounting difficulties within physics, physicists were able to transcend their preconceived philosophical, ideological, methodological, and psychological opinions that may or may not have been formed and conditioned by their cultural milieu.

Moreover, John von Neumann introduced an argument in 1927 that culminated in an apparently rigorous proof of the impossibility of any type of HVT in his 1932 book. That it actually fails to prove this impossibility was only pointed out with great clarity by John Bell in the mid-1960's (Cushing 1994, 131-133). A more concrete challenge to HVT, as Pauli, Heisenberg, and others pointed out, is how to handle the diffraction phenomenon ([19, pp. 107 – 118], letter on 07/02/35; and 119-122, letter on 07/09/35). That is, if there are two openings on a diaphragm and it is indeterminate through which opening each particle of an ensemble goes, the observed diffraction figure cannot be explained if we assume that each particle actually goes through only one determinate opening, and it is merely not known to us through which opening it goes, as a hidden variable theory may suggest.

Therefore, if we also take into account the stunning empirical success of QM, the scientific case for the acceptance of QM and rejection of HVT seems to be quite conclusive. Cushing himself mentions many of the aforementioned points. Then, why does he still insist on the weak Foreman thesis? I suspect that the reason is that Cushing holds a strong criterion for the elimination of under-determination between theories. He claims that, "There were logical and empirical grounds for rejecting certain attempts at hidden-variable theories ... but such consideration did not uniquely constrain the choice" (Cushing 1994, 214). In particular, he shows that BM can answer all the aforementioned challenges to HVT. It seems that his reasoning is that, since there are alternatives not excluded by logic and empirical evidence alone, the failure to explore them must have been based upon external (social, psychological, and ideological) factors.

However, if logic and experience are never adequate to determine theory choice, and if there are always external factors involved, then the only meaningful work is to discuss their relative significance in the acceptance and rejection of a scientific theory. In particular, given the mounting logical and empirical supports for QM and challenges to HVT, we should realize that it is the advocates of HVT that are more likely to be driven by social, psychological, and ideological factors. Moreover, it is dubious to say that the extra-logical and extra-empirical factors have to be social, psychological, or ideological. In the scenario from the late 1920's to the early 1930's, the physicists' choice of QM over HVT could be a practice of what Duhem called "good sense". Cushing thinks that "good sense" is vague and does not lead to agreement [20, p. 200]. Duhem himself acknowledges the fact that "these reasons of good sense do not impose themselves with the same implacable rigor that the prescriptions of logic do," which might lead to lengthy quarrels between the adherents of an old system and the partisans of a new doctrine, each camp claiming to have good sense on its side, each party finding the reasons of the adversary inadequate. [7, p. 217]

But unlike Cushing, Duhem argues,

> ...this state of indecision does not last forever. The day arrives when good sense comes out so clearly in favor of one of the two sides that the other side gives up the struggle even though pure logic would not forbid its continuation. [7, p. 218]

From this argument and from his treatment of some historical examples, we can imagine that, if he were alive today, Duhem would consider people who resisted quantum mechanics' radical break with classical mechanics in 1925 or 1926 reasonable, but the resistance afterwards unreasonable, although it can be done in a logically consistent manner.

Clearly, Cushing would not agree with this judgment. He would, and he actually does, appeal to the uncertainty of good sense. But he should at least admit that people who are in favor of QM may not have been motivated by external factors, but by good sense that is rational or reasonable. A good analogy to this situation is the idea of reasonable doubt in a court. In a trial, the jurists are asked to decide whether the defendant is guilty beyond reasonable doubt, not any logically possible doubt. Otherwise, no verdict could ever be rendered. Similarly, in a trial over theories, not every doubt is justified and should be addressed if we wish physics to go on. The Cartesian project to find something beyond any doubt is a fool's errand, and it is not reasonable to fixate on demanding absolute certainty in science. Of course, it is always possible that what is available to the jurists will turn out to be false or inadequate. When it turns out this way, we do not say that the jurists were irrational in rendering a verdict or that their verdict was externally motivated.

4. In 1952, David Bohm introduced his version of the hidden variable theory that, according to Cushing, can answer the aforementioned challenges to HVT. But it has been largely ignored. Cushing claims that this neglect is externally motivated. I shall only examine the case of Pauli's dismissal of this new development in this section.

In a 1952 paper, Pauli criticized Bohm's theory (1953). His criticisms are based upon physical and logical considerations, but Cushing thinks that they are not compelling. This lack of logical rigor by Pauli and his rejection of BM are driven by ideology [20, pp. 149 – 150]. It is true that

Pauli, in his later years, had developed a worldview that was compatible with his understanding of the philosophical implications of QM [1, 11, 13–15]. But at the end of this article, Pauli maintains that his judgment is based upon physical reasons "which have nothing to do with philosophical prejudices" [17, p. 42]. Even if we ignore Pauli's own denial to make Cushing's claim, the following points need to be shown: first, Pauli's worldview is only compatible with QM and is not compatible with BM; second, the physical reasons were not convincing to Pauli and he needed ideological reasons for his rejection of BM; third, he could not see past his own worldview, even if there were mounting theoretical and empirical challenges to it. I do not think that Cushing's judgment meets these burdens of proof. The Forman thesis, weak and strong alike, has a similar problem. Moreover, even if Pauli were partly driven by ideological factors, given the fact that Pauli did offer reasons based on physics to support his rejection, is it possible that the part of his judgment that was based on physics can be separated from his ideology, and so many other physicists who agreed with his judgment and rejected BM were convinced by the physical part of his judgment? For example, as Cushing mentions, Max Born maintained that Pauli's article "slays Bohm not only philosophically but physically as well"[4, p. 207](Cushing 1994, 150).

It is true, as Cushing points out, that Pauli and many others did not examine Bohm's theory as well as von Neumann's proof closely. But does this failure have to be driven by external reasons? I will not repeat the aforementioned arguments, but will only focus on a new situation in this second confrontation: by 1952, QM was not merely a promising theory anymore, as it was in the late 1920's, but a mature theory. Scientists have good reasons not to abandon easily such a mature theory. For modern science is not merely an enterprise based upon logic or upon an individual's heroic effort, but a communal project. Various theories within a large framework should be made compatible, if not consistent, so that there is a "harmonious community" of scientific theories. Young physics students need to be trained to master rules and techniques of the present paradigm so that they can work together in a scientific community. In particular, a mature scientific theory, especially one on the fundamental level, like the quantum theories of the 1950's, that have proven successful in many areas, is deeply intertwined with many other fundamental theories in a massive theoretical framework. To renounce it has rippling effects throughout the whole framework, greatly disrupting the communal work of a large group of scientists. Therefore, for science to develop, it is reasonable, even essential, to ignore an alternative even if there are some anomalies in the present framework and this alternative is a promising one. One can say that this is an internal "logic" of modern science. An analogy is that of a civil society in which a legal system has to be established. A citizen should not dismiss an article of law just because it is convenient to do so. Cushing dismisses a similar concern by Norwood Russell Hansen as sociological or psychological and at best "a sensible strategy for getting on with things" (1994, 155-156). But this strategy is built into the mechanism of modern science. The concerns this strategy is based upon are social and communal, but they are also internal and essential to how modern science is and should be done. Therefore, physicists are scientifically justified to ignore BM in particular and HVT in general.

5. Finally, Cushing makes a counterfactual argument in his book: if BM had been introduced earlier, we would not have to adopt the indeterministic worldview of QM (1994, xi and 174). In this sense, how we view the world is historically contingent. This is a reasonable claim. Cushing is also correct to point out an analogy between theory choice and natural selection (1994, 270, n35). Many biological mechanisms we have today can be traced back to historical accidents. This means that we or intelligent beings in a different environment – for example, a planet in a different solar system – could have functioned with drastically different mechanisms. Similarly, it is possible that those intelligent beings would develop a physics that is unintelligible to us. But I would like to point out that, when these contingent and contextual factors become reality, how things move forward does not have to be contingent anymore but has a life of its own that follows an internal, "a priori" mechanism of growth.

While adopting this diverse view, we have to be careful. Just as an opposition party might look better than the current administration because it does not have the burden for running everything but only needs to focus on a few faults of this administration, a rival theory like BM may only appear to be viable.

To support this diversity view, Cushing also appeals to Quine's dictum that "the world intrudes first as a surface irritation and remains thereafter as a constraint on our imagination (in constructing scientific theories)" (1994, 215). He uses this to offer a last defense of Bohm's theory because it demonstrates "the possibility of so dramatically different a view of the physical world" (ibid). But we can also use this line of defense to praise QM that, at its birth, offers us so dramatically different a view of the physical world. Besides, is Bohm's theory that holds on to certain classical concepts so radical as Cushing claims them to be? For example, Einstein did not like HVT precisely because they were not radical enough, which Cushing admits, although he once lists departing not radically from classical mechanics as a merit of BM [24, pp. 374–377], [8, pp. 57 – 58](also mentioned in Cushing 1994, 147). Second, the value of free imagination is limited. In fact, a genius in science and in practical matters is often someone who has the ability, the good sense, to pick the actual from the possible and the imaginable. There are also extra-logical, extra-empirical but reasonable and communal factors that constrain imagination and under-determination. Cushing often dismisses them as pragmatic, but they are part and parcel of how modern science is done. With an open mind, with the understanding that the world can be represented radically differently, let us practice in science Quine's "conservative" dictum: "where it doesn't itch, don't scratch" [22, p. 160].

Bibliography

[1] Bai, T. *Complementarity between the Rational and the Mystical – On Wolfgang Pauli's Worldview, His Exchanges with Carl Jung, and His Pessimistic Realism in QM*. 2006. Manuscript.

[2] Beller, M. *Quantum Dialogue: The Making of a Revolution*. Chicago: University of Chicago Press, 1999.
[3] Bohr. The philosophical writings of niels bohr. In Jan, F., Henry, J. F., Woodbridge, C. (Eds.), *Causality and Complementarity: Supplementary Papers*, vol. 4. Oxbow Press, 1998.
[4] Born, M. *The Born-Einstein Letters*. New York: Walker and Company, 1971. Translated by Irene Born.
[5] Carnap, R. Empiricism, semantics, and ontology. In *Meaning and Necessity*, 205–221. Chicago: The University of Chicago Press, 1956.
[6] de Broglie, L. *Physics and Microphysics*. New York: Pantheon Books, 1955. Translated by Martin Davidson.
[7] Duhem, P. *The Aim and Structure of Physical Theory*. Princeton: Princeton University Press, 1954.
[8] Fine, A. *The Shaky Game: Einstein Realism and The Quantum Theory*. Chicago: The University of Chicago Press, 1986.
[9] Forman, P. Weimar culture, causality, and quantum theory, 1918-1927: Adaptation by German physicists and mathematicians to a hostile intellectual environment. *Historical Studies in the Physical Sciences*, 3:1–115, 1971.
[10] Hacking, I. Experimentation and scientific realism. *Philosophical Topics*, 13:71 – 87, 1982.
[11] Heisenberg, W. Wolfgang pauli's philosophical outlook. In *Across the Frontiers*, 20 – 38. New York: Harper & Row, 1974.
[12] Hendry, J. *The Creation of QM and the Bohr-Pauli Dialogue*. Dordrecht and Boston: D. Reidel Pub. Co., 1984.
[13] Laurikainen, K. V. Wolfgang pauli and the copenhagen philosophy. In Lahti, P., Mittelstaedt, P. (Eds.), *Symposium on the Foundations of Modern Physics: 50 Years of the Einstein-Podolsky-Rosen Gedankenexperiment, Joensuu, Finland, 16-20 June 1985.*, 209 – 228. Singapore: World Scientific, 1985.
[14] Laurikainen, K. V. *the Philosophical Thought of Wolfgang Pauli*. Berlin and New York: Springer-Verlag, 1988.
[15] Laurikainen, K. V. *The Message of the Atoms : Essays on Wolfgang Pauli and the Unspeakable*. Berlin and New York: Springer, 1997.
[16] Pauli, W. Editorial. *Dialectica*, 2:308–311, 1948.
[17] Pauli, W. Remarques sue le problème des parameters cachés dans la mécanique quantique et sur la théorie de l'onde pilote. In George, A. (Ed.), *Louis de Broglie: Physicien et Penseur*, 34 – 42. Paris: Michel, 1953.
[18] Pauli, W. *Wissenschaftlicher Briefwechsel mit Bohr, Einstein, Heisenberg: 1919-1929.*, vol. I. Berlin and New York: Springer-Verlag, 1979.
[19] Pauli, W. *Wissenschaftlicher Briefwechsel mit Bohr, Einstein, Heisenberg*, vol. II, WB. Berlin and New York: Springer-Verlag, 1985.
[20] Pauli, W. *Writings on Physics and Philosophy*. Berlin: Springer-Verlag, 1994. Translated by Robert Schlapp.
[21] Quine, W. V. O. Two dogmas of empiricism. In *From a Logical Point of View*, 20 – 46. Cambridge, Mass: Harvard University Press, 1954.
[22] Quine, W. V. O. *Word and Object*. Cambridge, MA: The MIT Press, 1960.
[23] Serwer, D. Unmechanischer zwang: Pauli, Heisenberg, and the rejection of the mechanical atom 1923-25. *Historical Studies in the Physical Sciences*, 8:189 – 256, 1977.

[24] Stachel, J. Einstein and the quantum: Fifty years of struggle. In Colodny, R. G. (Ed.), *From Quarks to Quasars: Philosophical Problems of Modern Physics*, 349–385. Pittsburgh, PA: University of Pittsburgh Press, 1986.

[25] Stapp, H. The copenhagen interpretation. *American Journal of Physics*, 40:1098–1116, 1972.

[26] Wiggenstein, L. *On Certainty*. New York: Harper & Row, 1969.

The History of Quantum Theory and Quantum Curvature Interpretation[1]

Wu Xinzhong

Shanghai Jiaotong University

sju@sina.com

ABSTRACT. There is a frequency paradox in De Broglie's matter wave theory. Guoqiu Zhao constructs a new matter wave – Compton matter wave and solves this paradox. Zhao's quantum curvature interpretation clarifies the Copenhagen School's dense fog of quantum probability interpretation and analyzes different epistemological meanings of particle models in various theories of physics. Quantum curvature interpretation reconstructs a quantum homologue to Newton's particle mechanics using the Comptom matter wave presentation.

1 The Compton Matter Wave

De Broglie believes that the photon equation $E = h\nu$ and the material particle equation $E = mc^2$ are valid for photons and material particles. In a fixed reference system of a particle whose static mass is m_0, the inner frequency of a circle motion is $\nu_0 = m_0 c^2 / h$; but when a static observer sees the particle moving in v_0 velocity, its oscillation frequency decreases, namely

$$\nu_1 = \nu_0 (1 - v^2/c^2)^{1/2} = (m_0 c^2/h)(1 - v^2/c^2)^{1/2}$$

And then, the energy of moving particle is $E = m_0 c^2 / (1 - v^2/c^2)^{1/2}$, the corresponding frequency is $\nu_2 = \nu_0 (1 - v^2/c^2)^{-1/2} = m_0 c^2 / h (1 - v^2/c^2)^{1/2}$. It is obvious that v_1 and v_2 are different.

De Broglie derived the phase velocity $v_\omega = c^2/v$ from v_2 and $\lambda = h/mv = h(1-v^2/c^2)^{1/2}/m_0 v$, the phase velocity is different from the particle velocity v and it is greater than light velocity c. According to De Broglie, the phase velocity is greater than light velocity, but if phase waves can't convey energy and information, then there aren't any phenomena that violate the cause-effect law, this is a root of the quantum non-local phenomena.

In the view of a static observer, there is a particle of inner oscillation frequency v_1 moving in v velocity, the particle links with a wave of ν_2 frequency moving in v_ω velocity. Now a particle is regarded as a micro-clock whose phase is the same as the wave's phase, and the micro-clock moves in $v_\omega - v$ velocity relative to the phase wave. In other words, for any Galilean reference system, the phase of the

---

[1] The author is with Department of the History of Science, Shanghai Jiaotong University. Thank Chinese Ministry of Education 2007 youth research project funding of social science (Quantum mechanics interpretation and scientific realism, 07JC720016).

inner micro-clock of a moving quantum particle is equivalent to the phase of the wave in the particle's position at any moment.

In De Broglie's original phase wave theory, there are three contradictions: the inner oscillation frequency of a particle is incompatible with the quantum wave frequency, the particle velocity is different from the phase wave velocity, and the relativity formulas and Galilean reference system coexist in the process of analyzing problems. To solve these concept problems, Guoqiu Zhao introduced a new quantum mechanics representation, namely the Compton matter wave, and advanced the quantum curvature interpretation – geometry comprehension of the quantum probability.

Let us suppose that system A is the reference system of a static observer and system B is the reference system of a moving particle with v velocity. De Broglie supposed that the matter wave combining with the static particle in system B would be a vibration. The matter wave in system A results from moving particle oscillation. We can construct a real matter wave combining with the static particle according to the Compton matter wave. According to quantum mechanics, the wave frequency of a static particle is $\nu_0 = m_0 c^2/h$, and its Compton wavelength is $\lambda_0 = h/m_0 c$. It is obvious that ν_0 and λ_0 constitute the frequency and the wavelength of the Compton matter wave ϕ_0 of the static particle, the wave velocity of the Compton matter wave is light velocity c.

$$\phi_0 = a_0 \cdot exp\{(2\pi i/h)(P_0 \cdot r - E_0 t)\}$$

Here r is a displacement vector that stands for the motion direction of the particle, and $P_0 \cdot r = 0, P_0 = m_0 c (p_0 = m_0 c), E_0 = m_0 c^2$, P_0 is the Compton momentum of the static particle. ν_0 and λ_0 reflect the space-time size of the static particle. The Compton momentum means that a particle moves at light velocity in the imaginary proper time of Minkowski spacetime.

In view of system A, the particle is in uniform motion, and its mass changes form m_0 to m, $m = m_0(1 - v^2/c^2)^{-1/2}$. According to quantum mechanics, the wave frequency of the moving particle is $\nu_2 = mc^2/h$.

If the Compton wavelength of the moving particle is defined as $\lambda_c = h/mc$, then ν_2 and λ_c constitute the Compton matter wave ϕ of the moving particle in a system:

$$\phi = a \cdot exp\{(2pii/h)(P_c \cdot r - Et)\}$$

Here $P_c = m_c(p_c = m_c), E = mc^2$, P_c is the Compton momentum of the moving particle, c is wave velocity . ν_2 and λ_c reflect the space-time size of the moving particle . ϕ replaces for the space-time background information of the inner quantum wave of the moving particle.

Let us suppose the wave period of the static particle is T. When the particle is static, the measuring-time unit of a static observer in B system is τ_0, the static observer measures the wave-field period, and finds it is $T_0 = T/\tau_0$.

So wave frequency of the static particle is $\nu_0 = 1/T_0 = \tau_0/T$.

When the particle is moving, the static observer in A system gazes at the wave connecting with the particle again, its measuring-time unit becomes larger: $\tau_0' = \tau_0(1 - v^2/c^2)^{-1/2}$, then the wave period of the moving particle becomes $T_0' = T/\tau_0'$.

The wave frequency of the moving particle $\nu_2 = 1/T_0' = \tau_0'/T$, and so

$$\nu_0/\nu_2 = (\tau_0/T)/(\tau_0'/T) = \tau_0/\tau_0' = (1 - v^2/c^2)^{1/2}$$
$$\nu_2 = \nu_0(1 - v^2/c^2)^{-1/2}$$

This is consistent with the conclusion that results from the Planck formula and the matter wave ϕ_0 of the static particle, ν_2 is the field frequency of the Compton matter wave ϕ_2 of the moving particle. Obviously, the field frequency of the Compton matter wave of the moving particle increases, and the clock frequency of the moving particle decreases, not only two frequencies are consistent, but also they are cause and effect each other. It is because the frequency of the moving clock deceases so that the Compton matter wave frequency of the moving particle increases. This solves the frequency paradox of De Broglie naturally and always uses the Lorentz system in the reasoning process.

According to the Lorentz effect of length contraction resulting from the light wavelength of measuring-length unit increasing, we can also conclude that the Compton wavelength of the moving particle is shorter than that of the static particle. Certainly, the matter wave of the moving particle that De Broglie expected should be the Compton matter wave of the moving particle, but not the original phase wave. The Compton wave velocity is the light velocity c, the De Broglie phase wave velocity is $v_\omega = c^2/v$ where the phase wave stands for the wave image of the inner vibration but has no momentum, energy and information, like a light shadow moving at super-light-velocity.

Guoqiu Zhao connects the Compton wavelength with the particle circuit radius, and introduces the quantum curvature to stand for the spatial projection of the particle wave. Zhao finds that the quantum curvature resulting from the wave function in the normalization condition is proportion to the quantum probability, but quantum curvature includes more qualitative information, namely the phase information of the direction of the wave surface relating to the direction of the static Compton momentum (the inner momentum of the particle vibration).

2 Quantum curvature interpretation and the particle model

From the leading idea that the wave function stands for the space-time character of the micro-object in essence, we separate a curvature sector from the amplitude of the wave function. The curvature sector reflects the space-time character of the micro-particle. We call it the standard curvature: $R_n = \delta p_n/\hbar$. The relation between the standard curvature and the uncertainty principle is:

$$\delta P_n \cdot \delta x_n = \hbar, \delta x_n = 1/R_n$$

As we know, the standard curvatures in the various orbits of the hydrogen atom resulting from the uncertain relation is the same as the curvature sectors separating from the amplitude of the radial wave function, and $|\psi|^2$ are proportion to these curvatures, so we can form a new interpretation about the wave function, namely quantum curvature interpretation. The size of the quantum curvature stands for the particle character, and the space-time change of the quantum curvature stands for the wave character. Quantum curvature interpretation explains the wave-particle duality of the micro-particle reasonable.

In fact, it isn't an original creation of Guoqiu Zhao that explains the wave function as the curvature function, French mathematician Rene Thom advanced the similar idea in his work: "Structural stability and morphogenesis". Thom criticizes the particle model in quantum mechanics, and thinks that various controversies about the uncertain relation in quantum theory are based on a coarse and unfit model – the particle model: when micro-objects are filled in an unfit concept frame, chaos and paradoxes will happen. Thom advanced a new viewpoint regarding the wave function ψ as a local curvature in the super-camber, namely a super-camber form changing its topology type by the certain frequency. The total curvature of the wave patten relative to energy eigen functions is like the "standard curvature" in the electron energy level relative to the corresponding energy eigen value in the quantum curvature interpretation.

Quantum curvature interpretation doesn't agree with the particle model, and thinks that any concrete body can't be a particle. When we use some kind of field signal to observer the body, our knowledge about the body are relative to the field signal's nature, strength and propagating velocity. Newton mechanics neglects completely the influence of the motion and the observing signal on the body, then we can idealize the matter body into a particle. The limit of the observing signal velocity (light velocity c) is the root of the relativity effects, as soon as the discontinuity of the observing signal (action quantum h) is the root of the quantum effects. The space-time character of the micro-particle is discontinuous, the standard curvature of the feature camber of the micro-particle happens transitions from $R_n = 0$ to $R_n \neq 0$. The space-time has quantum features in the micro-world. If the relativity theory originates from the limit of the observing signal velocity, then the quantum mechanics originates from the discontinuity of the observing signal and "form-particle change". Whatever in the relativity or in the quantum mechanics, we failed to idealize the matter body into a particle because of the influence of the observing signal on the image of the matter body.

We adhere to Newton's force interaction principle. As Immanuel Kant tried to use some Aristotelian terms of physics and philosophy to understand a new Newtonian physics in his work Metaphysics Foundation of Nature Science, we try to understand the non-classical quantum phenomenon. The quantum curvature interpretation adheres to Newton's force interactions principle. According to Einstein's physics ideal, that regards the force as space-time geometry, makes use of Kant's philosophy category to rebuilds reasonable quantum mechanics that originated from the analysis mechanics: the curvature wave packets of the Compton matter wave replace the particle model of Newton's mechanics, the relative space-time background and the local quantum camber replace absolute space-time, the relative space-time is decided by the limit of the observing signal velocity, the local quantum camber is decided by the quantum action and reflects the space-time image of the micro-particle. Most popular formal systems and interpretations of quantum mechanics are constructed from an analysis mechanics that describes directly the whole mechanical system, but they lack a sort of quantum representation like Newton's particle mechanics that organize the whole system from the basic elements and their interrelations. So we are forced to discuss the quantum micro-phenomenon and the classical macro-phenomenon on the same level. We need to introduce some kind of complementarity in the quantum theory. The deep research on the measurement problem in the quantum curvature interpretation will reveal the secret of complementarity.

Bibliography

[1] Auletta, G. *Fountations and Interpretation of Quantum Mechanics.* World Scientific Publishing Co.Pte.Ltd, 2000.
[2] Max, J. *The Philosophy of Quantum Mechanics.* A Wiley-Interscience Pubilication, 1974.
[3] Unnikrishnan, C. S. Quantum correlations from wave – particle unity and locality: Resolution of the epr puzzle. *Annales la Fondation Louis de Brogile,* 25(3), 2000.
[4] Zeng, J. *New advance in quantum mechanics.* Beijing University Press, 2000.
[5] Zeng, J. *Quantum mechanics.* Science Press, 2000.
[6] Zhao, G. *Motion and field.* Metallurgical Industry Press, 1994.
[7] Zhao, G.-Q., Gui, Q.-Q., Wu, X., Wan, X. *A Novel Medicated Leaven of Physics.* Wuhan Press, 2004.

The Schröedinger's Cat Paradox, Entanglement and Decoherence[1]

Li Hongfang

Wuhan University

leehongfang@gmail.com

ABSTRACT. The recent experimental and theoretical progress in research into the Schrödinger's cat paradox have enriched our understanding of the relationship between the quantum world and the classical world and boosted the development of the philosophical issue of quantum mechanics. The objective realism as well as the positivism/subjectivism in interpretation of quantum mechanics is shown to be wanting, because the essential role attributed to the interaction between the measuring instrument and the measured object emphasized by the Copenhagen interpretation is maintained but at the same time corrected in the dynamic model of decoherence, where the environment is playing an essential role in the decoherence of the quantum states.

1 Introduction

The "Schrödinger's cat paradox", which was conceived by Schrödinger in 1935, remained an academic curiosity until the 1980s when it was proposed that a macroscopic object with many microscopic degrees of freedom could behave quantum mechanically provided that the object was sufficiently decoupled from its environment. In recent years, with the development of technology for quantum measurement, novel progress in demonstrating the macroscopic quantum behavior of various systems such as laser-cooled trapped ions, photons in a microwave cavity, C_{60} molecules and superconducting quantum interference device has been made. This progress has provided strongly experimental demonstrations for the understanding of the historical issue of whether quantum mechanical description of physical reality can be considered consistent. In the following sections we will first discuss the Schrödinger's cat "paradox" in Section 2. Then in Section 3 we will analyze three novel "Schrödinger's cat" experiments that provide us with a direct insight into quantum measurement and quantum decoherence. In Section 4 we will refer to the characteristic trait of entanglement of the cat and present a possible physical solution to the cat "paradox" using the notion of decoherence.

2 The Schrödinger's cat "paradox"

Whether quantum mechanics describes "just the phenomena" or "the reality behind the phenomena", in the first half of the 20th Century, under the influence of

---

[1] The author is with School of Philosophy Wuhan University, Wuhan 430072 China.

logical positivism/empiricism, the majority of physicists would presumably have given a positive answer. And even though neither N. Bohr nor W. Heisenberg can be reckoned a logical positivist, their Copenhagen interpretation has been couched in the empiricist language of "phenomena". [1] In particular, the emphasis placed on the essential role played by the measurement arrangement has fostered the idea that quantum mechanics is just dealing with phenomena to be observed in the course of measurement. It is therefore incorrect to regard a certain property of quantum objects as a property of the quantum object itself; rather, it is an attribute which must be assigned to both the quantum object and the experimental arrangement. Since the choice of the experimental arrangement is purely a matter of human intention, the properties of quantum objects cannot be considered objective. That is, the human intention influences the structure of physical reality.

To illuminate the conflict between our ordinary experience of observation / measurement and the Copenhagen interpretation, Schröedinger skillfully devised a diabolical device in which the microscopic quantum indeterminacy of the atomic world which governs radioactive decay might be translated into macroscopic indeterminacy, accordingly a cat put in this device is whatever simultaneously dead and alive, viz. being in a quantum coherent superposition of states:

$$|\Psi> = \alpha| \downarrow> \otimes |\text{dead cat}> | + \beta| \uparrow> \otimes |\text{alive cat}> \quad (1)$$

Where $(| \downarrow>$ and $(| \uparrow>$ denote interior states of an atom that has and has not radioactively decayed, and α and β denote probability amplitudes, $|\alpha|^2 + |\beta|^2 = 1$.

According to Schrödinger, provided that the discussion of quantum measurement of the Copenhagen interpretation is valid, it should be also valid for macroscopic objects made of microscopic particles, thus inferring, if a "macroscopic cat" is put in a quantum coherent superposition of alive and dead states, whether the cat is dead or alive will not be an objective reality which is independent of observers, but dependent of a observer's measurement. Or, according to at least von Neumann, the cat has been suspending an ambiguous state simultaneously dead and alive until someone opens the device and looks at its content, there is a sudden reduction of the wave function:

$$|\Psi> \xrightarrow{\text{observation of human eyes}} |\text{dead cat}> or |\text{alive cat}> \quad (2)$$

The situation defies our sense of reality because we only observe live or dead cats in our real world, and we expect whether the cat is alive or dead should be independent of our observation. Accordingly, the consistency and universality of quantum mechanics will suffer from severe challenge unless there is a certain mechanism in quantum mechanics to eliminate the coherent superposition of alive and dead states. This is Schrödinger's cat "paradox". [7]

3 Experimental Progress in "Schrödinger's cat"

Physics is ultimately an experimental science and it takes new experiments to drive us to new understanding. With the development of the technology of quantum measurement in recent years, Schrödinger's cat-like states of matter have been generated in meso-scopic systems or systems that have both macroscopic and microscopic features and their decoherence from quantum superposition to statistical mixtures and classical behavior have been observed, which effectively prompts our ability to understand quantum measurement.

In 1996, Monroe et al[4] first experimentally demonstrated the basic principles of Schrödinger's cat using a cooled single atom. They used laser pulses to make the atom oscillate harmonically as a combination of wave-packets representing two different electronic states and produced a superposition of two "coherent-state wave packets". Because the two wave packets are separated by a mesoscopic distance of more than 80nm, which is large compared with the size of the individual wave packets ($\approx 7nm$), this is a "Schrödinger's cat" at the single atom level, viz. an atom was prepared in a quantum superposition of two spatially separated but localized positions.

In the same year, Haroche et al[3] realized another remarkable experiment of "Schrödinger's cat" using a microwave cavity. They sent a Rydberg atom to run through a high Q microwave cavity in which a few photons was trapped in advance, so the atom would interact with a radiation field with classically distinct phases produced by the photons and generate a bigger-size superposition of quantum states. The meso-scopic superposition is also the equivalent of an "atom + measuring apparatus" system in which the "meter" simultaneously points at two different directions what qualifies as Schrödinger's cat in a box with an atom in a linear superposition of its excited and ground states.

In order to check the coherence of the superposition and to study how it is transformed with time into a mere statistical mixture, the researchers have further probed the "cat state" with a second atom which acts as a "quantum mouse" and goes across the cavity after a delay, accordingly have realized a quantum measurement without opening the lid of the box, but letting a sub-atom "mouse" walk to the nose of the "cat" and look what has happened. The experimental results show that the time for the coherent combination |live > plus |dead to decohere (because of unavoidable interactions with the environment) can be measured and is found to be shorter for more complicated quantum cats, viz. when the number of photons become larger, the decay becomes faster and faster, which is an objective decoherence process deriving from quantum entanglement between the photo field state and its environment.

In other words, Haroche's quantum-cat is observed in an indirect measurement by the quantum-mouse in a coherent quantum non-demolition (QND) experiment in which the phase of the quantum-mouse wave function is sensitive to the state of the quantum-cat, but does not decohere it, which extends the sense of quantum measurement in a way not directly anticipated in the original Copenhagen interpretation. So, this experiment has provided us a direct insight into the transition from a quantum superposition to a statistical mixture at the heart of quantum measurement by allowing controllable studies of quantum measurement and quantum decoherence.

In 2000, Friedman et al[2] achieved a quantum superposition of truly macroscopically distinct states, which is closer to the Schrödinger's metaphor, in which a superconducting quantum interference device (SQUID) can be put into a superposition of two current states (consist of billions of paired-up electrons moving in perfect harmony and without resistance) in the ring corresponding to the indeterminate state of "Schrödinger's dead-and-alive cat", which "narrows the gap between theoretical ideas and reality" and indicates that quantum mechanics can describe assuredly physical behavior at all scales. Thus, quantum theory does not break down when the system becomes more complex. Of course, besides superconduct, supercurrent and radiation, this type of large coherent quantum state is well isolated from the inside environment, and the common macro-systems do not have this peculiarity, so macro-scopic quantum coherent states cannot be seen everywhere.

4 Entanglement and Decoherence of Schrödinger's cat

The experimental progress which has been achieved in understanding the Schrödinger's cat paradox has further promoted our understanding of some of the basic concepts in quantum mechanics, such as entanglement. According to Schrödinger, entanglement is one characteristic trait of quantum mechanics, the one that forces its entire departure from classical lines of thought. So entanglement has an inextricable relationship to quantum measurement. It is crucial for us to understand non-locality in quantum mechanics. And the real cat state should include the internal states corresponding to the collective states |alive cat $>$ and |dead cat $>$, which may denote using $|D_j>$ and $|L_j>$, where $j = 1, 2, 3, \cdots, N$, refer to the number of internal microscopic particles that make up of the macroscopic cat. Accordingly, the complete state of macroscopic cat should be wrote as

$$|\Psi>= \alpha|\text{alive cat}> \otimes \prod_{j=1}^{N} |L_j> +\beta|\text{dead cat}> \otimes \prod_{j=1}^{N} |D_j> \qquad (3)$$

This is a quantum superposition state in which the macroscopic collective states |alive cat $>$ and |dead cat $>$ get entangled with the corresponding internal states $\prod_{j=1}^{N}|L_j>$ and $\prod_{j=1}^{N}|D_j>$. If we only consider the dead and alive states without the explicit inclusion of the internal states of the cat, the information of internal degrees of freedom need be removed. The rule is to construct the density matrix $|\Psi><\Psi|$ of the composite system and "trace out" all internal degrees of freedom of the complete state of the cat. In this way we shall obtain the density matrix of the measured system:

$$\begin{aligned}\rho &= tr_{int.}|\Psi><\Psi| \\ &= |\alpha|^2|\text{alive cat}><\text{alive cat}| \\ &+|\beta|^2|\text{dead cat}><\text{dead cat}| \\ &+\alpha\beta^*|\text{dead cat}><\text{alive cat}| \prod_{j=1}^{N} <L_j|D_j>,\end{aligned} \qquad (4)$$

where the diagonal elements $\rho_{cd} = |\alpha|^2|\text{alive cat}><\text{alive cat}|+|\beta|^2|\text{dead cat}><\text{dead cat}|$ denote the dead state and alive state of the cat respectively, and the off-diagonal elements (interference term) $\rho_{cnd} = \alpha\beta^*|\text{dead cat}><\text{alive cat}| \prod_{j=1}^{N} <L_j|D_j>$ denote the superposition of dead and alive states. Because the off-diagonal elements involve the scalar products of different coherent states of internal degrees of freedom $\prod_{j=1}^{N} L_j|D_j>$, where $|<L_j|D_j>| \leq 1$, therefore, when the internal degrees of freedom that make up of the cat is to infinite $N \to \infty$, $\prod_{j=1}^{N} < L_j|D_j >$ is to zero, which may rapidly eliminate the off-diagonal elements and lead to the quantum decoherence of superposition states. [6]

Therefore, we may write out formally the coherent superposition of dead and alive states in terms of the rule of quantum mechanics. Yet strictly speaking, only referring to macroscopic objects, their quantum coherent features will not exist. The decoherence of Schrödinger's cat shows us a spontaneous dynamic process in which the internal states of degrees of freedom of the macroscopic cat get entangled with its collective states, viz. the states of the measured system. This tends to eliminate the interference term because the internal states of degrees of freedom can be orthogonal. It is irrelevant whether there is a human observer or even a real measuring instrument on the scene. And it is just the large number of internal degrees of freedom of the macroscopic cat that makes the process of decoherence become very momentary. Before human's eyes observe, it has finished for a long time. This is also why we open the box and always look at a classical cat that accords with our experience perfectly rather than a quantum cat of superposition as quantum mechanics predicts.

In sum, the recent experimental and theoretical progress in research into the cat paradox have enriched our understanding of the relationship between the quantum world and the classical world and boosted the development of the philosophical issue of quantum mechanics. The naively objective realism as well as the positivism/subjectivism in interpretation of quantum mechanics is shown to be wanting, because the essential role attributed to the interaction between the measuring instrument and the measured object emphasized by the Copenhagen interpretation is maintained but at the same time corrected in the dynamic model of decoherence, where the environment (e.g. the external air molecule or the internal degrees of freedom of macroscopic object) is playing an essential role in the decoherence of the quantum states. [5] [8]

Bibliography

[1] De Muynck, W. M. Towards a neo-copenhagen interpretation of quantum mechanics. *Foundations of Physics*, 34(4):717–770, 2004.

[2] Friedman, J. R., et al. Quantum superposition of distinct macroscopic states. *Nature*, 406:43–46, 2000.

[3] Haroche, S., et al. Observing the progressive decoherence of the 'meter' in a quantum measurement. *Physical Review Letters*, 77:4887–4890, 1996.

[4] Monroe, C., et al. A 'Schrödinger cat' superposition state of an atom. *Science*, 272:1131–1136, 1996.

[5] Omnès, R. *The Interpretation of Quantum Mechanics*. Princeton University Press, 1994.

[6] Sun, C.-P. New progress of fundamental aspects in quantum mechanics. *Progress in Physics*, 21(3):317–359, 2001.

[7] Wheeler, J. A., Zurek, W. H. *Quantum Theory of Measurement*. Princeton: Princeton University Press, 1983.

[8] Zurek, W. H. Decoherence and the transition from quantum to classical. *Physics Today*, 44:36–44, 2002. "Decoherence , Einselection and The Quantum Origins of The Classical", arXiv: quant-ph/0105127 v2 11 Jul.

The Descriptive Object of the Schrödinger Equation and Wave Particle Unification in Quantum Mechanics Curvature Interpretation[1]

Zhao Guoqiu
Wuhan Engineering Institute
zhao66@126.com

Gui Qiquan
Wuhan University
guiqq@sina.com

ABSTRACT. The curvature interpretation of quantum mechanics acknowledges that the micro-universe object is not mass point. Though the "image" of the micro-universe object is simply not accessible, it can be formed by using the matter wave length. The degree of the thus formed "image: cannot be ignored when discussing the atomic problem. The mass point abstraction principle in classical mechanics is not applicable to the atomic world (or can not be followed without discrimination). The wave function in quantum mechanics is the curvature wave, the degree of curvature of which indicates the corpuscular property and the change of curvature indicates the wave property. The mass point model has been modified greatly by curvature interpretation. The wave-particle dualism in the curvature model harmoniously has been unified. If quantum phenomena in the atom are transformed into the subject studied for classical mechanics in macroscopic experience, we must change the discontinuous distribution of energy into a continuous distribution, change the discontinuous mutual effect between energy levels into sequential effects, and change a pure state into a mixed state. This is the task of quantum measurement. The association of the increase in quantum entropy (thermodynamic property) with quantum measurement reflects the nature of taking the enduring action in measurement.

1 Wave-particle unification in quantum mechanics curvature interpretation

What is curvature interpretation in quantum mechanics? What is the basic train of thought for curvature interpretation?

In quantum mechanics curvature interpretation, the wave function in quantum mechanics describes the "self" of the micro-universe object and the variable rule of the "surface curvature". The wave function is the curvature wave, the degree of curvature indicates the granularity, and the curvature change shows

---
[1] The authors are with Wuhan Engineering Institute, Wuhan 430080 China, and Wuhan University, Wuhan 430072 China, respectively.

the wave property. In quantum mechanics curvature interpretation, the annotation system can be set up through "all-optical transformation" of the electron, "quantization forming of image" and "transformation of the image and point".

The basic train of thought for quantum mechanics curvature interpretation is to acknowledge that the microcosmic object is not the mass point. The abstract principle of mass point in classical mechanics is not applicable to the atomic world and cannot be indiscriminately imitated. The microcosmic object of the electron in the atom into mass point is an abstraction, so the mass point has a virtual reality while the wave is real. In quantum mechanics, the equivalent structure with a phenomenal meaning corresponds to the mass point embodied in the theoretical result in the wave function mode.

1.1 Transformation of the image into light

In the macro-universe, the object of experience is apparent to the naked eye. The "image" of the object is set up by bio-instruments of the eye-brain system through observed information which increases the extension function of the bio-instruments. That is, the "image" of the object is converted in a person's brain. In the micro-universe, however, the electron "entity-in-self" can be differentiated finally only by atomic luminescence. Atomic luminescence is the result of electronic transition, but we only know the optical frequency and strength sent or absorbed in the electronic transition. It is a discontinuous spectral line.[1, 2, 5]

In order to form the electronic "image" in the atom, we apply the bending of the curved surface to carry out an analogy on the change of the luminous intensity and set up the following coincidence relationship:

1. if the luminescence is zero, corresponding to the plane, the curvature is zero;
2. if the luminescence is weak, corresponding to the curved surface, the curvature is "small";
3. if the luminescence is strong, corresponding to the curved surface, the curvature is "big".

For a description of the micro-universe object, the mass point model can be replaced with the "curvature model". Thus, the luminous intensity at the electronic transition in the atom can be related to the geometric "image" of the formed electronic "phenomenon entity" at some energy level. Because the luminous intensity at different energy levels of the atom differs, the "curvature" related to the electron is also changeable. Therefore, the electronic "shape" or "image" is also changeable. As for the electron of the continual transition in the atom, its "image" is changeable as it moves; this is very different from the macroscopic phenomenon, though the phase circle corresponds to the eigen state matter wave in the atom. It is a changeable form and the electronic "image" should be provided by it. The curvature wave is also defined by it and corresponds with the interference of the wave at the macroscopic phenomenon level. We call the above model a "curvature model". Actually, the "curvature model" is simply a "field theory" model.

1.2 Quantitative construction of the electronic "image" in the atom

1. Enlightenment of the wave function of the hydrogen atom

   The radial wave function of the hydrogen atom is generally as follows

   $$R(r) = \alpha B_0 e^{-\rho/2} \rho^l L_{n+l}^{2l+1}(\rho) \tag{1}$$

   where, $\alpha = 2/na_0$, $\rho = \alpha r = 2r/na_0$, $B_0 = -b(2l+1)!(n-l-1)!/[(n+1)!]^2$.
   It can be changed by calculation of the above equation

   $$R(r) = R_n 2 B_0 e^{-R_n r} (2R_n r)^l L_{n+l}^{2l+1}(2R_n r) = R_n G(r) \tag{2}$$

   where, $R_n = 1/na_0$, a_0 is the Bohr radius, n is the quantum number of energy level, and R_n has the dimension of curvature (curvature of phase circle).

   R_n is the wave amplitude (R_n^2 is just the electronic "surface curvature"), and the wave function of the hydrogen atom can be regarded as the curvature wave.

2. Construction of the electronic "image" in the hydrogen atom and the relation between the "image" and the curvature R_n

   The De Broglie matter wave length of electron on energy level n in hydrogen atom

   $$\bar{\lambda} = \bar{\lambda}/2\pi = na_0 \tag{3}$$

   $\bar{\lambda}_n$ can be regarded just as the radius of a circle with the perimeter λ_n, R_n is just the curvature of the phase circle. It is the R_n of equation (2). Therefore, a curvature R_n corresponds to the electron as defined by de Broglie's matter wave for each energy level n of the hydrogen atom which we call curvature. $r_n = na_0$ is called the curvature radius. It gives a basic image of the electron on each energy level in the hydrogen atom and reflects the "structure" information of the electron. The wave function of the hydrogen atom is actually a curvature wave with an amplitude of R_n. The simplified form of the wave function $R_n \cdot G(r)$ or $R_n \cdot G(x)$ indicates that there is a curvature on each space-time point corresponding to the electron which represents a correspondence of strength of weakness of luminescence to transition frequency of the electron in the experiment phenomenon. By means of R_n or $R_n \cdot G(r)$ (or $R_n \cdot G(x)$), it is possible to analyze and recognize the movement status of the electron in inner space. The size of the "reference curvature" corresponds to the status vector module of Von Neuman, and Hilbert space has a practical physical significance. The status vector is just a curvature vector. A curvature radius r_n and a curvature R_n defined by matter with a wave length λ_n can be separated from the amplitude of the wave function in any other matter wave.

   The above analysis of the hydrogen atom is universally significant.

3. The transformation of image and mass point

   In the atom, we established a basic "image" in inner space for the electron by using the De Broglie matter wave length λ_n on each energy level and represented it by curvature R_n. It is possible to prove the shape we have established for the electron in the atom. And when discussing the atom, its "shape" is not negligible and does not conform to the principle of abstraction for macroscopic particles. For making the particle abstraction a certainly, this particle will be virtual and can exist on any point in the "shape" without movement track while the wave is real[4]. Obviously, the less the "image" of the electron, the larger the curvature R_n, and the more possible it becomes to find the "point particle" in the shape. The larger the shape of the electron, the less the curvature R_n, and the less possible it becomes to find the "point particle" in the image. When the reference image of the electron on each energy level is different, their occurrence probabilities are also different. The curvatures on each time-space point change. Therefore, the occurring probabilities also change. The probabilities and the curvatures in inner space can be inter-conversed. The cloud image of electron in the atom is a geometrical image made by the curvatures and represents the entire image of the electron on different energy levels. This is referred to as "transformation of image and mass point".

The "transformation of image and mass point" in the curvature interpretation of quantum mechanics admits the effectiveness of the interpretation of critical probability. The interpretation of the curvature can eliminate a basic contradiction in the probability interpretation and bring the rational part of the probability interpretation into it. Compared to the probability interpretation, the physics theoretically described in the curvature interpretation is real with a more profound and more complete cognition: the wave function is essentially the curvature wave, the size of the curvature indicates the granularity and the change of the curvatures reveals the wave property. The wave property and granularity are unified in the same model, and the wave-particle dualisms in the microcosmic object itself are unified harmoniously. The curvature interpretation of quantum mechanics is logically consistent with philosophy, physics, the experimental phenomenon and the mathematics of the annotation system.

4. Indeterminacy of the electronic "image" and electronic position.

From the indeterminacy relation, the physical significance of δx_n is the position indeterminacy of the mass point electron. This indeterminacy quantity is just equal to the "curvature radius" of the electronic "image" in inner space. The indeterminacy quantity just reflects that the electronic "image" in the atom is not neglectable, and the principle of abstraction for mass point is not applicable or this basic idea can not be followed without discrimination. The indeterminacy relation has a realistic background.

2 Descriptive objects of the Schroedinger equation

2.1 Interference of waves in classical mechanics

If several rows of waves are transmitted simultaneously in space, the vibrations at some points in space will be constantly strengthened, while at other points, they will be constantly weakened or completely counteracted. This phenomenon is called interference of waves. Generally, when waves overlap with different amplitudes, frequencies and phases at any point, the situation becomes very complicated. The overlapping waves sent by two wave sources with the same frequency, or the same vibrating direction, or the same phase or constant phase difference will result in the wave interference phenomenon.

2.2 Reanalysis on superposition of the wave in quantum mechanics, wave interference and the Schrödinger equation descriptive object

In quantum mechanics, the matter wave equation – the Schrödinger equation is set up analogously with the wave equation in classical mechanics–it is also a linear wave equation. If $\Psi_1, \Psi_2 \cdots \Psi_n$ is the solution of the equation, $\Psi = \Psi_1 + \Psi_2 + \cdots + \Psi_n = \sum_{n=1}^{n} \Psi_n$ is also the solution of the equation (it is $\Psi = \int \Psi(x)dx$ for the continuous spectrum). This is the principle of the superposition of the wave in quantum mechanics. The probability wave and energy level transition concept, no matter in the quantum mechanics or quantum field theory, is the recognized concept. Superposition of the matter wave is the superposition of the probability wave, and the matter wave interference is the interference of the probability wave. [3]

In interpreting the quantum mechanics curvature, the wave is the description of the change in the particle "image", which may correspond to the experiment by optical phenomena. Of course, it refers to a new understanding of the concept of space in the micro-universe. In inner space, the space is real, the wave is dynamic, and the particle is virtual and trackless. The curvature wave is a kind of inner space structure wave that can be easy to understand. It integrates the entity attribute of the point particle in the macroscopic experiment with the matter wave, and its propagation needs no medium. In the atom, the electron transits from one energy level to the other, absorbs or emits a photon, and the state changes abruptly. Information among eigen states is not related. In addition, the mathematic analysis indicates that there interference must occur during the superposition among eigen states in the atom:

$$|\Psi|^2 = |\Psi_1 + \Psi_2|^2 = \Psi_1^2 + \Psi_2^2 + \Psi_1^*\Psi_2 + \Psi_1\Psi_2^* \tag{4}$$

where $\Psi_1^*\Psi_2 + \Psi_1\Psi_2^*$ is is the interference item.

When applying the Schroedinger equation to the hydrogen atom, it is necessary to assume that the energy E_n of the electron is quantized, which conforms to the interpretation of the actual physical meaning and exactly reveals that the energy level separation in the atom is the physical essence. It is obvious that the concepts of energy level separation and quantum transition in an atom cannot be given up, according to the requirements in Schrödinger equation. There is a physical process in the coherence of the matter wave in the atom.

In our opinion, both in and out of the atom, the discontinuous number of matter wave state and continuous number of matter wave state are different in physical essence, though they are the objects described in Schrödinger equation. It embodies that from the inside to outside of the atom, the energy is changed from discontinuous distribution to continuous distribution, interaction is from discontinuous to continuous, and object from quantitative change to qualitative change

It is obvious that to change the quantum phenomenon in the atom to a classical mechanics research object in the macroscopic experience, we must change the discontinuous distribution of energy into a continuous distribution. This does not mean we have to cut the quantum infinitely. When we change the orderly quantum transition into a disorderly quantum transition, the non-continuity of the energy and the interaction disappear. When we change the discontinuous interaction into a continuous action, it changes from a pure state into a mixed state changes the real body wave and the virtual particle into a real particle and a virtual wave. This is the task of quantum measurement. People can connect the increase in quantum entropy (thermodynamic property) with quantum measurement, which exactly embodies the essence of the continuous action in the measurement.

Based on this, the automatic decoherence of the macroscopic object in the decoherence interpretation of the quantum considers the influence of the external and internal environments. Actually it introduces a kind of continuous action and changes a pure state into a mixed state.

Bibliography

[1] Rene, T. *Catastrophe Theory: Ideology and Application*. Shanghai Translation Publishing House, 1989.
[2] Sakata, S. *Collection of Thesis on Science and Philosophy*. Beijing: Knowledge Press, 1987.
[3] Xue, D. *Four Speeches about Wave Dynamics*. Beijing: Commercial Press, 1965.
[4] Zhao, G. *Between Physics and Philosophy*. Beijing: China News United Publishing, 2007.
[5] Zhao, G., Gui, Q., et al. *A New Light on Physics*. Wuhan: Wuhan Press, 2004.

An Afterthought on the Interpretation of Quantum Decoherence[1]

Zhao Guoqiu

Wuhan Institute of Engineering

zhao66@126.com

ABSTRACT. The entanglement state formed by a measured system (S) and a macro-instrument (M) is logically inconsistent in the theoretical system of decoherence interpretation. The application model of decoherence interpretation needs to be re-established.

1 The basic idea of the interpretation of quantum decoherence

The physicist John Bell pointed out that, "In the theory of quantum measurements, the disappearance of quantum coherence is the cornerstone of philosophic discussion".[1] The basic idea of the interpretation of quantum decoherence can be summarized as follows:

1.1 The macro-object has quantum coherence

As a universal theory, quantum mechanics describes a macro-object and even the cosmos, that is, they can be written as a pure wave function satisfying the Schr?dinger equation. A macro-object, therefore, has quantum coherence.

1.2 The external or internal environments can induce the decoherence of a macro-object

Due to the influence of the external or the internal environment, macro-objects generally can decohere instantaneously. So the coherent superposition of a macro-object cannot be seen in real life. The external environment can be molecules and atoms in the air, or photons in radiation, or even include cosmic microwave background radiation. And the internal environment has a great internal degree of freedom with the random motion of a macro-object. A macro-object can entangle with either the external or internal environment, thus resulting in the loss of coherence.[2]

---

[1]The author is with Wuhan Institute of Engineering, Wuhan 430080, P. R. China.

1.3 No decoherence without a system of internal dissipation

Coupling to the environment, the energy will be dissipated from a macro-object. Generally, decoherence happens with dissipation, and without internal dissipation, the system will not decohere. If a macro-object and its environment (the external or internal degrees of freedom) are not coupled or coupled weakly, then there isn't dissipative effect, and quantum interference of the system will remain. This can be used to illustrate superconduction and superfluid. Though they have many internal degrees of freedom, they are still kept in a macroscopic quantum superposition state. In conclusion, Brownian fluctuations and dissipation are the sources of decoherence and the former is the primary factor.

2 Interpretation of decoherence applied to a model of quantum measurements

Correlation between an instrument and a measured system

Von Neumann's measurement theory has pointed out: the operation of quantum measurement is to "read out" the state of a measured system (S) from the state of an instrument (M).[4] If the quantum mechanics are used for describing the "read-out" process, there must be correlation between the instrument (M) and the measured system (S). If $|\Psi\rangle = \sum_n c_n |n\rangle$ is the initial state of the measured system (S), and $|e_n\rangle$ are the initial state of the measuring instrument (M), an entanglement state between the system (S) and the instrument (M) will be formed by quantum measurement, which can be described as a wave function of the total system (S+M) (initial states of factors) as follows:

$$\Psi_z(0)\rangle = \sum_n c_n |n\rangle |e\rangle \qquad (1)$$

Based on decoherence theory, this quantum entanglement tells us that as for non-factorable final state,

$$|\Psi(t)\rangle = \sum C_n(t) |n\rangle |e_n\rangle$$

$|e_n\rangle$ is the final state of the instrument, $n = 1, 2, 3 \cdots \cdots k$.

Once the instrument (M) is found to be in $|e_k\rangle$, we will soon know the total system (S+M) is in the component $|k\rangle |e_k\rangle$. That is, the total wave function collapses to $|k\rangle |e_k\rangle$, and the measured system (S) must be in $|k\rangle$.

Note that in formula Eq.1, the wave functions of both the system (S) and the instrument (M) are pure quantum states. Following the basic train of thoughts of decoherence interpretation, above operations of decoherence interpretation can be simply concluded as follows:

System (S) and the instrument (M) both in pure quantum states ($|n\rangle, |e_n\rangle$) → establish a quantum entangled state between system (S) and the instrument (M) → macro-instrument (M) state $|e_n\rangle$ auto decohere. → find the instrument (M) in the $|e_k\rangle$ state → the measured system (S) decohered to $|k\rangle$ state (mixed state).

This is an application of decoherence interpretation to the above model of quantum measurement. As a kind of applied model, its idea of Eq.1 is clear but logically inconsistent, and the difficulty of wave packet collapse still remains:

First, since macro-objects including measuring instrument (M) can be decohered instantaneously, once an instrument (M) is "born", it will evolve quickly from "a pure state" to "a mixed state", and the evolution is irreversible.

Second, due to the first reason, the instrument (M) has already decohered to a mixed state before the quantum system (S) can be measured. Thus, in any actual measurement, the instrument (M) cannot exist in a "pure state". As the interaction of (M) and (S), the ideal state of Eq.1 does not make any actual physical significance, but it does have a mathematical significance.

Third, for formula Eq.1 to have an actual physical significance, the macro-instrument (M) cannot be decohered before it is coupled to the measured system (S). Thus we have to make a painful decision to allow the "instrument cat" to be neither dead nor living, but this state of affairs conflicts with the view that macro-objects decohere quickly in the interpretation of decoherence. Otherwise, the macroscopic instrument (M) can be decohered quickly, but before coupling to the system (S) it must return to a pure quantum state, and then turn into a mixed state with the measured system (S). This obviously conflict with the principle of quantum measurement and experimental fact that a pure state will be evolved irreversibly into a mixed state.

Fourth, if we know the instrument (M) is in $|e_k\rangle$ state, the total wave function (S+M) will collapse to $|k\rangle|e_k\rangle$ state. Thus it can be concluded that the system is in $|k\rangle$ state. Because Eq.1 is logically inconsistent and M has decohered before measuring of instrument (M), the time for decoherence of macroscopic instrument (M) has no contribution for wave packet collapse of measured system (S), thus, the difficulty of the wave packet collapse is not eliminated.

In light of the above four reasons, we think that the decoherence manipulation described in Eq.1 is just mathematically constructed "for all practical purpose"[1], and does not represent any actual physical process or make any physical significance. The very trouble to us is that the difficulty of the wave packet collapse is still retained in this model.

An analysis of the automatic decoherence of a macroscopic cat also faces the same destiny. In the Schröedinger's cat cage, the "cat" behaves as the "instrument (M)" which drives decoherence of the disintegrative atoms. Its destiny is not better in the current auto decoherence model. The poor cat is just waiting, neither dead nor living. Not until the disintegrative atom system (S) is placed into the cage, does it start to decohere and return to the macroscopic world with the atoms. Otherwise, after the cat automatically decoheres independently, it will return to the dead and alive superposition state immediately while the disintegrative atoms are placed into the cage, and then it will decohere with the atoms. Neither case accords with the basic principles of quantum mechanics.

Obviously, the current operation mode in the interpretation of quantum decoherence cannot interpret consistently the problem of quantum measurement.

3 Afterthought on the auto decoherence interpretation of quanta entanglement: the basic new train of thought

3.1 Von Newman assumed that "pure quantum state" of macro object only has the symbolic meaning in decoherence theory, once the macro object is born, it will decohere momentarily to be transformed to the mixed state. Any actual measuring instrument (M) only appears in mixed state in quanta measurement.

3.2 Essentially, the quanta measurement is the macro instrument (M) to provide a continuous interactive mechanism for measured system (S), eliminating the discontinuity between eigen states, making the coherence disappears and transforming the pure state to the mixed state.

3.3 The mixed state instrument (M) is to identify the mixed state measured system(S), and this identification has the randomness.

3.4 The decoherence of measured system (S), the pure state is transformed to the mixed state, which reflects the micro quantum world and macro classical world, the theory describes the transformation of physical actual structure.

Take micro particle of energy level 2 as example, set up decoherence model. If $|A\rangle, |B\rangle$ is the branch of energy level 2 pure quantum state of the micro particle, and Ψ_A, Ψ_B is the branch of corresponding mixed state after measuring. We think that the $|A\rangle$ transformed to Ψ_A and $|B\rangle$ transformed to Ψ_B in measuring is a real physical process. Therefore, there is quanta correlation between $|A\rangle$ and Ψ_A, $|B\rangle$ and Ψ_B. The quanta measurement only provides the environmental change for measured system (S) from saltation of the state to continuous state. Decoherence in the measured system (S) is the result of self state entanglement before and after measuring. Based on the above though, we make out the self entanglement state as:

$$\Psi = \Psi_A \otimes |A\rangle + \Psi_B \otimes |B\rangle \qquad (2)$$

Eq.2 is similar to "through Rabi rotation, space state Ψ_B to describe the atom to what path and Ψ_c linked with interval state $|2\rangle$ and $|3\rangle$ of the atom to form the so-called entanglement state.

$$\Psi_T = \Psi_B \otimes |2\rangle + \Psi_C \otimes |3\rangle$$

It is accordingly found out that under condition of no change of atomic momentum, after making the internal state mark on space movement, and the interference fringe between Ψ_B and Ψ_C disappears."[3, 4]

In Eq.2, state $|A\rangle$ and $|B\rangle$ before measuring is regarded as an internal state of micro object (pure quantum state), Ψ_A, Ψ_B is the mixed state after measuring, and continuous action of instrument (M) to system (S), and regarded as internal power in transformation of $|A\rangle \to \Psi_A, |B\rangle \to \Psi_B$. But the discontinuity of original pure state function is changed by this internal power, and the transformation from pure state to the mixed state is realized in self entanglement.

An Afterthought on the Interpretation of Quantum Decoherence 225

Above model can be applied to analysis on double-slit experiment and Schrödinger cat. The double-slit provided the branch of energy level 2 of an electron wave function. The electron wave function $|A\rangle, |B\rangle$ before measuring is of the pure quantum state, it can be regarded as internal state of electron (interval between double-slit is similar to creation of saltation zone), and state Ψ_A, Ψ_B after measuring is the mixed state of electron. Making the self entanglement state as Eq.2 and making the internal state mark can realize the electronic decoherence. The quanta measurement only provides a transformation power of continuous action and inside system (S).

As for Schrödinger cat, we can also make the alive cat $-|A_{alive}\rangle$ and dead cat $-|A_{dead}\rangle$ of pure state be transformed to alive cat $-\Psi_{alive}$ and dead cat $-\Psi_{dead}$ of mixed state respectively, the internal environment effect and energy dissipation only provide the continuous action mechanism for system (S) from internal pure state transformed to the mixed state. Thus, the self entanglement state of cat decoherence can be indicated as:

$$\Psi = \Psi_{alive} \otimes |A_{alive}\rangle + \Psi_{dead} \otimes |A_{dead}\rangle \qquad (3)$$

In Eq.3, $|A_{alive}>, |A_{dead}>$ is the pure quantum state before cat decoherence (can be regarded as internal state of cat), $\Psi_{alive}, \Psi_{dead}$ is the mixed state of cat, $\otimes$ indicates the correlation provided for continuous action mechanism of pure state transformed to the mixed state. To educe the probability distribution by Ψ in Eq.3 is maked with internal pure state, which can realize the instant auto decoherence of the cat[7]. There is no wave packet collapse, but self entanglement and automatic decoherence of the measured system occur because of the intervention of continuous action in measuring.

In a word, the quanta measurement theory in the curvature interpretation of quantum mechanics[5, 6] can be described as follows:

1. The wave packet collapse is not needed.
2. Each wave function branch exists in parallel and corresponds with macro world. In the micro world, such as in the atom, the branches of the wave function of different energy levels exist in parallel. The energy between energy levels is discontinuous, and the action is discontinuous and saltative which creates an independent coherent wave source. But the eigen function branch itself corresponding to each energy level is continuous, of single value and in periodic function. In this way, the eigen function of each energy level corresponds with the macro world. The macro particle information of the micro object is included in the amplitude (fiducial curvature) corresponding to the eigen state.
3. The quanta measurement process is just the process of eliminating the saltation or eliminating the independent coherent wave source. In self entanglement and automatic decoherence, the vast coherence wave disappears, and the macro "local" conformation particle is generated. The process of "particle collected by wave" and "particle emerged to wave" long-cherished by De Broglie can finally be interpreted here.
4. In measuring, instrument (M) has randomicity in the identification of the eigen value. The continuous action of instrument (M) on measured system (S) forces a large number of eigen states transformed simultaneously to macro world in self entanglement. And it is random for instrument (M)

to collect the eigen value of any eigen state. But the eigen value of eigen state nearby atomic nucleus can be identified easily, and the probability is also high, or vice versa. As the eigen value of the eigen state of some energy level is actually collected by the instrument (M), the probability is transformed to 1. It is unknown why the eigen value of this eigen state will be measured at a certain time. What is known is only that the statistical rule of probability functions.

5. Persons and measuring instrument (M) belong to s same macro world.

The instrument (M) is the macro object, the impact of the internal and external environment can be used for automatic decoherence, and the movement of the instrument (M) is macro deterministic. Human beings and the macro instrument (M) belong to the same macro object. The people only perceive the continuous action in experience, and the human observation movement is also deterministic. Therefore, human beings and the measuring instrument (M) belong to the same macro world. The readings indicated on the instrument (M) are just the human cognitive readings. The measuring process is of the process of forcing the quantum probability transformed to macro classical probability through the intervention of continuous action. The human beings apperceive the experience. The curvature interpretation of quantum mechanics does not require multiple worlds; it only requires the macro and micro worlds.

Bibliography

[1] Bell, J. S. On wave packet reduction in the coleman hepp model. *Helv, Phys, Acta,* 48:93 – 98, 1975.

[2] Li, H. Decoherence interpretation of quantum mechanics and philosophy. *JOURNAL OF DIALECTICS OF NATURE,* (2), 2005.

[3] S. Dure, T. N., Rampe, G. ? *Nature,* 395:33, 1998.

[4] Sun, C. *Quanta decoherence problem, New progress of quantum mechanics (1).* Beijing: Beijing University Press, 2000.

[5] Zhao, G. *Interpretation on interaction actuality between physics and philosophy and curvature of quantum mechanics.* China News United Publishing, 2005.

[6] Zhao, G., Gui, Q. Descriptive object of schrödinger equation and wave particle unification in curvature interpretation of quantum mechanics. In 13th *International Convention on Logic, Methodology and Scientific Philosophy.* Beijing, 2007.

[7] Zhao, G., Gui, Q., et al. *A New Light on Physics.* Wuhan Press, 2004.

Hierarchy, Form, and Reality[1]

Chen Gang

Huazhong University of Science and Technology

mgchen@hust.edu.cn

ABSTRACT. Scientific progress in the 20th Century has shown that the structure of the world is hierarchical. A philosophical analysis of the hierarchy will bear obvious significance for metaphysics and philosophy in general. Jonathan Schaffer's paper, "Is There a Fundamental Level?", provides a systematic review of the works in the field, the difficulties for various versions of fundamentalism, and the prospect for the third option, i.e., to treat each level as ontologically equal. The purpose of this paper is to provide an argument for the third option. The author will apply Aristotle's theory of matter and form to the discussion of the hierarchy and develop a form realism, which will grant every level with "full citizenship in the republic of being." It constitutes an argument against ontological and epistemological reductionism. A non-reductive theory of causation is also developed against the fundamental theory of causation. Due to the limitations of size for this volume, this is a shorter version of the full paper.

Acknowledgements: This research for this paper was supported by a joint scholarship from the Royal Institute of Philosophy and UCCL in London. The support enabled me to visit Oxford for three months in early 2006. The research also received years of support from The Center for Scientific Progress and Human Spirit, HUST and the GPSS MAP Fund (GPSSMAP03).

1 Schaffer on Fundamentalism

Jonathan Schaffer's recent paper in NOUS, "Is There a Fundamental Level?" (2003), systematically reviews the works of philosophers and scientists. It provides the context for the problems and an entry point for research in the field. As the title of the paper suggests, the focus of his attention is on 'fundamentality'. According to Schaffer, fundamentalism consists of three theses: 1) the thesis of hierarchy, i.e., the world is hierarchical, stratified into levels; 2) the thesis of fundamentality, i.e., there is a bottom level which is fundamental; 3) the thesis of primacy, i.e., the bottom level is primarily real, other levels are only derivative. [9, p. 498]

Schaffer leaves the first thesis intact and does not pay much attention to the third thesis, but he thinks that the second thesis is the source of the problem. Therefore, Schaffer's key question is, "is science actually in the process of discovering atoms" or "does science really practicing atomism" - that is, is science finite or infinite?[9, p. 502] From the current state of research in science, we

---

[1]The author is with Department of Philosophy, Huazhong University of Science and Technology, Wuhan, 430074, China.

do not know if a quark or super string is the fundamental building block. But the history of science, as Schaffer admits, "is a history of finding ever-deeper structure".[9, p. 503] Whenever we find a fundamental building block, atoms or quarks for example, soon we always find that they have parts and an internal structure. Therefore, there seems to be a trend or a trajectory in science that is descending toward the infinite. However, the tendency or trajectory, like induction, is by no means a logical proof. Although I believe that there will never be a complete microphysics, scientific inquiry in this dimension will never come to an end. I do agree with Schaffer that we should remain agnostic about this issue.

With the thesis of fundamentality in doubt, Schaffer proposes three options as outlets: 1) that a certain version of fundamentalism can be re-formulated without presupposing fundamentality, that is, a fundamentalism without fundamentality; 2) that there might be evidence for a fundamental something else, such as a fundamental supervenience base which consists of more than one level; 3) that we treat each level as equal and grant each one "a full citizenship in the republic of being". After a detailed discussion of options 1 and 2 by examining the four versions of fundamentalism (physicalism, the Humean idea of atomism, epiphenomenalism and atomism), Schaffer comes to the conclusion that all four versions of fundamentalism face fatal difficulties for options 1 and 2, leaving option 3 as the most desirable. At the end of his paper, Schaffer shows the possible benefits and prospects for the third option.

The following quick comments are in order: 1) The thesis of primacy is based on the thesis of hierarchy and the thesis of fundamentality. If the thesis of fundamentality does not hold, the thesis of primacy loses ground. If we cannot identity the fundamental level, no level is primary. Schaffer's work has established a good ground for us to tackle the thesis of primacy. 2) Although Schaffer has successfully spelt out the difficulties for options 1 and 2, as two variances or revisions of fundamentalism, and the benefits and prospect for option 3 as a version of non-reductionism, it will be difficult to prove that there are at the utmost three options and that at least one option is viable. After all, we have to remain agnostic about the possibility of an infinite descending. Consequently, Schaffer does not provide a decisive argument against the thesis of fundamentality. Therefore it might be relatively easier simply to provide a positive argument for option 3. 3) The two possibilities on the thesis of fundamentality impact the thesis of primacy in different ways for reductionism and non-reductionism. If there is an infinite descending in the hierarchy, it poses a serious problem for fundamentalism but no threat to option 3; however, if there is indeed a complete microphysics, it does not prove that fundamentalism or reductionism is the only viable option. There might still be room for non-reductionism because it does not require an infinite descending. Non-reductionism is compatible with both possibilities on the thesis of fundamentality. 4) Fundamentalism is a reductionist interpretation of the hierarchical worldview. There might be a non-reductionist interpretation of the hierarchy. That is, we accept the thesis of hierarchy and remain agnostic on the thesis of fundamentality but deny the thesis of primacy. 5) Schaffer seems to think that only the second thesis of fundamentality begs the question. He takes fundamentality as the focus of the discussion. I think it is the thesis of primacy that should be the focus of discussion. Our key question is not whether there is a fundamental level, but the relationship between two adjacent levels, namely, whether the relationship is reductive or not. Hence we move our attention from the thesis of fundamentality to the thesis of primacy.

The purpose of this paper is to provide an abstract argument for Schaffer's option 3. First, I will extrapolate Aristotle's theory of matter and form in the light of modern science, that is, apply Aristotle's theory to the discussion of hierarchy and develop form realism, which effectively grants every level with "full citizenship in the republic of being". This is, at the same time, an argument against physicalism and atomism. Secondly I will address the problem of causation and provide a theory of causation according to form realism which constitutes an argument against epiphenomenalism. Finally I will shift the focus from ontological reductionism to epistemological reductionism and provide an argument against Hume's idea of atomism.

2 A Theory of Form Realism

I would like to start my argument at the middle level in the hierarchy of existence, i.e., the meso-cosmic objects of sensible magnitude. Let's take a chair for example. It has two components: the wood beams (matter) and the design of a chair (form). When I bought a chair from IKEA, it was hardly a chair. It came as a set of parts tightly packed in a flat box. It is not yet a chair. It was not in the form of a chair. You cannot sit on it. It became a chair only when I finished the assembly. As Aristotle points out, matter is the potentiality, form the actuality.[1, De Anima 412a10] The word "actuality" means "complete reality".[1, Metaphysics 1047a32] The chair, like a statue or a saucer, is a composite of matter (the wood beams) and form (the structure of the chair) according to Aristotle's hylomorphism.[1, Metaphysics 1013a25] Now the chair is an existence distinct from the existence of a box of wood beams. A chair is not identical to a box of wood beams. The wood beams arranged in a certain form become a chair.

Aristotle is absolutely right when he interprets "matter" as "which in itself is not a this"; "form" as "essence, which is that precisely in virtue of which a thing is called a this".[1, De Anima 412a6-9] A chair is a chair not because it is made of wood beams. Since a chair can be made of steel or plastic; we can also build a table or a house using wood beams. A chair is a chair because it is in the form of a chair. We can find ample examples for Aristotle's thesis. The essence of Coca-cola is its formula of secret ingredients, water, salt, and sugar. Form differentiates music from noise, water from ice, a diamond from graphite, etc.

If we apply Aristotle's theory of matter and form to the hierarchy of reality and keep going downward, we reach something unexpected. What is a wood beam? It is made of wood cells. A wood beam is the foot for the chair because its wood cells are arranged in the form of a foot; a wood beam is an arm for the chair, because its wood cells are in the form of an arm. If we keep going downward further and further, we get the diagram:

What implications can we draw from the diagram? On the ladder of the downward analysis, at each level, matter can always be further analyzed into forms and sub-matter. All that remains on the ladder are forms. The forms on the steps of the ladder are distinctive from each other. Matter almost resolves into forms, though not completely if we assume that the descending is finite. Matter does not "vaporize" and disappear. It only "melts" down the ladder,

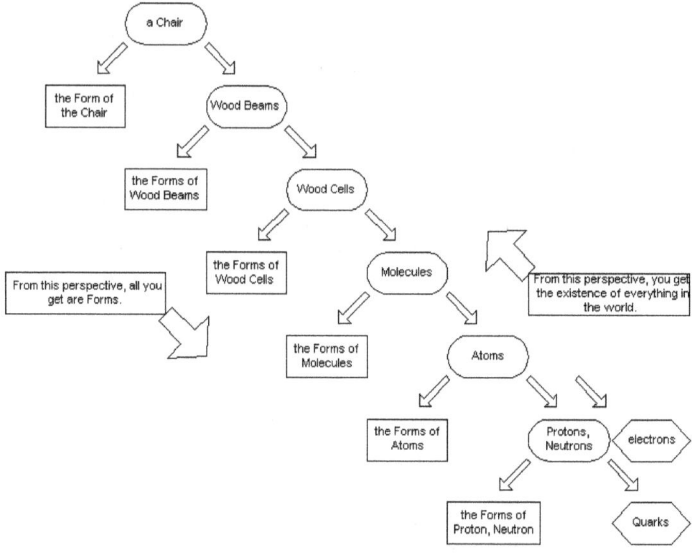

Figure 1: The Ladder of Reality.

with some residue: electrons and quarks or super strings, assuming they are not penetrable. If we think that only matter is real, then what exists in the world is nothing but electrons and quarks or super strings. The whole world is nothing but the arrangement or properties of electrons and quarks. This is the conclusion in favor with the fundamentalists. However, if forms are not real, only matter matters, then we can refuse to pay the bills for electricity, the telephone, Internet service, etc., since nothing material flows into my house via the wires from those services. Music would be the same thing as noise, since both are vibrations of air. Software piracy would be perfectly legal, since we only duplicate the magnetic patterns on our own floppy discs. It does not sound right. It runs against most of our intuitions. Only by admitting the reality of forms, and looking upward, i.e., by going through the ladder of upward combination, can we reconstruct and acknowledge the existence of everything in the world. This is the conclusion of non-reductionism. If form is real, then a chair is not identical to a box of wood beams, because a chair is the wood beams plus something extra, i.e., the form of a chair. The reality of forms as something extra also explains why a system is more than the totality of its parts. Fundamentalists seem to think that the arrangement at the higher level is not important. They usually give primacy to the objects or particles at the fundamental level. They tend to hold that the essential or permanent properties are attributes of the atomic objects or particles. Fundamentalists did talk about the shape of atoms, but they failed to realize that there are distinctive forms at each level. That form at higher level, the form of a chair for example, is not identical or reductive to the forms at lower levels, e.g., the forms of wood beams, and that form at each level can give rise to the emergence of a new set of properties.

3 How Does Causation Happen?

Epiphenomenalism as presented by Schaffer in the context of metaphysics means that all causal powers inhere at a fundamental level. Epiphenomenalism seems to receive support from science. A professor in physics might tell us that all causation happens via one or more of the four fundamental forces: 1) strong interaction via gluon, 2) weak interaction via neutrino, 3) electro-magnetic interaction via photon, and 4) gravitational interaction via graviton as the agent. Does this mean that causation happens only at the fundamental level? We should not instantly jump to that conclusion. New progress in science may prove that the four fundamental forces are not fundamental. With the thesis of fundamentality in trouble, a similar and related question also perplexes epiphenomenalism: "What is the fundamental level at which causation really happens?" "Has physics found the final causal agent and the lowest level of causation for us?" Again, the possible answers to the questions could be manifold: 1) If descent is finite, that is, if we have actually found the fundamental causal agent and the lowest level of causation, it is doubtful that causation happens only at the lowest level. We may choose to prove either causation happens only at the fundamental level, or it happens at multiple levels. 2) If the descent is infinite, that is, if we have not yet found the final causal agent and the lowest level of causation, can we say that our knowledge about causation is zero? Epiphenomenalism then is in real trouble! Scientists, especially physicists, never hesitate to provide causal explanations at macro-levels. We already have self-complete causal accounts at many levels on the ladder of reality. It is amazing that each one of them is unique and coherent. Therefore, causation must have happened at higher level. All we need to do is to explain how causation happens at the macro-level, especially how causation can happen at multiple levels at the same time and why causal account at each level could be unique and self-complete.

Now with form realism in hand, we face a different set of questions: Does form have any causal power? Is form involved in causation? Can we say that forms at macro levels play no role in causation? If form is real, it is better off to assign some causal power to it. It seems against intuition to say that form has causal power; however, it is obviously not right to say that forms (especially forms at macro levels) play no causal role. Still, take a chair for example. A chair is solid only if the design of the chair is good, the material is reasonably solid (which is determined by forms at lower level), and the chair is used in the right way (which means the form at upper level). If the structure of a chair is impossible, no matter how solid its wood beams are, it will crash. If a chair is made of tofu or cheese, no matter how perfect the design is, it will not hold its shape. Even a perfect chair cannot be abused. It follows that forms at each level are involved in causation. There are some cases that are less radical. A chair with a relatively poor design may still hold a reasonable weight if it is made of material that is perfectly solid. A chair made with a relatively poor material may still be stable if it has a perfect design. What I mean is that the causal roles of forms on three adjacent levels are complementary to each other. What we get is the collective effect of them all. If we can show that forms are involved in causation, shall we still stick to the intuition that forms have no causal power?

Some philosophers[3] believe that only fundamental laws in physics are real laws, while Ceteris Paribus laws in special sciences, e.g., economics, sociology, biology, etc., are not. In the context of our hierarchical discussion, the question is which level within the physical domain is the fundamental level on which the fundamental laws reside - the level of sensible objects, the level of atoms, or the level of sub-atomic particles? Since there is no fundamental level, the laws of atoms may also be subjective to the interference from sub-atomic level. We have to say, laws in physics and laws in the special sciences are the same. Boyle's PV Law, which assumes that T (Temperature) remains the same, is one typical example for this. Therefore I agree with Peter Lipton[7] that, most or even all laws are Ceteris Paribus by nature. They cannot hold without idealization or approximation.

4 Epistemological Non-Reductionism

What is the relationship among scientific disciplines? Some further discussion in the context of epistemology is surely necessary. Is there a reductive relation between concepts and theories of various disciplines like physics and chemistry? Hume's idea of atomism is an example of classical epistemological reductionism. The theoretical reductionists like those in the Vienna Circle believe that all scientific theories can be derived from physics. From a form realism perspective, this assumption becomes at best doubtful. Scientists in different disciplines have their own layers of reality. Their major concerns are the forms at their own levels. Their theories are designed to describe forms at different levels. Forms at different levels are distinct; there is no reduction between them. Naturally there is no reduction between theories of forms at different levels. If we put the argument by example, the solidity of a chair is determined by forms at three levels collectively. It is not determined by micro forms only. It follows that, the theory of macro forms is different from the theory about micro forms. Therefore, there is no logical derivation between them.

We have to admit that the reductive approach, as a scientific method, is quite powerful in some cases. Occasionally scientists may go deeper into an adjacent level for a better understanding of some macro properties and features. That is the typical practice in physical chemistry or chemical biology. Some macro properties can be reduced to micro properties, and some macro properties can be explained by micro forms. For example, the color of a chair, if it is made from the same material or painted in one color, is identical to the color of its wood beams, and the color is a macro manifestation of certain micro features such as wavelength and frequency, which is determined by micro forms. However, there are many macro properties which cannot be reduced to micro properties and cannot be explained by micro forms alone. That is, some macro properties are mainly determined by macro forms, e.g., the solidity of a chair is mainly determined by the design of a chair, and this kind of macro properties cannot be derived from the description of micro forms. The discovery of DNA provides another fine example. Heredity as a macro feature can only be partly explained by the molecular structure of DNA. Many hereditary characteristics cannot find corresponding DNA segments. There is no one to one correspondence between hereditary characteristics and DNA segments. We have not yet reached a full understanding of the mechanism by which DNA segments control the formation of proteins. Heredity as the description of macro forms, therefore, cannot be derived from the structure of DNA alone. We can find plenty of examples in other contexts. For instance, the World Cup Soccer Championship is the property of a team, not for any individual player. Can we imagine an account of social change in terms of properties of chemical elements?

5 Conclusion

In this paper I have briefly examined Schaffer's paper on hierarchy and fundamentality and tried to provide an argument for his third option of non-reduction. I have applied Aristotle's theory of matter and form to the discussion of hierarchy and reached a kind of form realism, i.e., an ontological non-reductionism, which is also an argument against physicalism and atomism. Based on form realism, I have also argued against epiphenomenalism and epistemological reductionism. The only remaining issue is the economic concern, that is, the violation of the principle of parsimony leads to the unnecessary proliferation of reality. As Schaffer points out, the economic concern is only a secondary concern. It should not be over-weighted. Form realism as a theory in metaphysics, if it is valid, should apply to other issues in philosophy. Further exploration into its application to other issues is, however, beyond the scope of the current paper.

Bibliography

[1] Aristotle. *The Basic Works of Aristotle*. Random House, modern library paperback edition edn., 2001. Ed. by Richard McKeon.

[2] Davidson, D. Mental events. In *Essays on Actions and Events*. Oxford: Clarendon Press, 1980.

[3] Earman, J., Roberts, J., Smith, S. Ceteris paribus lost. *Erkenntnis*, 57:281–303, 2002.

[4] Kim, J. Supervenience as a philosophical concept. In *Supervenience and Mind*. Cambridge, UK: Cambridge University Press, 1990.

[5] Kim, J. *Mind in a Physical World*. The MIT Press, 1998.

[6] Kim, J. Making sense of emergence. *Philosophical Studies*, 95:3–36, 1999.

[7] Lipton, P. All else being equal. *Philosophy*, 74:155–168, 1999.

[8] Liu, C. Explaining the emergence of cooperative phenomena. *Philosophy of Science*, 66(Proceedings):92–106, 1999.

[9] Schaffer, J. Is there a fundamental level? *NOUS*, 37(3):498–517, 2003.

From Determination, Indetermination, Extrinsic Under-determination, to Intrinsic Under-determination : A Suggestion about the Curvature Interpretation of Quantum Mechanics[1]

Wan Xiaolong

Huazhong University of Science And Technology

hwanxl@yahoo.com.cn

1 Generalization

In the hundred year main stream history of the modern philosophy of science, analytical metaphysics has moved toward the forefront of science. The philosophy of quantum mechanics is a typical illustration. In passing through an eighty year dispute, it still has survived until now in those main interpretations that not only have promulgated a certain nature of quantum mechanics respectively, but moreover have together formed three problems in the interpretation of quantum mechanics (measurement, probability, and correlation) and three questions in metaphysics (causality, reality, and individuality).

1.1 six main groups of QM interpretation

There are six groups of interpretations of quantum mechanics survived after so many years of research: Copenhagen Interpretation (+von Neumann's Approach), Assemble Interpretation (Minimal), Bohmian Theory, Many-Worlds and Many Minds, Modal Interpretations , and The Role of Decoherence. Each of this six groups of interpretations exposed some features in Quantum mechanics, but all of them based on a mainstream in understanding of quantum measurement during seventy years: firstly, Born's Probability Rule, then von Neumann's measurement theory, including his interpretation rule (eigenstate - eigenvalue correlation) and his famous Projection Postulate ([2, pp. 211–212]), and gave different answers about the "Determination" vs. "Indetermination".

---

[1] The author is with Huazhong University of Science And Technology, Wuhan 430074 China.

	Ontology	epistemology	methodology
Aristotle Physics	multiple multiple	Natural position	Elementary math Natural observe
General Modern Physics	Single quality	To saves the phenomena	differential equation measurement
Classic Physics	Classic particle	Agree with Reality -one to one	the same as above
Quantum Mechanics	Single - quality not "atom"	To saves the but phenomenon but Constructionism	differential equation, but no mechanic function

Table I:

1.2 Six Main Problems in the Philosophy of Quantum Mechanics

Three problems in the interpretations of quantum mechanics: Measurement, probability, and correlation. Three metaphysics questions: causality, reality, and individuality. Quantum correlations include: EPR argument and aggregate with identical particles.

1.3 Three features in QM

After above half century's study, we now know the main features in quantum mechanics:

Quantum state: non-locality and entanglement; Quantum measurement: causality+ statistics; Quantum probability: probability amplitude +interference. The core feature in QM is "probability". About quantum probability, we realize:

1. In physics, we obey Born's Probability Rule
2. In methodology, we interpret the Probability: single system or ensemble, un-classic or classic.
3. In epistemology and ontology, we must ask: how , where and what the nature and reality of the probability?

Is QM a revolution theory? Non. See Tab.I A Conclusion of Section 1
The quantum mechanics is a revolutionary theory to the classical mechanics, but a faithful one to the general thought of modern physics. Thus, we will study the nature of the "quantum probability" not only in the interpretational history of QM, but also in the whole history of modern physics.

2 Comparing Six Interpretations

According to Peter Mittelstaedt in 1998 [1, pp. 6–7], those mainly interpretations are still facing a dilemma, when regard the quantum mechanics as an universal theory, between the increasing the explanation power and the reducing the explanation request attached.

Copenhagen -Neumann	Minimal -Assemble	Bohmian	Many -Worlds	Modal	Decoherence
Complementarities Self-referential character	Positivist attitude	Realistic And ordinary reality	Realistic But Ontology extravagant	Quantum Individuality	Interference phenomena
2	1	5	4	3	3

Table II:

2.1 Explanation Power

In Tab.II, we arrange the 5-1 as most explanation power to lest one.

2.2 Explanation Request Attached

In Tab.III, we arrange the 5-1 as **Explanation Request Attached** to lest one.

Copenhagen -Neumann	Minimal Assemble	Bohmian	Many -Worlds	Modal	Decoherence
Classical Instrument +CPP	Calibration Pointer objectification Probability reproducibility	Revised formulation, quantum potential +CPP	Preferred basis	Value state +CPP	Approximate and idealization
4	3	5	4	4	4

Table III:

2.3 "Determination" vs. "Indetermination"

In Tab.IV, there are six interpretations interpret different nature of quantum probability. **Conclusion of Section 2**
There are no single interpretation which has most powerful explanation and lest explanation request attached at the same time, vise visa. There are 3x2=6 kinds solves to understand the feature of quantum probability :(Indetermination +Under-Determination+ Determination) x (Single system +Assemble)

Copen- hagen- Neumann	Minimal -Assemble	Bohmian	Many- Worlds	Modal	Decoherence
				Modal	Decoherence
Indeter- mination	Indeter- mination	Strict Deter- mination	Under Deter- mination (Extrinsic)	Under Deter- mination (Extrinsic)	Indeter- mination
Single system	Assemble	Single system	Single system	Single system	Assemble

Table IV:

3 A Review about Curvature Interpretation CCI

The Curvature Interpretation of Quantum Mechanics is especially focus on Born's Probability also. It believes the success of the Born's Probability rule as well as of the Heisenberg's uncertainty principle, and needs not revise anything in von Neumann's formula, but wants to eliminate the puzzles in quantum mechanics by the way of deeply understanding her methodology in sight of whole history of modern physics.

3.1 A mainstream in understanding of quantum measurement

Every interpretation of quantum mechanics must begin with a discussion of measurement. There is a mainstream in understanding of quantum measurement during seventy years: firstly, Born's Probability Rule, then von Neumann's measurement theory, including its interpretation rule (eigenstate - eigenvalue correlation) and its famous Projection Postulate.

CCI begins with Born's Probability, but unlike most other works, it nor wants to reduce Born's Probability as classic one (as Carnap), neither focus on its epistemological nature(as van Fraassen, Popper, etc).Rather, it realizes, as an universal theory of QM, that Born's Probability is about single system and QM is a Under-Determination theory, but it doesn't totally agree with van Fraassen's way (his Under-Determination is cause +statistics, it is Extrinsic Under-determination) , it is Under-Determination Intrinsic one, it realize the Quantum Probability itself is do Under-Determination.

3.2 The cause of Born's Probability puzzle: "particle" supposes

At the beginning of quantum mechanics, the physicist exactly regarded microphysical objects such as electrons as point-particles. Unfortunately electrons are not point-particles according recent examinations. In the early history of modern physics, Galileo used "atoms" to replace Aristotle's "five elements", and then in analytic mechanics, "mass-point" with a differential equation is very useful and a core method.

From the two maps below, we can see: **Probability $\propto$ Curvature**

3.3 Probability $\propto$ Curvature+

In CCI school at Wuhan, China, the opinion is probability amplitude[3, pp. 111–112] $\propto$ Curvature. But Prof. Zao think probability $\propto$ Curvature [3, pp. 113–114].

Now, we suppose Probability $\propto$ Curvature +(as a function of Curvature)in Hilbert space (at first, we do not study the details here) , then we get:

Probability is not only a rate of two quantities, but also presents a real physical quality,

Curvature, it comes from when we abstract something which is not particle as particle, we have to give this "particle" intrinsic probabilistic, for no continuousness of action and limited speed of propagation in QM.

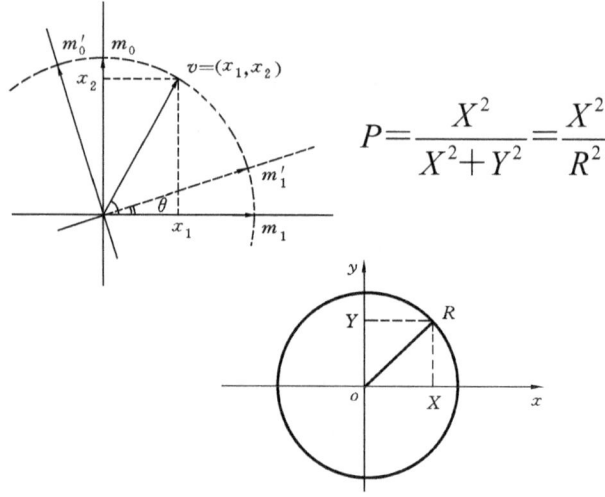

Figure 1:

Prof. Zhao defies the Curvature+(1/R) from Bohr's radius [3, pp. 98–99]. But I rather suggest Dr. Wu Xinzhong's concept that this 'R' as the 'radius of particle' must be the length of Compton Wave [3, pp. 120–121]. We easy to get the Compton Waves of many kinds of " element particle" in the different situations. We regard the distribution of Born Probability as the distribution of Curvature + in the score of the length of Compton Wave. In other word, the normalized condition of Born Probability is the length of Compton Wave.

Probability vs. Curvature:

1. We aren't easy to understand negative Probability, but we are very easy to understand negative Curvature.
2. We can understand the physical nature of probability amplitude: it is do element function of Curvature. Probability amplitude is not a physical quality, but Curvature is.
3. We can thoroughly unify the Complementarity or wave-particle dilemma: the degree of Curvature is particle-ness, the change of Curvature is wave-ness.

Conclusion of Section 3

1. CCI can give wave function a reality, single system, and under-determination interpretation. But Curvature interpretation is not a h.v.theory, rather as a Intrinsic Under-determination of Copenhagen amended.
2. Prof. Zhao want to use his Curvature interpretation now to solve the puzzles about measurement, two-slit experiment, and the relations between space-matter, but I only try to increase comprehensibility about the question of identity and Individuate by CCI.

4 Conclusions: The possibilities and impossibilities of Curvature Interpretation

If Curvature Interpretation is correct, then Quantum correlation is a statistics correlation as well as a causal correlation. The negative correlation of fermions aggregate represents negative curvature with quantum inner space, the positive correlation of boson aggregate represents positive curvature with quantum inner space, and the zero correlation of classic particle aggregate shows du-seperatibility between space and matter in classic physics. A kind of the identical particles is the same of all inner properties and outer relations only in their matter feature, but they will still be distinguishible from their space-features.

Curvature Interpretation is not a mature interpretation and rather a suggestion now, but if it works, it will solve the dilemma between the increasing the explanation power and the reducing the explanation request attached. As a minimal interpretation, it increase lest explanation request attached, but give wave-function a most powerful real interpretation as Bohmian Theory, and don't revised Born-von Neumann's formula as former.

In the other hand, even though CCI exposes the physical nature of quantum probability, the distribution of curvature in quantum space is still dependent on the Born probability. Thus, it is not a determination interpretation, but rather an intrinsic under-determination one. It follows Born-von Neumann's mainstream measurement, learned much from Many-Worlds, Modal Interpretations, and even Decoherence, as a new Copenhagen revised.

Bibliography

[1] Mittelstaedt, P. *The Interpretation of Quantum Mechanics and the Measurement Process.*

[2] van Fraassen, B. *Quantum Mechanics: An Empiricist View.* Oxford: Clarendon Press, 1991.

[3] Zhao, G.-Q., Wu, X.-Z., Wan, X.-L. *A Novel Medicated Leaven of Physics(in Chinese).* Wuhan Press, 2002.

D

Science and Society

Logic and Higher-Order Cognition[1]

Wu Shuxian

National Center for Nanoscience and Technology

wushuxian@tsinghua.org.cn

ABSTRACT. Human cognition has levels of cognitive psychological perspective. Higher Order Cognition (HOC) is a multi-faceted and complex area of thinking which refers to the mental processes of reasoning, decision making, creativity etc. Logic, as the science of thinking, has a widespread and pervasive effect on the foundations of cognitive science. Both Human Cognition and Logic have a relationship to Language. In this paper, I analyze the contributions of Logic in the main areas of HOC and find that the crucial element of HOC is reasoning. Moreover, the interface of logic and experiment is to mutual benefit. So, logical theory, with its historical development and further strengthening, is the most effective approach to HOC research.

1 Introduction

Thinking refers to the general process of considering an issue in the mind, and logic is the science of thinking. Thinking and logic have been the subject of speculation for a long time. More than two thousand years ago, Aristotle introduced a system of reasoning or of validating arguments called the logical syllogism; late-nineteenth- and early-twentieth-century logicians such as Hilbert, Russell etc., developed formal systems and mathematical models of reasoning. These fundamental developments in logical theory have had perhaps a more widespread and pervasive effect on the foundations of cognitive science than any other contributions from philosophy or mathematics. Indeed, it has been found that valid arguments can be justified by showing that they have correct logical forms, even if one does not understand the meaning of the words in the arguments. One can easily be led into error by relying on one's intuition about what is true instead of using careful logical deduction in complicated situations.

Cognitive science, in some sense, is concerned with mechanism, with how humans reason. The central hypothesis of it is that thinking can best be understood in terms of representational structures in the mind and computational procedures that operate on those structures. So, formal logic provides some powerful tools for looking at the nature of representation and computation. Propositional and predicate calculus serve to express many complex kinds of knowledge, and many inferences can be understood in terms of logical deduction with inference rules such as *modus ponens*.

---

[1] The author is with National Center for Nanoscience and Technology, Institute of Science, Technology and Society, Tsinghua University, Beijing 100084 China.

Obviously human cognition operates on different levels. Fundamental cognitive operations such as perception, attention, etc. are lower-order cognitive abilities. Memory and creativity are higher-order cognitive abilities. Higher-order cognition is based on low-order cognition and influenced by lower order cognition in the process of evolution. Some information people define "higher-order cognition" as an information processing phenomena in which the meta-cognitive factors of monitoring and control play the fundamental role. This term is practically synonymous with complex cognition [5].

Higher-Order Cognition involves the ability to understand and implement the steps necessary to solve problems, attack new areas of learning, and think creatively. Dr. Levine[2] stated, "Higher order cognition is the pathway to complex thinking. It enables students to grapple with intellectually sophisticated challenges, integrate multiple ideas and facts, undertake difficult problems, and find effective and creative solutions to dilemmas whose answers are not immediately obvious." This description makes it apparent that higher order cognition is not a single entity but a multi-faceted and complex area of thinking. Higher order cognition is composed of some interrelated processes: concept formation, problem solving, reasoning, decision making, creativity, etc.

For the sake of economy, I focus here on the interplay between logic, reasoning and decision making.

2 Logic and Reasoning

One of the marvels of human nature is that people are endowed with the ability to reason. Even though there is more to intelligence than reasoning, it is a crucial aspect of any understanding of human intelligence. Logic is the intellectual discipline which studies the nature of reason and ways of improving our use of reason.

As we know, there are two main forms of reasoning: deductive reasoning and inductive reasoning. The former can be contrasted with the latter in terms of necessity versus probability. A conclusion is reached through a process of reasoning called deductive reasoning, which is the logical technique in which particular conclusions are drawn from more general principles. It is concerned with which beliefs are licensed or entailed by other beliefs by necessity. In inductive reasoning a conclusion is usually expressed implicitly or explicitly in terms of a probability statement. This form is concerned with certain beliefs supporting or being supported by other beliefs. Francis Bacon proposed induction as the logic of scientific discovery and deduction as the logic of argumentation. In fact, both processes regularly are used together in the empirical sciences. By the observation of particular events (induction) and from already known principles (deduction), new hypothetical principles are formulated and laws induced. Generally, people employ both inductive and deductive reasoning to arrive at beliefs; but the same argument that is inductively strong or powerful may be deductively invalid.

An appealing feature of using logical reasoning in cognitive research is that it makes it possible to evaluate, or validate, the correctness of the thinking process on the basis of its forms rather than its content. Particularly, a valid proof clearly demonstrates that whenever the premises of an argument are true, its conclusion is also true. A typical case might be:

All A are B
All B are C
Therefore, all A are C.

From the cognitive view, we need to understand the differences between valid and invalid reasoning. What are the valid principles of reasoning? What makes a principle of reasoning valid or invalid? As models of valid reasoning, formal systems have important uses within mathematical logic and computer science. Johnson-Laird (1991), who developed important models of human cognition and logic, have argued that postulating more mental models makes for better predictions about the way people actually reason. When people reason, they must go "beyond the information given". They attempt to deduce the consequences or implications of a rule, set of premises, or statements using warrants by logic or by information that is either given in the problem or assumed to be true within the discourse. Clearly, reasoning abilities are not static. They are developed through experience and rendered easier to perform through exercise.

The ability to reason in a rational way with incomplete or inconsistent information is a major challenge, and its significance is obvious. Much of human reasoning is approximate in nature. In this perspective, non-classical formalisms are much better suited than classical logic for handling tasks with the uncertainty of human cognition. Fuzzy logic, model logic and default logic are all very useful and fruitful in this area. Some [4] propose a novel approach to quantum logic which is different from von Neumann's traditional approach.[11] The logic incorporates probabilistic reasoning in order to deal with uncertainty on the outcome of measurements and dynamic reasoning to cope with the evolution of quantum systems. It has the advantage of making quantum logic an extension of classical logic.

The divide between logic and human reasoning is enshrined in Frege's famous doctrine of "Anti - Psychologism", which claims that human reasoning practice can never tell us what is a correct conclusion. Do the facts matter? Is logic not experimental? Indeed, anti – psychologism is still defended to-day with a great deal of dogmatic fervor, but experimental facts about human reasoning are coming to light these days. The boundary is delicate. Logic approaches human reasoning with its purposes of its own, while a logical theory is useless if it was totally disjoint from actual reasoning of people.

3 Logic and Decision Making

Decision making is the process of choosing a preferred option or course of action from among a set of alternatives. People make decisions all the time, knowingly or unknowingly. The decision-making process often has three stages: information-gathering, information processing (thinking), and the act of choosing. It is clear that thinking is the central stage in decision-making. Good reasoning can lead to success, while bad reasoning can lead to catastrophe. There are many reasoning forms in decision making: inductive reasoning, deductive reasoning, conditional reasoning, etc. The most well-known conditional argument form is referred to as the *modus ponens* (MP). It has the structure:

$P \rightarrow Q$
P
$\therefore Q$

It is valid whether we want it or not, but where is the necessary opposition to practice? More importantly, if our observed practice diverges from some logical norm, what does that mean? This leaves rich space for cognitive science research. A barrier thesis like Frege's Anti-Psychologism may have worked in keeping logic at a safe distance from other communities.

The art of decision making is shared by such disciplines as mathematics and economics, artificial intelligence, psychology and philosophy. The study of decision making addresses two accounts: the normative analysis and the descriptive analysis. The former is concerned with the nature of rationality and the logic of decision making; the latter, in contrast, is concerned with people's beliefs and preferences as they are, not as they should be. Experimental evidence though indicates that people's choices are often at odds with the normative assumptions of rational theory. In light of this, some scientists focus on methods of improving decision making, bringing it more in line with normative routine.

Most decisions are made under conditions of uncertainty. In such decision contexts, the decision maker has to consider both the desirability of the potential outcomes and their probability of occurrence. Reasoning under uncertainty is a key issue in many areas. From a logical point of view, uncertainty basically concerns formulas that can be either true or false, but their truth value is unknown due to incompleteness of the available information. Traditionally, decision making under uncertainty (DMU) relies on a probabilistic framework. This proposal was made by economists in the 1950's and justified on an axiomatic basis by Savage. Bayes's theorem is a very helpful model that provides a method for evaluating changing probabilistic values.

Expected utility theory has dominated the analysis of decision making under risk. It has been generally accepted as a normative model of rational choice. It assumes that all reasonable people would wish to obey the axioms of the theory, and that most people actually do, most of the time. The assumption of rationality means that the people make their choice to maximize something of value, whatever that something may be. But, the findings of Tversky and Kahneman as well as studies of syllogistic reasoning, suggest that human beings are less than perfectly rational creatures. For most decisions, there is no one perfect option that will be selected by all people. Then, how can we predict the optimal decision for a particular person? Simon [7] showed the condition of "bounded rationality" for humans in making decisions. He suggested the term "satisficing" to describe the first minimally acceptable option. Daniel Kahneman [1] also argued for the definition of rationality and present evidence, challenging the assumption in new ways.

Formal rules and mental models do not exhaust the domain of possible theories of reasoning. Yingrui Yang and Johnson-Laird [14] suggest a new theory. Some psychologists have argued that human reasoners never make logical mistakes. Others have argued that apparent errors can be explained as a rational performance of a different task from the one the experimenter intended. Their view is that people are rational in principle but err in practice-for example, by the ability of the participants to recognize their errors once they are explained to them. Theories of reasoning need to account for logical competence and explain which errors occur and why they occur. The model theory bases rationality on the principle of validity-that is, the ability to grasp that a counterexample to an inference refutes it. This accounts for error in terms of the limitations of working memory, which imply that reasoners are unlikely to cope with fully explicit models. A sensible compromise is for reasoners to take into account what is true but to treat what is false as though it did not exist. Truth is more useful than falsity, but if reasoners rely only on the truth, they will occasionally succumb to the illusion that they grasp a set of possibilities that is in fact beyond them.

It is instructive to compare logic with rationality. They have several properties in common. Both characterize part of human cognitive nature; both are formally rigorous; and both are proved to be limited. Given these similarities, we might find that logic is the best synonym for rationality in this sense.

4 Summary and perspective

The nature of human cognition is embodied. In recent years, real – life links with actual experiments have also started emerging. This trend towards a more realistic modeling of information in reasoning. Moreover, since abstract theory influences actual behaviour, the interface between logic and the empirical cognitive science today appears in many ways, and is much more diverse than the traditional distinction ever allowed us to see. In my view, logical theory already follows practice ¬¬¬¬¬¬ – when the reasoning is running, we can go and read off the results.

As I have demonstrated above, Logic has a profound impact on HOC research. Different higher level mental processes can be transformed into logical reasoning. Although there is no logic that captures all complex entailments, many proposed systems formalize a lot of domains. After all, people tend to make decisions more rationally and precisely, not contrariwise. Logical theory, with its historical developments and further strengthening, will be the most effective approach to future HOC research. I view this as a broader agenda for logic.

Bibliography

[1] Kahneman, D., Tversky, A. (Eds.). *Choices, values, and frames*. Cambridge University Press, 2000.

[2] Levine, M. D. *Higher order cognition, Developmental Variations and Learning Disorders*. Cambridge, MA: Educators Publishing Service, 1999.

[3] Lohman, D. F. Reasoning abilities. In Sternberg, R. J., Pretz, J. E. (Eds.), *Cognition and intelligence: identifying the mechanisms of the mind*. Cambridge University Press, 2005.

[4] Mateus, P., Sernadas, A. Reasoning about quantum systems. In Alferes, J. J., Leite, J. (Eds.), *Logics in artificial intelligence, JELIA 2004 proceeding*. Springer, 2004.

[5] Necks, E., et al. Higher-order cognition and intelligence. In Sternberg, R. J., Pretz, J. E. (Eds.), *Cognition and intelligence: identifying the mechanisms of the mind*. Cambridge University Press, 2005.

[6] Nielsen, T. D., et al. Symbolic and quantitative approaches to reasoning with uncertainty. In *ECSQARU 2003 proceeding*. Springer, 2003.

[7] Simon, H. A. *Models of bounded rationality*. Cambridge: MIT Press, 1982.

[8] Solso, R. L. *Cognitive psychology*. Peking university press, 2005.

[9] Sternberg, Robert, J. *Cognitive psychology*. Thomson/Wadsworth, 2000.

[10] Van Benthem, J. *Logic and reasoning: do the facts matter?*, vol. 88. Studia Logica, 2008.

[11] Von Neumann, J., Oskar, M. *Theory of games and economic behavior.* New York: J. Wiley, 1944.

[12] Wilson, R. A., Frank, C. K. *The MIT encyclopedia of the cognitive sciences.* Shanghai Foreign Education Press, 2000.

[13] Yang, Y., Johnson-Laird, P. N. How to eliminate illusions in quantified reasoning. *Memory & Cognition*, 28(6):1050–1059, 2000.

[14] Yang, Y., Johnson-Laird, P. N. Illusions in quantified reasoning: How to make the impossible seem possible, and vice versa. *Memory & Cognition*, 28(3):452–465, 2000.

The Problem with the Deflationary Account of Self-consciousness[1]

Zhou Yuncheng

Tsinghua University

yczhou@tsinghua.edu.cn

In this paper, I examine the deflationary account of self-consciousness as formulated by Bermudez[1]. I claim that the reason why the deflationary account harbors the paradox of self-consciousness is that it identifies the wrong problem to account for–what to account for is not a how-question but a why-question. I also point out a full understanding of linguistic abilities requires an appropriate conception of language. I propose a robust conception of language to elaborate linguistic abilities that are necessitated for Bermudez's argument for non-conceptual self-consciousness. The upshot is that, with the robust conception of language, Bermudez's effort to dissolve the paradox of self-consciousness is disintegrated.

1 Deflationary Account of Self-Consciousness

The account of self-consciousness I examine in this talk is the deflationary account formulated by Bermudez[1]. It roughly goes as follows. In order to give an explanation of self-consciousness, we need to explain what it is to be capable of thinking I-thoughts, the thoughts that are immune to error through misidentification. The capacity to think thoughts that are immune to error through misidentification amounts to linguistic competence as to the use of the first person pronoun. Once we have established a theory to account for the use of the first person pronoun, we arrive at a solution to the problem of self-consciousness. So the account of self-consciousness is reduced to a theory of linguistic competence as to the use of the person pronoun.

According to [1], however, a deflationary account like this is unacceptable because it relies on circular reasoning – explanatory circularity and capacity circularity. Explanatory circularity arises from the phenomenon that the required theory of linguistic competence cannot be given unless an explanation of the capacity to think self-conscious thoughts is available, which is initially the very thing to account for. Capacity circularity surfaces when there is no foundational account to explain how a person acquires the capacity for self-conscious thoughts or that for mastery of the first person pronoun. These two forms of circularity constitute what Bermudez terms "the paradox of self-consciousness".

---

[1]The author is with Department of Foreign Languags, Tsinghua University, Beijing 100084 China.

2 How to Dissolve the Paradox of Self-Consciousness

Bermudez's strategy[1] for dissolving the paradox of self-consciousness is to try to keep the account of self-consciousness from being contaminated linguistically or conceptually. To illustrate, Bermudez shows us several cases in which some form of primitive self-consciousness is devoid from linguistic or conceptual elements. Included in these cases are pre-linguistic infants and some highly developed animals. He offers an argument that consists in rejection of the Conceptual Requirement Principle (CRP) and endorsement of the Priority Principle (PP).

> The Conceptual Requirement Principle (CRP): The range of contents that one may attribute to a creature is directly determined by the concepts that the creature possesses. [1, p. 41]
>
> The Priority Principle (PP): Conceptual abilities are constitutively linked with linguistic abilities in such a way that conceptual abilities cannot be possessed by nonlinguistic creatures. [1, p. 42]

It is claimed that in these cases pre-linguistic infants and some highly developed animals do not possess linguistic abilities but are capable of self-conscious thoughts or at least of self-awareness. According to PP, there would be no conceptual abilities without linguistic abilities. So the self-consciousness displayed in pre-linguistic infants and some highly developed animals is of a different form of self-consciousness, which is termed primitive or non-conceptual self-consciousness. The existence of non-conceptual self-consciousness suggests the invalidity of CRP and reveals the fact that self-consciousness seems to be a far more pervasive phenomenon that is shared by human beings and other highly developed animals. Once we have an account of non-conceptual self-consciousness, we can hope for the discovery of a genesis of self-consciousness. As a result, the paradox of self-consciousness is likely to be dismissed.

3 Bermudez's Failure to Dissolve the Paradox of Self-Consciousness

It is obvious that the above argument presupposes a common sense understanding of linguistic abilities. It is assumes for example, without explanation, that pre-linguistic infants and some highly developed animals are without linguistic abilities. This assumption seems to me to be problematic in that the implied conception of language is not clear enough for the required reasoning.

An appropriate theoretical conception of language is crucial to a full understanding of PP. As a matter of fact, the background story for the proposed PP is what Bermudez calls the Thought-Language Principle, which says, "The only way to analyze the capacity to think a particular range of thoughts is by analyzing the capacity for the canonical linguistic expression of those thoughts."[1, p. 13] Obviously, the canonical linguistic expression of I-thoughts is nothing but the first person pronoun. What is needed is to establish a semantic theory to account for the use of the first person pronoun. However, as Bermudez argued, it is not sufficient to have semantic rules which regulate solely the use of the first person pronoun because the semantics alone do not give us an adequate account of the

The Problem with the Deflationary Account of Self-consciousness 249

mastery of the first person pronoun. Bermudez suggests we provide pragmatics in addition to semantics in order to explain how it is possible for a person to be capable of using the fist person pronoun. The required account is what comes to be token-reflexive rules for the use of the first person pronoun. It is this very account, however, that harbors the problem of explanatory circularity. As seen from Bermudez [1, p. 14–16], a series of token-reflexive rules are given to account for the use of the first person pronoun, but no matter how they are revised to accommodate the theoretic need, the problem of explanatory circularity does not go away.

Why not? The diagnosis I have in mind is that the proposed token-reflexive rules appear to be a nonstarter to account for the use of the first person pronoun. We do not have a meta-language at our disposal to state the process of how it is possible for a person to be capable of the use of the first person pronoun. The conjecture of the so-called scientific language is of no use here. Decades ago, Quine taught us that translation into any known language would presuppose semantics of its own. There is no scientific language that is neutral to everything to be expressed.

So the problem is not that we choose a wrong meta-language to state the token-reflexive rules for the use of the first person pronoun. Rather we have identified the wrong problem to approach. Instead of providing an account of how it is possible for a person to be capable of the use of the first person pronoun, we need to consider why it is the case that a person could be capable of the use of the first person pronoun. With this task on the table, what we need is not a pragmatics but a meta-semantics along with semantics. According to Kaplan[4], semantics gives us the rules to associate expressions with semantic values, while meta-semantics tells us why the association of semantic values with expressions is possible. In other words, meta-semantics tells us what is the something that could make the association possible. We need a theoretical account to explain the contingent fact that an expression is associated with a semantic value of some appropriate sort. A good candidate for a model of such a theory would look like Chomsky's model for a theory of syntax.

The theoretic resources we would draw from Chomskyan generative project centers on the methodology of theorizing concerning the use of the first person pronoun. A syntactic analysis of the use of the first person pronoun is a functional account which aims to explicate what role the first person pronoun plays in a syntactic construction. Notice that when we talk about the first person pronoun we do not refer to the expression of the first person pronoun. Rather we want to isolate the first person element that plays the role that the first person pronoun is supposed to play in the utterance of a sentence.

We could even generalize the theorizing to any representational system. We argue that syntactic analysis of a representational system will reveal the functional role played by the first person element in the system. A system like this is claimed to be modularized but not encapsulated, a typical case of which is the computational system of human language. Chomsky called it I-language. But I propose to use a robust conception of language to characterize the representational properties in a system. The conception of language is robust because it provides a way to generalize linguistic abilities to the effect that everything is linguistic that could be syntactically or semantically computed.

I suppose that this robust conception of language is what is required of a correct understanding of linguistic abilities in PP and what is implied in the Thought-Language Principle. This is clearly seen in Bermudez's citation of Dummett [3]:

> We communicate thoughts by means of language because we have an implicit understanding of the workings of language, that is, of the principles governing the use of language; it is these principles, which relate to what is open to view in the employment of language, unaided by any supposed contact between mind and mind than via the medium of language, which endow our sentences with the sense that they carry. In order to analyze thought, therefore, it is necessary to make explicit those principles, regarding our use of language, which we already implicitly grasp.

What Dummett means by "the principles governing the use of language" is certainly not the specific semantic rules employed to regulate use of language. Although Dummett's conception of language is different from ours, what he has set in the agenda to analyze thought regarding the principles of how language works has something in common, in spirit, with our robust conception of language. That is, both aim to give a principled account to explain why language works as it does. This is further corroborated in the following passage from Dummett [2]:

> Here, an implicit grasp of certain general principles, naturally represented by axioms of the theory, has issued in a capacity to recognize, for each sentence in a large, perhaps infinite, range, whether of not it is well-formed, a capacity naturally represented as the tacit derivation of certain theorems of the theory. To each of these theorems corresponds a specific practical ability, i.e. the ability to recognize of a particular sentence whether it is well-formed or not; but this is not true of the axioms. A knowledge of certain axioms, taken together, issues in a general capacity, in this case to recognize of any sentence whether or not it is well-formed; and the ascription to the speaker of an implicit knowledge of those axioms is based on the confidence that he has a general capacity which embraces all the specific abilities which correspond to theorems derivable from that set of axioms.

The theory of language which Dummett was after is to construct an axiomatic system in which the general capacity of a language user guarantees any specific abilities he has for use of any sentence because those abilities, as theorems in the system, could be derived from the set of axioms which is embodied as the general capacity. So Dummett's axiomatic system could be assimilated to the working mechanism of human language, which is similar to our concern with the robust conception of language.

The robust conception of language has the implication that pre-linguistic infants and some highly developed animals, possessed of a representational system that involves the first person element, have linguistic abilities. If pre-linguistic infants and some highly developed animals are not without linguistic abilities in terms of the robust conception of language, then they are likely to have conceptual abilities according to PP. The representational content available to them is not exempt of the conceptual content. Thus, there seems to be no possibility for non-conceptual content to be all that is needed for representational content. Therefore, the rejection of CRP is implausible. If CRP cannot be rejected, Bermudez's effort to dissolve the paradox of self-consciousness is susceptible to failure.

4 The Key to the Problem of Deflationary Theory

A possible rebuttal is that there is no need to adopt the robust conception of language to characterize the linguistic abilities required in PP. Linguistic abilities might be explained with many other conceptions of language as well. The answer I have in mind is this. Whatever conception of language is to be employed to account for linguistic abilities, one question that cannot be evaded is what it is that language uses in the way it does. An account of how language is actually used does not seem to provide the answer to the question. What is needed is a principled account that explains why language works as it does. In other words, the theory that is needed is a meta-theory of linguistic use. Kaplan told us that the meta-theory of linguistic use can be done in two ways: the semantic account of linguistic use and the syntactic account of linguistic use[5]. When we do semantics on the level of meta-theory, we try to find out "the principles governing the use of language". When we do syntax on the level of meta-theory, we aim to give a functional explanation with syntactic analysis.

On the ground of the above theorizing, what justifies the belief that pre-linguistic infants and some highly developed animals do not have linguistic abilities? The answer does not lie in the description of behavioral performance because the description of behavioral performance does not give us a meta-theoretic account of the intrinsic properties of the performance system. Instead, we are supposed to explore why the pre-linguistic infants and some highly developed animals behave as they do with regard to self-consciousness. Any theory that could tackle this sort of why-question would take the form of either syntactic calculation or semantic calculation as long as what is to be figured out is a representational system. There seems to be no doubt that the performance system possessed by pre-linguistic infants and some highly developed animals is a representational system. If the working mechanism of a certain performance system contains syntactic or semantic calculation as its core configuration, then the performance system is said to be possessed of linguistic abilities.

It seems that linguistic abilities so characterized may be overreaching. In this sense, we even say a computer has linguistic abilities by virtue of its operative system. It is not wrong to attribute linguistic abilities to a computer, but what a computer has is derived linguistic abilities whose origin could be traced back to the person who designed the computer. The point I emphasize is this. Whether linguistic abilities or derived linguistic abilities, they constitute the intrinsic properties of a representational system. Our job is to isolate the intrinsic properties rather than to locate their origin.

Bibliography

[1] Bermudez, J. L. *The Paradox of Self-consciousness*. MIT Press, 1998.
[2] Dummett, M. What is a theory of meaning. In Evans, G., McDowell, J. (Eds.), *Truth and Meaning*. Clarendon Press, 1976.
[3] Dummett, M. *Truth and Other Enigmas*. London: Duckworth, 1978.
[4] Kaplan, D. *Themes from Kaplan*. Oxford University Press, 1989.
[5] Kaplan, D. Reading 'on denoting' on its centenary. *Mind*, 114:933–1003, 2005.

A New Image of Science[1]

Qiu Huili

Ordnance Engineering College

no email

ABSTRACT. The recent rise of interest in the history and perspective of science has a strong interdisciplinary flavor. This paper briefly reviews the current approaches to science studies. Then the paper lays out five mayor themes of situated and embodied cognition. According to these themes, the paper analyses the new image of science in terms of situated and embodied cognition.

1 Current approaches to science studies

Science is studied in very different ways by historians, philosophers, sociologists, and psychologists. Not only do researchers from different fields apply markedly different methods, they also tend to focus on apparently disparate aspects of science.

For several decades, the philosophy of science was dominated by the logical empiricist approach. The logical empiricists emphasized the logical structure of science rather than its psychological and historical development. In 1962, Kuhn published his influential *Structure of Scientific Revolutions* which championed a more historical image of science. Kuhn charged the logical empiricists with historical irrelevance. While philosophers of science have increasingly employed historical ideas, the historiography of science has taken a more sociological turn, paying increasing attention to the social context of science. Historians have thus found common cause with sociologists such as Barnes and Bloor. A school called sociology of scientific knowledge (SSK) developed in the early 1970s and resulted in a thoroughgoing sociological contextualization of science. It examined, for example, how internal scientific standards and experimental evidence fail to provide for scientists' beliefs and how the beliefs and knowledge claims of scientists are influenced by their social context. In the mid-1970s, another SSK approach emerged which was called "Laboratory Studies". Laboratory studies are studies of science and technology through direct observation and discourse analysis at the source where knowledge is produced, in modern science typically the scientific laboratory. The study of laboratories has brought to the fore the full spectrum of activities involved in the production of knowledge. It shows that scientific objects are not only "technically" manufactured in laboratories but also inextricably symbolically and politically construed.

---

[1]The author is with School of Humanities and Social Sciences, Ordnance Engineering College, Shijiazhuang 050003, China.

In this paper, I will outline a new image that views scientific communities from the perspective of embodied and situated cognition (ESC). Although Kuhn and Hanson did use psychological ideas, they did not fully employ cognitive ideas to explain science. Although a few isolated voices have been heard within ESC advocating the study of science, for the context of a field that is still largely unexplored in explaining science, I shall show that the ESC approach provides a helpful perspective on the investigation of science in multiply embedded and varied sites.

2 Five major themes in embodied and situated cognition

ECS has – as one of its most salient and attractive features – a strong interdisciplinary flavor. Contributors have come from backgrounds as diverse as robotics [2], cognitive anthropology [4], cognitive psychology [1], and developmental psychology [9].

First, the most central theme of ESC is that cognition is for action. Intelligence is shown not in detached thought but in adaptive behavior. According to ESC, the function of the mind is to guide action, and cognitive mechanisms such as perception and memory must be understood in terms of their ultimate contribution to situation-appropriate behavior.

Second, cognition is situated. One of the enduring slogans of ESC is the "power of the situation" over human behavior. It has been shown that the framing of cognition in terms of the body and its environment provides not only limiting but also enabling constraints for cognition [7].

Third, cognition is embodied. Cognition is now broadly understood as part of an overall functional and motivational system. In this light, the mind is no longer seen as passively reflective of the outside world, but rather as an active constructor of its own reality. In particular, cognition and bodily activity intertwine to a high degree.

Fourth, groups extend the cognitive powers of the individual. One of the most conceptually significant claims of situated cognition is the idea that people rely on the environment to facilitate and structure cognition of fact. We often directly manage the environment to aid cognitive tasks [5]. Thus on the one side, the physical environment can actively participate in cognitive processes by cueing, aiding, prioritizing, or otherwise structuring the processes. On the other side, cognition is shared and not limited to the individual. Communication processes allow shared representations to be constructed and cognitive effort to be distributed among individuals.

Fifth, situated cognition and symbolic thought constitute a dual-process model. In humans, situated cognition (based on implicit action-oriented representations) coexists with a more explicit, symbolic style of processing [8]. In this mode, we use language and symbolic structures effectively as tools.

3 A different image of science

Science provides arguably the best example of a higher cognitive activity. Scientific knowledge is a result of a series of cognitive activities that has been oriented by human beings. Therefore, we have good reason to explain science, a special kind of cognitive activity, from an ESC perspective.

3.1 Science is an action-oriented practice

According to the point of ESC, action has an important cognitive meaning in all kinds of cognitive activities. In this light, if we want to get an inspirational comprehension of science, we need to pay attention to the behaviors that relate to science, such as how does it work out and what kinds of operations happen in the cognitive process, how do researchers arrange their experiments, and how are conclusions about the observed object derived? Scholars in this field need to focus on the actions in the process of knowledge production and pay attention to how producers guide or adapt their actions in the local situation in order to get object-relative knowledge.

When we take action as an important factor in thinking about science we in fact give practice a philosophical meaning. Laboratory Studies show how the factors themselves are reconfigured, not as a result of the political strategies of specific agents but as the outgrowth of specific forms of practice. In fact, most of our knowledge is a working out of actions as an aspect of activities that have practical, not theoretical objectives; and it is this knowledge, itself an aspect of action, to which all reflective theory must refer.

3.2 Science is situated

ESC points out that situations and particularly contexts, including the relationship of the individual to working partners and interlocutors, are among the most important regulators of cognition in scientific research.

The laboratory is often the situation in which scientific research happens. In the laboratory, scientists drive out recognition of the object by observing and interpreting the results of their actions. Then they decide on new actions to be executed on the environment. This means that the researcher's conception may change according to what they "see", which itself is a function of what they have done. We may speak of a recursive process, an interaction of making and seeing. This interaction between the researcher and the environment strongly determines the course of how experiments are designed and conducted. That means, the subtlest of situational cues may directly influence these supposedly fundamental and automatic cognitive processes of the researcher. This is a typical feature of scientific research. It has led to the concept of "situated knowledge". This notion of situated knowledge offers a differently motivated understanding of scientific practice, allowing for a rather broad understanding of the human-thing relationship in which humans and things are intricately woven into one another.

3.3 Science is embodied

ESC tells us that the mind is no longer seen as passively reflective of the outside world, but rather as an active constructor of its own reality. In this sense, there is no such thing as the mind separate from the body. Thinking is not the product of some disembodied mind located somewhere outside the physical, but it is part of an active relationship between embodied humans and the world.

So we have good reason to say that all kinds of knowledge are embodied, and that knowledge is created in the unity between subjects and objects that is the direct result of having a body. A good example of this kind of embodied knowledge is an experiment operation. The plan and rules of the experiment, strategies and complex physical movements combine in ways that mean a researcher often has no time to think through how to play. They must feel the object in their bones. That is, the principles of the experiment are embodied, and the full meaning can only be expressed in actions. They propose that their physical experience of the world - their spatial awareness, their bodily movement, and the way they manipulate objects - provide the pattern for how they reason about the world.

3.4 Science is a kind of collective knowledge

Thinking in terms of ESC focuses on the process of acquiring information about objects rather than the static possession of knowledge. The physical environment, individual and communication processes all actively participate in cognitive processes. So we can say that knowledge is collected from individual, environment and communication processes.

In the laboratory, the cognitive task determining the location of the object is performed by a collective, an organized group, and, moreover, under the circumstances, could not physically be carried out by a single individual. In this sense, collective cognition is ubiquitous in science. In many laboratories there are some tasks that are clearly cognitive and, in the circumstances, could not be carried out by a single individual acting alone. Completing the task requires coordinated action by several different researchers and different artifacts. So ESC invites us to think differently about common situations. Rather than simply assuming that all cognition is restricted to individuals, we are invited to think of some actual cognition as being distributed among several individuals. The individual bits of knowledge are acquired and coordinated in a carefully organized system operating through real time. It is particularly enlightening to think of the whole facility as one big cognitive system which has to consider the human-machine interactions as well as the human-human interactions. So parts of the cognitive process take place not in anyone's head but in an instrument or on a chart. The cognitive process is distributed among humans and material artifacts.

3.5 Science is a kind of language-dependent thought

According to ESC, it is more enlightening to think of the scientist plus the external visual representation as an extended cognitive system that produces a judgment about the fit between an abstract model and the world. The visual representation is not merely an aid to human cognition; it is part of the system engaged in cognition. What scientists have inside their skins are representations of a few general principles together with bits and pieces of prototypical models. They also possess the skills necessary to use these internal representations to construct the required external representations that are then part of an extended cognitive system.

In fact, language itself is an elaborate external scaffold supporting not only communication but thinking as well. This distributed view of language implies that cognition is not only embodied, but also embedded in a society and in a historically developed culture. The scaffolding that supports language is a cultural product [3]. There is no "language of thought". Rather, thinking in language is a manifestation of a pattern matching brain trained on external linguistic structures.

4 Conclusion

Science must be understood as a concerned dwelling in the midst of the dynamic interactions between mind and world, things and hand, symbol and artifacts–intertwined to a high degree. It does not have the de-contextualized cognition of isolated things. For too long we have tried to build our understanding of science out of nouns, to construct a science around entities. Perhaps it is time to construct our understanding of science in verbs and around the dynamics of interactive processes.

Bibliography

[1] Barsalou, L. W. Language comprehension: archival memory or preparation for situated action? *Discourse Processes*, 28:61–80, 1999.

[2] Brooks, R. *Cambrian Intelligence*. Cambridge, MA: MIT Press, 1999.

[3] Clark, A. *Being There: Putting Brain, Body, and World Together Again*. Cambridge, Mass: The MIT Press, 1997.

[4] Hutchins, E. *Cognition in the wild*. Cambridge, MA: MIT Press, 1995.

[5] Kirsh, D. The intelligent use of space. *Artificial Intelligence*, 73:31–68, 1995.

[6] Norenzayan, A., Schwarz, N. Telling what they want to know: participants tailor causal attributions to researcher interests. *European Journal of Social Psychology*, 29:1011–1020, 1999.

[7] Semin, G. R. Agenda 2000: Communication: language as an implementational device for cognition. *European Journal of Social Psychology*, 20:595–612, 2000.

[8] Smith, E. R., Decoster, J. Dual process models in social and cognitive psychology: conceptual integration and links to underlying memory systems. *Personality and Social Psychology Review*, 4:108–131, 2000.

[9] Thelen, E., Smith, L. *A dynamic systems approach to the development of cognition and action*. Cambridge, MA: MIT Press, 1994.

ACT-R, a Computational Cognitive Architecture Model and its Methodological Reflection[1]

Qin Yulin
Zhejiang University
yulinq@yahoo.com

Cao Nanyan
Tsinghua University

ACT-R (Adaptive Control of Thought - Rational) is a theory and a computational model of human cognitive architecture. As a theory, it integrates cognitive psychology research in memory, learning, reasoning, problem solving, as well as perception and motion, and proposes systematic hypotheses on information processing in the functional components of human cognition and the interaction among them to generate human cognitive behavior. As a computational model, it offers a computer software platform to develop the models that can quantitatively simulate and predict human behavior (such as reaction time and accuracy) across a wide range of cognitive tasks. Combining with brain imaging research, it is growing toward a human cognitive architecture model with a neural basis and has been applied in education, industry, the military and medical research. After briefly introducing ACT-R, this article discusses the characteristics of ACT-R methodology, such as its modular structure and its integration of history as well as the difference between ACT-R and traditional information-processing theories.

1 What is ACT-R?

ACT-R (Adaptive Control of Thought - Rational) is a well-known theory of human cognitive architecture and computational model developed by John Anderson at Carnegie Mellon University and the ACT-R community across the world. In North America, there are currently eight laboratories (such as the Air Force Research Laboratory) and 23 universities with researchers involved in ACT-R. In Europe, the number of laboratories and universities is nine in Germany, seven in the United Kingdom, and six in other countries[2]. The publications in ACT-R community cover 48 issues, such as psychological refractory period, graphical user interface, skill acquisition, mathematical problem solving, emotion and motivation and so on, across six categories: ACT-R theory, Perception and Attention, Learning and Memory, Problem Solving and Decision Making, Language Processing, and Other (including fMRI study of the neural basis of ACT-R)[3].

---

[1] The authors are with Department of Psychology and Behavior Science, Zhejiang University, Hangzhou, Zhejiang 310028 China, and School of Humanities and Social Sciences, Tsinghua University, Beijing, 100084, China, respectively.
[2] From http://act-r.psy.cmu.edu/people/
[3] From http://act-r.psy.cmu.edu/publications/

ACT-R is an endeavor toward answering the ultimate scientific question–how can the human mind occur in the physical universe in an empirical way instead of as philosophical speculation [4]. It is an integrated psychological theory of mind [5]; it is an artificial intelligence model of human cognitive architecture; and it is also a software platform for developing computational models to quantitatively simulate and predict human behavior for a wide range of cognitive tasks. Therefore, it has various applications. It is evolving towards a system which can perform the full range of human cognitive tasks: capturing in great detail the way we perceive, think about, and act on the world, with the aspect concerns the relation between the function (cognition) and the physical structure (brain). This is the central issue of this article and will be discussed in detail.

As an example of the integrated theory of mind, Anderson et al. [6] developed an integrated theory of list memory. It shows the strength of the ACT-R theory is that it offers a completely specified processing architecture that serves to integrate many existing models in the literature, such as backward and forward serial recall, set size effects in the Sternberg paradigm, length-strength effects in recognition memory, the Tulving-Wiseman function, serial position, length and practice effects in free recall, and lexical priming in implicit memory paradigms.

As a software platform, ACT-R simplifies the processes of developing a computational model for simulating/predicting human performance of certain cognitive task to the process of listing the declarative and procedural knowledge used in those tasks. As a result, for example, a graduate master's degree student could use ACT-R to develop a computational model to simulate a driver's behavior ([10], following Salvucci[14]). In China, more than half the drivers are new on the road, on average with only about a year of driving experience. By successfully simulating the differences between new drivers and experienced drivers, the model reveals the possible causes of a new drivers' poor performance in some critical periods and therefore offers ways for improving road safety training programs.

2 Modular structure of ACT-R

An essential assumption in ACT-R is that 'The human mind is what emerges from the actions of a number of largely independent cognitive modules integrated by a central control system'. It is an expression of the production of the functional needs of the human mind and the physical constraints of the human brain [4, p. 45].

There are three kinds of modules in ACT-R:

1. Perception modules (input), such as the visual module and the aural module;
2. Action modules (output), such as the manual module and the vocal module;
3. Inner modules, such as the goal module (keep problem solving goal and subgoals to control the process), the imaginal module (mental representation of problem state), the declarative module (for declarative knowledge), and the procedural module (for procedural knowledge).

The unit of declarative knowledge is the chunk, which is a piece of knowledge or facts such as 3+4=7. The unit of the procedural memory is a production rule, which is a condition-action pair, such as 'IF the goal is to add 3 with 4, the math fact 3+4=7 has been retrieved. The manual module is available. THEN press number key '7' and delete current goal'.

Each module, except for the procedural module, has its own buffer, which can hold one chunk of declarative knowledge. Most of the information processing within and across modules can be in parallel, but the cooperation of the modules can only be down through production rules: if the condition part of a production rule matches the current chunks in the buffers of the modules, this production rule will be fired and then the action part of the production will change the chunks in the buffers of some modules and the situation of some other modules. Then this new situation of modules and their buffers causes new production fires and so on to generate cognition in the brain. Only one product can be fired at a time in ACT-R. And only one chunk in a buffer forms the center cognitive bottle-neck.

The brain regions corresponding to these modules have been gradually identified by fMRI study. The left prefrontal region corresponds to declarative memory retrieval (For example, see Anderson et al. [9]).

3 Physical restrictions sharpen the architecture

In addition to the module structure, a good example of physical restriction sharpening the architecture in ACT-R is the principle that the mind is adapted by the environment. The 'R' in ACT-R reflects this kind of rational analysis base. Even though there are about 100 billions of neurons in the human brain, the capacity of our memory is still limited. But the mind has to meet various demands in our life and the explosion of information in our environment. However, as shown in Anderson and Schooler (1991), the repetition of information in our environment drops quickly. If we do not need a piece of information for a while, it is unlikely we will need it in the future. As a mirror to the environment, in ACT-R, if the declarative memory has not been needed for a while, it will be treated as a low probability that we will need it in the future. This is the base of the sub-symbolic level of declarative memory.

A third source of evidence of physical restriction sharpening the architecture in ACT-R is the effort to map modules in ACT-R to brain regions and using ACT-R models to predict the activity in these brain regions. ACT-R has been improved by this process and some further possible modifications have been discussed (Anderson and Qin, [8]).

4 Integration in ACT-R

John Anderson believes that the unique human ability is 'to acquire nearly arbitrary competences (e.g., driving a car, solving an algebraic equation) that were not anticipated in our evolutionary history' (see Rahman [13]) by 'organizing novel combinations of behavior' [4, p. 46]. As discussed previously, ACT-R is an integrated psychological theory of the integrated mind. The method which has been used to develop ACT-R itself is also based on this type of integration. For example, the modular structure is on symbolic lines, but the sub-symbolic level of declaration memory has strong inferences of connectionism. The history of ACT-R is also a history of integration, from a symbolic declarative model, HAM

(Human associative memory, Anderson and Bower [7]), to ACT* (Anderson [1]) by combining the symbolic level of procedural memory and the neural inspired sub-symbolic levels, to ACT-R 2.0 (Anderson [3]) by combining his own rational analysis (Anderson [2]), to modular structure in ACT-R 5.0 (Anderson, et al. [5]) by combining perception-motor modules of EPIC (Meyer and Kieras [11]), to the current version of ACT-R 6.0 with brain mapping of modules through fMRI study (Anderson [4]).

5 Other methodological reflections

From the methodological aspect, ACT-R uses the hypothesis-testing method to explore the mechanism of human cognitive behavior. Similar to the traditional information processing theories of human cognition, ACT-R maintains clear boundaries between the objective and subjective worlds. Knowledge is thought of as the internal representation of the outside world. However, there are clearly differences between ACT-R and the physical symbolic system proposed by Newell and Simon [12]. The physical symbolic system theories put emphasis on the general ways in which a system shows intelligence whether that system is a machine, a human being, or anything else. ACT-R emphasizes the uniqueness of human intelligence. It is the model for the cognitive architecture of the human brain.

ACT-R has a goal module to perform goal directed cognitive tasks, but there is no special way to represent human intention, emotion, or motivation. However, ACT-R 6.0 is an open system, and it offers an approach for adding new modules. Some researchers did try to add an intention module, an emotion module, or a motivation module to simulate affective cognition. Some other researchers want to adjust the value of related parameters or to simulate the individual differences among those parameters.

Bibliography

[1] Anderson, J. R. *The Architecture of Cognition*. Cambridge, MA: Harvard University Press, 1983.

[2] Anderson, J. R. *The AdaptiveCharacter of Thought*. Hillsdale, NJ: Erlbaum, 1990.

[3] Anderson, J. R. *Rules of the Mind*. Hillsdale, NJ: Erlbaum, 1993.

[4] Anderson, J. R. *How Can the Human Mind Occur in the Physical Universe?* New York: Oxford University Press, 2007.

[5] Anderson, J. R., Bothell, D., Byrne, M. D., Douglass, S., Lebiere, C., Qin, Y. An integrated theory of mind. *Psychological Review*, 111:1036–1060, 2004.

[6] Anderson, J. R., Bothell, D., Lebiere, C., Matessa, M. An integrated theory of list memory. *Journal of Memory and Language*, 38:341–380, 1998.

[7] Anderson, J. R., Bower, G. H. *Human Associative Memory*. Washington, DC: Winston & Sons, 1973.

[8] Anderson, J. R., Qin, Y. Using brain imaging to extract the structure of complex events at the rational time band. *Journal of Cognitive Neuroscience*, 32:1323–1348, 2008.

[9] Anderson, J. R., Qin, Y., Jung, K.-J., Carter, C. S. Information-processing modules and their relative modality specificity. *Cognitive Psychology*, 54(3):185–217, 2007.

[10] Cao, S. Modeling the car drivers' behavior and the effect of drivers' experience

with computational cognitive architecture ACT-R, 2007. Unpublished thesis for Master's degree, Department of Psychology and behavior science, Zhejiang University.

[11] Meyer, D. E., Kieras, D. E. A computational theory of executive cognitive pprecesses and multiple-task performance. part 1. basic mechanisms. *Psychological Review*, 104:2–65, 1997.

[12] Newell, A., Simon, H. A. Computer science as empirical inquiry: Symbols and search. *Communications of the ACM*, 19:113–126, 1976.

[13] Rahman, S. Celebrating a decade of tics. *TRENDS in Cognitive Science*, 11(4):137–139, 2007.

[14] Salvucci, D. D. Modeling driver behavior in a cognitive architecture. *Human Factors*, 48(2):362–380, 2006.

A New Scheme for Embodied Cognition: from Chaos to Order[1]

Ma Yongjun

Henan University

Mayongjun22@126.com

ABSTRACT. This paper introduces the "order out of chaos" scheme of Ilya Prigogine, a type of self-organizing world model for exploring cognitive processes. At present, new research approaches focus largely on embodied cognition. Researchers consider generally that embodied cognition can only form in interaction with the environment, but the cognitive process is worthy of further exploration. With Ilya Prigogine's inspiration, cognition can be considered an 'order out of chaos' self-organization process. When the system communicates nonlinearly with the outside, maybe it is gradually far from an equilibrium state. In the non-equilibrium state, random micro-fluctuation probably expands rapidly into macro-fluctuation, which leads to the emergence of self-organization. As Ilya Prigogine says, 'order through fluctuation'. The scheme is not inconsistent with the basic characteristics of current cognitive thinking. Therefore, I do not seek to be different by introducing the structure but I seek rather to further explore the ways in which to adapt 'order out of chaos' to understand the cognitive process and open a path on the basis of embodied cognitive science.

1 Move towards Embodied Cognition

The classical cognitive view of the mind conceptualizes cognitive functions in terms of a computer metaphor and views the human brain as a formal system for operating and processing symbols, whatever cognition and intelligent activities can be calculated on the basis of the Turing algorithm. Although this research program is still prevalent, it faces many insurmountable difficulties on the theoretical and operational levels.

Embodied cognition emphasizes interaction with the environment, which is an unprecedented transformation in research approach. Such a transformation is the significance of Kuhn's paradigm, which changed the computer metaphor into an interaction metaphor, the isolationist analysis into relational analysis, the symbolic or encoded representations into sensory-motor representations etc. Embodied cognition studies have led the science of cognition to a new process and formed gradually a number of general methods and concepts. R. Brooks summarized four key concepts about embodied cognition: (1) "situate-edness"; (2) "embodiment"; (3) "intelligence"; and (4) "emergent" [1, 133–186]. All of

---

[1]The author is with College of Philosophy and Public Administration, Henan University, 85 Minglun Street, Kaifeng, Henan, China.

these conceptions stress that cognition systems must communicate with the environment. They also indicate that the cognitive system is an open system. So embodied cognition science coincides unconsciously with the new paradigm of the physical world in its own development process, which is Ilya Prigogine's "order out of chaos" scheme which I am just about to introduce in this paper.

2 Discussion of the Embodied Cognitive System

These features of embodied cognition can only form in interaction with the environment in the cognitive process, but the cognitive process is worthy of further exploration. As far as I know, the nonlinear cosmic scheme can provide new approaches for the exploration of cognitive science.

2.1 Summary of the "order out of chaos" complex system

The Open System

According to dissipative structure theory, an isolated system will descend from order to disorder with the increase of entropy. As for an open system, it can exchange material and energy with the environment and obtain negentropy to counteract its own increasing entropy. An open system, therefore, can evolve from chaos to order. All systems treated in synergetics may be considered open systems and thus fulfill a necessary condition for self-organization [3, p. 49].

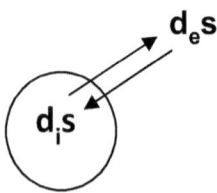

Figure 1: The exchange of entropy between the outside and the inside[4]

Far from Equilibrium Systems

When systems are far from equilibrium systems, stability is no longer the consequence of the general laws of physics. We must examine first the way a stationary state reacts to different types of fluctuations produced by the system or its environment. In some cases, the analysis leads to the conclusion that a state is "unstable"–i.e. in such a state that certain fluctuations, instead of regressing, may be amplified and invade the entire system, compelling it to evolve toward a new regime that may be qualitatively quite different from the stationary state corresponding to minimum entropy production. [5, p. 140]

Fluctuation

Fluctuation means that the system deviates from the stable state. If a system is in a state of equilibrium, random fluctuation is called micro- fluctuation. In systems far from an equilibrium state, random fluctuation is called macro-fluctuation. As for far from an equilibrium system, random micro-fluctuation

A New Scheme for Embodied Cognition ...

may magnify rapidly, which can drive a system to make a transition from an unstable state into a new state of order.

Self-organization

Order is really produced "spontaneously" from disorder in a self-organization process when an open system exchanges materials, energy and information with the outside. The process itself is a state of disorder and non-equilibrium. In this state, the fluctuations in one or a small number of factors may determine the changes for the whole system. In other words, self-organization is a way of forming order out of chaos.

Ilya Prigogine's Cosmic Scheme

It has been shown that demonstrations of impossibility, whether in relativity, quantum mechanics, or thermodynamics, have shown that nature cannot be described "from the outside" as if by a spectator. Description is a dialogue, a communication in which the subject is constrained and demonstrates that we are macroscopic beings embedded in a physical world. [5, p. 299–300]

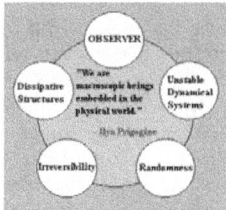

Figure 2: Macroscopic Beings Embedded in a Physical World

2.2 Choice of Embodied Cognitive System

Nonlinear Outcome of Nonlinear Universe Evolution

Ilya Prigogine's world scheme has enlightened cognitive research. The Prigogine paradigm has produced a revolutionary change in the way we view the world. According to Prigogine, the classical physics model only happens in a limited number of occasions which may be deterministic and reversible; however, the course of nature contains many primary elements which are random and irreversible. There are, for example, a great of nonlinear systems in the world. HUman beings place themselves in a complicated material universe. As macroscopic beings embedded in a physical world, it is consonant with organic evolutionism/Darwinism that human beings form and evolve their cognitive ability in response to the constant and dynamic changes going on in the world. Human beings gradually form their cognitive ability by interaction with their external environment through millions of years. The human brain is the outcome of that long process of adaptation to the universe. According to organic evolution, it is impossible that the nonlinear, non-equilibrium, complicated universe would create a linear, mechanical brain. The human brain system is the nonlinear outcome of a nonlinear process of evolution in the universe.

The Neurophysiology of the "Chaotic" Brain

Recently, the theory of chaos and research into neural networks has been linked to each other. The research indicates that the output figure that we get when we combine chaos and the model of the human neural network from an electroencephalogram resemble each other extremely closely. The output graph which a chaotic neural network describes is similar to an EEG (electroencephalogram). The 'chaos' of brain waves has its base in neurophysiology. The nerve cells produce physiological brain wave activity. There are over ten billion neurons in the human brain. Neurons have a highly nonlinear organization with connections to each other through synapses. The linkages among nerve cells forms an enormous and complex neural network which is the center of information processing. There are massive numbers of neurons (10^{11}) and even more neural connections (10^{12}) in the human brain. They produce signals of brain waves which display chaotic behaviors. Modern brain science has viewed the working process of brain as a dynamic system which is complex, multilayered, and chaotic [2, p. 35]. At present, many labs engage in this kind of study. Although modern science research is still primarily focused on revealing superficial brain functions, it is beginning to lay the scientific foundations for the study of the nonlinear brain.

The 'Order out of Chaos' Scheme of Embodied Cognition

Ilya Prigogine argues that the open system relies on material flow, energy flow and information flow from the outside world. So the cognitive system needs to communicate constantly with the outside. This kind of communication is not a linear or changeless process. Visual cognition at different times is not a linear response by changeless neurons to an identical object shape. Rather it represents the random synergetic activity of billions of neurons. The same individual sees an ambiguous figure; the subject has a different response at different times. Sometimes the subject sees a duck and sometimes a rabbit. Similarly, different individuals see pictures in dissimilar ways. Simple visual cognition is indeed simple, but abstract cognition is more complex and chaotic than visual cognition. It can be characterized as nonlinear, random and emergent. According to the dynamics mechanism of 'chaos', the brain does not produce linear cognition as the information it has access to increases. Apparently, the system becomes more and more chaotic as the amount of information available increases. The system communicates in a nonlinear mode with the outside. As its communications with the external world intensify, gradually the system verges far from a state of equilibrium. In a state of non-equilibrium, random micro-fluctuation probably expands rapidly into macro-fluctuation. As macro-fluctuation kicks in, which a situation develops where self-organization spontaneously begins to take place–as Prigogine says, 'order though fluctuation' [5, p. 177–189].

3 Conclusion

In my opinion, the cognitive system is, in fact, consistent with such an open non-equilibrium model of self-organization. Cognition is such a self-organizing process, whose characteristics are a state of non-equilibrium, randomness, from chaos to order. In fact, the 'order out of chaos' cognitive scheme that I introduced is not inconsistent with the basic characteristics of cognitive science (embodiment, situated-ness, emergent, etc.). These features are coincident because the characteristics of embodiment cognition can be displayed only in such a system. Therefore, it does not seek to be different by introducing structure, but it further explores ways of realization for the cognitive process in a non-linear mode.

Bibliography

[1] Brooks, R. *Cambrian Intelligence: The Early History of the New AI*. Cambridge, MA: MIT Press, 1999.
[2] Fei, D.-y. The form work of creation studies. *Dialectics of Nature*, 22(10):35, 2006.
[3] Hermann, H. *Principles of Brain Functioning*. Berlin Heidelberg: Springer-verlag, 1996.
[4] Prigogine, I. Structure and fluctuations. *Time*, 1977. URL http://nobelprize.org/nobel_prizes/chemistry/laureates/1977/prigogine-lecture.pdf.
[5] Prigogine, I., Stengers, I. *Order Out of Chaos*. Bantam Books Inc, 1984.

Toward a Naturalistic Account of Self-Knowledge[1]

Jerry J. Yang

National Taipei University of Technology

jyang@ntut.edu.tw

ABSTRACT. I challenge people's belief about each of us enjoys "privileged access" to one's own inner states from a Naturalist perspective. I first clarify in what sense we enjoy such "privileged access." I further prove that the recognition of such special access does not entail the kind of epistemic value that many people think one's self-knowledge is entitled to.

Since Descartes' Meditations, it has been widely believed that the essence of one's self-knowledge gained via introspection is different from that of others' knowledge toward one's inner states through their empirical observation. The essential difference between the two kinds of knowledge does not lie only in the epistemic routes by way of which each kind of knowledge is acquired, but also in their epistemic values, meaning self-knowledge enjoys a superior epistemic status as opposed to other types of knowledge given that one's self-knowledge is only open to oneself while other-knowledge can be publicly shared. People thus believe that each of us enjoys "privileged access" to one's own inner states, an access not shared by others. Such special access warrants the superior status of one's self-knowledge as far as its epistemic value is concerned. I challenge the above-mentioned belief from a Naturalist perspective. I first clarify in what sense we enjoy such "privileged access." I further prove that the recognition of such special access does not entail the kind of epistemic value that many people think one's self-knowledge is entitled to.

One does know the content of one's own mental states by looking into, i.e. by introspecting, one's mind. Granted, this epistemic route by way of which one obtains the knowledge about one's inner states is only open to oneself. By contrast, to understand how one feels other people have to rely on empirical observation such as seeing. The term "privileged access" hence refers to the direct and immediate access in terms of which one finds out about what's going on in one's mind. Although people have different views about the connotation of this concept[2], there is still an agreement among them. They all agree that the essence of one's introspective self-knowledge is different from that of other-knowledge, i.e. other peoples knowledge about one's own mental states. The difference is not only exhibited in the cognitive routes by way of which each kind

---

[1] The author is with Ellery Eells Memorial Center for the Philosophy of Science & Professional Ethics, General Education Center, National Taipei University of Technology, No.1, Sec. 3, Chung-Hsiao E. Rd., Taipei, 10608, Taiwan.

[2] The most classic paper on this topic arguably is written by William Alston in [1]. People who are interested in the topic might find the following book useful in [4]

of knowledge is obtained; it is also shown in the degrees of certainty each kind of knowledge is endowed with.[3] For people tend to believe that one's grasp of the content of one's mental states enjoys an epistemic privilege, i.e. a higher degree of certainty than others' grasp of the very same content.

There is no doubt that any inferential knowledge based on empirical observation is easily subject to error, and hence it carries lower degree of certainty. My knowledge about my own mental states, however, must carry higher degree of certainty as I am the only person who posses such non-inferential justification for my knowledge about my inner world. The phrase "epistemic asymmetry" therefore refers to the type of warrant I have for my introspective self-knowledge which is unavailable to others. And further, we may conclude that it is because of such "epistemic asymmetry" that results in the epistemic privilege, i.e. a higher degree of certainty one's introspective self-knowledge is endowed with. To challenge the aforementioned view we should first inquire how one achieves one's introspected self-knowledge. Upon reflection, we find that introspected self-knowledge is constituted by one's consciousness or awareness of one's mental states. [4] We now follow Rosenthal's distinction by dividing mental states into two groups, i.e. phenomenal states on one hand, and intentional states on the other in [7]. Phenomenal states refer to the type of mental states that exhibit phenomenal properties or characters. Sometimes we say they refer to whatever appear in one's mind when one exercises one's senses such as sensations, feelings, mental images, etc. By contrast, intentional states refer to the type of mental states with intentional properties or characters. Sometimes we say they refer to the kind of mental states that involve or concern objects or states of affairs in the world such as beliefs, desires, thoughts, etc. We thus call one's consciousness of one's phenomenal states, one's phenomenal consciousness; and, one's consciousness of one's intentional states, one's intentional consciousness.

It is not that easy to distinguish the difference between phenomenal consciousness and intentional consciousness. A possible difference is that, at least some, if not all, of the contents of phenomenal consciousness are not representational while all the contents of intentional consciousness are representational.[5] It becomes obvious what the concept of self-knowledge with epistemic privilege refers to once we make out well the distinction of the above-mentioned two kinds of self-consciousness. Self-knowledge with epistemic privilege must refer to the kind of self-knowledge which is based on phenomenal consciousness. I call this kind of self-knowledge phenomenal self-knowledge. On the other hand, self-knowledge based on intentional consciousness does not own any epistemic privilege. I call it intentional self-knowledge. Granted, a lot of our beliefs or concepts are prone to errors, most likely as a result of our misjudgment about things, and hence self-knowledge concerning intentionality is subject to mistakes. By contrast, no one would understand better than I do toward my inner feelings or sensations inside my body even though a third person might have a good grasp of those

---

[3] A helpful discussion can be found in McKinsey [6]. Reprinted in [5].

[4] In this article, "consciousness" and "awareness" is used interchangeably. Accordingly, to say "one is conscious of ..." is equivalent of saying "one is aware that..."

[5] I have to admit that the point I made here could be controversial. For some might think that all of the sensations have representational content, and indeed some of the phenomenal contents are also representational. But I do not intend to get involved in the dispute of the topic in the literature here. I think my point exhibited here is useful enough for the sake of my argument.

feelings or sensations. Thus, *prima facie*, phenomenal self-knowledge is the kind of self-knowledge that is entitled to epistemic privilege. If so, to reassess the soundness of the claim of epistemic privilege we have to first disclose the process one's phenomenal consciousness is formed while placing an emphasis on one's sensations and feelings.

One's consciousness of one's phenomenal states is a sort of "information-based" perception. Precisely speaking, one's phenomenal consciousness is a sort of self-awareness based on sensory information. The process of attaining such self-awareness is a process of building an "informational link" between oneself and the phenomenal state one is in. This has been confirmed by the reports and discoveries in neuroscience. Accordingly, certain sensory information must be presented to oneself when one is conscious of the phenomenal state one is currently in. The information in question is provided by the sensory part of our nervous system, which is in charge of detecting and analyzing stimuli from the external and internal environment of the organism. The reason that the sensory system is able to provide sensory information is the fact that there are many cells, named receptors, which are located at the interface between the nervous system and the outside world, such as one's skin, that can respond to sensory stimuli. These receptors would send signals to the nerve cells in the spinal cord before the signals are transmitted to the cerebellum via the ascending pathways in the sensory system. After the cerebellum, known to be responsible for higher cognitive functions, receives the signals, the brain then issues commands to trigger one's actions. [6] The term "informational link" hence refers to the neuron-physiological fact that if one is aware of being in a certain state, one must receive certain information from that state. Whenever one is aware of oneself in being aware of the state one is in, certain sensory information about oneself is presented to him. This informational link is built thanks to the cooperation of several "parties" which include at least the sensory receptors, the ascending pathways, and the nervous system in the spinal cord and brain. The details of such cooperation were already revealed by the neuron-physiological basis of ones self-awareness, of one's phenomenal states.

However, if we agree that there must be an informational link between oneself and the state one is in, then it is reasonable to claim that the activity of being aware of oneself is a matter of receiving the sensory information of one's state. Being aware of one's moving hands in space is to receive the kinesthetic information about the movement of one's limbs. Being aware of one's crossed legs is to receive the proprioceptive information about the position of one's legs. The activity of being aware of oneself, thus, can be reduced to an activity of receiving (sensory) information from one's sensory system. One reacts or responds accordingly after such sensory information is registered in one's consciousness. The journey of the information starts with the motion of sensory receptors and ends with the command from the brain via the transmission of the neural system in the spinal cord. Given that all the transmission of the information or signals occurs within one's body through an inside route in which no second person can intervene, the sensory information about the state one is in that one receives is hence private, i.e. no second person can receive this information better than one does because no one other than oneself can share the transmission of such information. I call the kind of self-awareness "proprietary self-awareness" (or

---

[6]The reports and discoveries from neuroscience can be seen in [2] and [8]

"proprietary self-consciousness") and the kind of self-knowledge based on such self-awareness or self-consciousness "proprietary self-knowledge."

One may object to the view that the sensory information is private by pointing out that it is not causally impossible for one's nerves be wired over to my brain so that the information of one's phenomenal states is delivered to my brain. In that case, one might continue to argue, the information one has about one's phenomenal states is no longer private because I can be aware of one's phenomenal states in the same way that one would have been if one had not been rewired. Granted, the information that travels through pathways internal to one's body, as in one's sensations, could be sent via other pathways into the brain and body of another person such as mine, which is so far as is known not causally impossible.

To the objection just made, I agree that it is not only conceivable, but causally possible that two people's brains can be wired to each other in this way. I agree that implies the information going through one person's mind can be shared by another, i.e. the information is a shared message and hence is not private any more. But we have to be clear about exactly what concept of causal possibility is involved. Whether it is causally possible for something to happen depends on two factors: the laws of nature, and the antecedent conditions. First, it depends on whether its happening is consistent with the laws of nature (or more narrowly, the causal laws of nature). These laws may be expressed by conditional propositions, e.g. of the form "if certain antecedent conditions hold, then a certain event will occur". The second factor, determining whether something is causally possible, is what the relevant antecedent conditions are. It appears that the challenge was made with the following idea in mind. That is to say, the possibility of the imagined rewiring of nerves is logically consistent with any causal laws including the neuron-physiological principles or any electrochemical laws that might be relevant to the transmission of information along the pathways. If that is the case, then we are speaking of possibility in this sense:

(1) It is causally possible that P, if P is logically consistent with all the causal laws.

If the challenge posed by the possibility of rewiring nerves is taken to be causally possible in this sense, then it is not a logical feature of this information that it is private. The claim that the information one receives about the state one is in is private, thus seems to be in jeopardy; and, therefore we cannot interpret the concept of proprietary self-awareness in this way because the information one has about the state one is in, i.e. one's self-awareness is not proprietary in that it is in this sense causally possible for it to be had by another person. That is not the kind of causal possibility I have in mind when I claim that no second person can intervene in the transmission of sensory information from one's own body.

When I pointed out that, other than oneself, no person can share whatever information one receives regarding the state one is in, the claim is made based on the way that our nervous system and the transmission pathways are actually arranged. Scenarios that are merely logically possible are not relevant to this claim, and neither are those which are causally possible in the sense just considered. Hence, the kind of possibility I have in mind is relative to the actual facts of a situation. To speak of possibility in this sense means:

(2) It is causally possible that P, if P is actually consistent with all the causal laws and with the actual conditions that determine whether the antecedents of those conditionals expressing such laws do or do not hold.

If someone in Taiwan claims "it is causally possible for watermelons to grow in January", he is not using the term "causally possible" in sense (2), for given the actual January climate of Taiwan, it is not causally possible for watermelons to grow. It is not causally possible for watermelons to grow because the antecedent conditions for them to grow (e.g. climate, temperature, and soil condition, etc.), do not obtain. One instead is using the term "causally possible" in its primary sense, (1), i.e. that it's happening is not logically inconsistent with any causal law. The climate might possibly change in the future so that January is warm. The statement about watermelons is true if the term "causally possible" is used in the sense (1), for it would not be difficult for watermelons to grow if the antecedent climatic conditions were other than they actually are.

It is clear that it is not a logical truth about watermelons that they do not grow in January in Taiwan; it is only a contingent truth about watermelons that they cannot grow in January in Taiwan. Similarly, it is not a logical truth about our sensory information that it is not shared; it is only a contingent truth about the actual transmission of our sensory information. Given the way our bodies are actually arranged, there are no actual antecedent conditions that allow for a second person to intervene in the process of transmission and in turn share our sensory information. Hence, the challenge involving the possibility of rewiring one's nerve can only prevent us from claiming that the private information one receives from one's sensory system is a logical truth about one's self-consciousness of one's phenomenal states; it does not prevent us from using the concept of private information as the de facto feature of one's self-consciousness. Therefore, it is justified to claim that the information one has about the phenomenal state one is in is de facto private, i.e. one's self-consciousness or self-awareness is de facto private. Therefore, we can still call one's self-consciousness or self-awareness de facto private or "proprietary"; and one's self-knowledge based on such self-consciousness or self-awareness de facto private or "proprietary." I now conclude that the de facto private information we have about the states we are in is all that is required for our self-awareness to be "proprietary."

I have recognized the fact that the information we receive about the state we are in is de facto private. Upon reflection, we can see that the status of the information in question being private is guaranteed by the private transmission route of such information. That is, the reason that our sensory information is private is because the transmission route is private. I believe that it is the private transmission route that explains why we enjoy the privileged access to the state we are in. The widespread belief among philosophers is that, as noted in the very beginning, the epistemic asymmetry that one has for the mental state one is in, which is unavailable to others, results in the epistemic privilege that one's introspective self-knowledge is entitled to. The epistemic asymmetry as we have pointed out is warranted by the privileged access one has toward one's own mind. Nevertheless, it might be generally true that one can know which state one is in with more certainty than others can, given that no second person can intercept the signals by cutting in during the process of the transmission. But the fact that we have private access to our own states, which no outsiders have, does not entail that we have better access to our own states, that we know them better, or with greater reliability, than people other than ourselves. In other words, the private access we have to our own states does not award us the kind of epistemic privilege or advantage over others that some people thought we would have, enabling us to know which state we are in with more certainty or reliability than others. In conclusion, it might be contingently true that one's knowledge toward the content of one's phenomena states is more certain than others' knowledge toward the very same content; it nevertheless is not necessarily true for one to claim that one's introspected self-knowledge of one's phenomenal states enjoys a higher degree of certainty that other-knowledge is not entitled to.

Bibliography

[1] Alston, W. Varieties of privileged access. *The American Philosophical Quarterly*, 8(3), 1971.

[2] Beatty, J. *The Human Brain: Essentials of Behavioral Neuroscience*. Sage Publications, Inc, 2001.

[3] Descartes, R. Meditations on first philosophy. In Cottingham, J., Stoothoff, R., Murdoch, D. (Eds.), *In The Philophical Writings of Descartes*, vol. II. Cambridge University Press, 1993.

[4] Gertler, B. (Ed.). *Privileged Access: Philosophical Account of Self-Knowledge*. Ashgate, 2003.

[5] Ludlow, P., Martin, N. (Eds.). *Externalism and Self-Knowledge*. CSLI Publications, 1998.

[6] McKinsey, M. Anti-individualism and privileged access. *Analysis*, 51:9–16, 1991.

[7] Rosenthal, D. Two concepts of consciousness. *Philosophical Studies*, 94:3, 1986.

[8] Tyldesley, B., Grieve, J. I. *Muscles, Nerves and Movement: Kinesiology in daily living*. Blackwell Science Ltd, 1996.

The Ontogenesis of Language: Evidence from the Early Grammatical Development of Chinese[1]

Yang Xiaolu

Tsinghua University

xlyang@tsinghua.edu.cn

1 Introduction

Language development constitutes an important area in cognitive science, intriguing not only linguists and psychologists but attracting also the attention of philosophers interested in epistemology, i.e. where knowledge comes from and how it develops. It is believed that findings concerning language development shed light on the issue of the origin and development of human knowledge.

The study of language development has long been driven by the nature-nurture debate, which is deeply rooted in the dichotomy between nativism and empiricism in philosophy since it was first outlined by Plato and Aristotle. The centre of the debate concerning language origins and development focuses on whether early grammatical development is rule-governed or experience-driven. For generative nativists like Chomsky (1981) and others [1, 7, 12], children's grammar is rule-governed and, from early on, young children have access to abstract syntactic categories as part of an innate linguistic system of principles in the form of a Universal Grammar, which includes both lexical categories such as nouns and verbs and functional categories. The empiricist view, on the other hand, argues that early grammatical development is usage-based and data-driven. There is no endowed linguistic knowledge at birth and early sentences are fairly simple and are free of abstract syntactic categories [4, 6, 8–11].

The present study looks at the nature of early syntax by looking at the status of verbs in the language of two Chinese-speaking children under the age of two. In the next section we will briefly introduce the Verb Island Hypothesis that assumes no abstract knowledge of verbs as a general syntactic category. We will then present our findings that challenge the Verb Island Hypothesis. We will discuss our findings and draw some conclusions in the last section.

2 The Verb Island Hypothesis

Early verbs, according to Tomasello [8], develop along different paths. Verbs operate as "individual islands of organization." [8, p. 257]. Young children do

---

[1]The author is with Department of Foreign Languages and Literatures, Tsinghua University.

not yet possess verbs as a general abstract grammatical category, but rather they pick up verbs individually one by one. So, young children become productive in the use of verbs in a limited and incremental way and newly learned structures seldom 'transfer across verbs.' Early multi-word combinations are simple, mostly derived on the basis of general cognitive processes such as symbolic integration ([8]; see also the term 'structure combining' in [9]).

Following [5], [8] suggests that early verbs should not form abstract, coherent categories if they do not appear as the arguments of other predicates. Verbs that appear after modals verbs or in serial verb constructions, for instance, are arguments of modals and other predicates. Verbs with a present or past tense morphology are also signs of verbs as arguments of higher predicates. Only when children develop the use of verbs in this way can one say that verbs are possessed as a syntactic category.

One question that immediately arises now is whether the Verb Island Hypothesis accounts for early verbs in Chinese, a language that is well-known for its lack of overt verb morphology. If what Tomasello and his colleagues suggested is universally correct, we would expect similar results in the L1 acquisition of verbs in Chinese, i.e. the early use of verbs would be simple and limited, with different verbs behaving in different ways.

3 The present study

3.1 Methodology

In the present study, we used transcripts of two Chinese-speaking children, CY and ZHZ, from the CELA (Chinese Early Language Acquisition) project. Both kids were born and now live in Beijing. The two kids were visited weekly or biweekly for one hour each visit over an extended period of time until they were two year olds. CY was observed from 00;10 and ZHZ from 01;04.[2] We collected 53 one-hour audio-video recordings from CY and 24 one-hour audio-video recordings from ZHZ, which were digitized and transcribed.

We extracted and analyzed multi-word sentences containing verbs from CY's and ZHZ's transcripts. If verbs in early Chinese follow the same developmental pattern as that reported for the L1 acquisition of English verbs, we would expect that: a) there is no sign of CY's and ZHZ's early verbs as a general and coherent word class; b) their sentences containing verbs in the observed periods should be relatively simple, with no or few cases where verbs occur as arguments of higher predicates; c) their early verbs should be limited in their use, i.e. occur mostly in limited sentence patterns.

3.2 Findings

Altogether CY produced 880 multi-word combinations containing verbs and ZHZ 693 verbs. During the observed period, one argument sentences and two argument sentences were common in the two kids' transcripts. They were quite productive in expressing AGENT and THEME. Over 25% of CY's early sentences contained

---

[2]Figures like 00;10 indicate the age of the child. 01;04;11, for instance, stands for one year, four months, and eleven days.

AGENT and 17% of ZHZ's did so. All AGENTs appeared in the subject position. THEME appeared in about 50% of CY's sentences and 40% of ZHZ's sentences.

Both kids tended to be very sensitive to the word order of Chinese. There were very few cases of incorrect positioning of agents and themes at this stage. They consistently put the agent in the subject position of their one-argument and two argument sentences, as already mentioned. Themes mostly occurred in the post-verbal position, as arguments of transitive verbs, mostly action verbs with a few exceptions of psychological verbs like *xihuan* ('like') or verbs of existence like *you* ('have'). But there were also cases when the subject of the sentence was a theme. In such cases, verbs were either unaccusatives or ergatives. The position of themes thus invites one to think that CY and ZHZ were aware of different verb types.

To answer whether verbs occur as arguments of higher predicates, we checked whether our kids' verbs co-occurred with functional categories such as negators, modals, aspect markers, focus adverbs and whether there were sentences with two verbs. Our analysis shows that both the kids used negators and focus adverbs in some sentences and these words were almost all placed correctly before verbs. Besides, both produced serial verb constructions, particularly CY. About one third of the verbs in CY's data entered into the serial verb construction. In addition to serial verb constructions, our kids' early verbs also feature a complexity in the VP-internal structure, as evident in their V-V compounds, V-reduplication, and V-*yi-xia* structures.[3]

It is also found that the co-occurrence of verbs with functional categories such as negators and focus adverbs was not restricted to a small number of verbs: with a limited number of verbs: 39 out of 135 verbs in CY's transcripts co-occurred with negators and 35 out of 162 verbs for ZHZ. When they started using the negator (i.e. the 1;5-1;7 period), they already used it with different verbs (with 6 different verbs for CY and with 7 different verbs for ZHZ) and the number of co-occurrences increased as they became older. This is also the case with V-V compounds and V-reduplication: before 2, V-V compounds and V-reduplication were not uncommon. Around 15% of different verbs appeared as the second verb in the compounds for both kids and a similar number of verbs also appeared in V-reduplications.

Though English-speaking children are reported to be quite limited in the use of almost all their early verbs [4, 8], i.e. using them only in one sentence pattern, the picture we get with our Chinese-speaking children appears to be a different one. We have found that early verbs in the two Chinese-speaking children's production were not limited to only one sentence pattern.[4] CY was particularly productive in her early use of verbs. Except for the period before 1;5, the number of her verbs that entered into two or more sentence patterns was similar to or even more than that of verbs employing one sentence pattern (22 vs 31 in the 1;5-1;7 period, 38 vs. 30 in the 1;8-1;9 period and 62 vs. 41 in the 1;10-1;11 period). ZHZ's data also indicate that there was no lack of verbs that entered into two or more sentence patterns (38 in the 1;8-19 period and 57 in the 1;10-1;11 period)

---

[3]V-V compounding is a very important and very productive process in Chinese Chinese. Its complexity lies in the interaction of semantics and syntax and RVCs (Resultative Verb Compounds) particularly raise a lot of significant issues concerning the argument structure of the verbs involved and as a composite (Li 1990, Cheng and Huang 1994).

[4]Our 'sentence pattern' is defined as any change in argument types, functional categories, internal VP structure, etc.

before 2.

4 Conclusion

In the present study, we analyzed multi-word combinations containing verbs before 2 by two Chinese-speaking children, CY and ZHZ. Contrary to Tomasello's [8] findings that his English-speaking subject's early verbs are characterized by "concreteness, particularity, and idiosyncracy" (p.264), our data demonstrate some generality and systematicity in young Chinese-speaking children's use of early verbs. We have seen that quite a number of different verbs appeared as arguments of higher predicates: negators and focus adverbs consistently appeared pre-verbally. Many different verbs appeared in serial verb constructions and in V-V compounds. It was also common for the two kids to use verbs in two or more patterns during the observed period. We would take such findings as evidence that early verbs in Chinese do form a general and coherent syntactic category and that early child Chinese is quite productive and complex.

Our findings are consistent with findings of some recent studies of the acquisition of Chinese, such as [2, 3, 13–15]. These studies all provide evidence against the experience-driven account of early syntax. Though the evidence is far from conclusive, we believe that cross-linguistic research of early grammar will shed light on the ontogenesis of language, which will eventually inform us of the ontogenesis of human knowledge.

Bibliography

[1] Crain, S. Language acquisition in the absence of experience. *Behavioral and Brain Sciences*, 14:597–650, 1991.

[2] Ji, S., Yang, X. Children's acquisition of the nominal de in mandarin. Zhengzho: the International Conference on Forms and Functions in Chinese, 2006. April 21-23.

[3] Lee, T. Productivity in the early word combinations of cantonese-speaking children. In Shi, F., Shen, Z. (Eds.), *The Joy of Research – A Festschrift in Honor of Professor William S.Y. Wang on His Seventieth Birthday*. Tianjin: Nankai University Press, 2004.

[4] Lieven, E., Pine, J., Baldwin, G. Lexically-based learning and early grammatical development. *Journal of Child Language*, 24:187–219, 1997.

[5] Nino, A. On formal grammaticall categories in early child language. In Levy, Y., Schlesinger, I., Braine, M. (Eds.), *Categories and Processes in Language Acquisition*. Hillsdale, NJ: Erlbaum, 1988.

[6] Pine, J., Lieven, E., Rowland, G. Comparing different models of the development of the english verb category. *Linguistics*, 36:4–40, 1998.

[7] Pinker, S. *Language Learnability and Language Development*. Cambridge, Mass: Harvard University Press, 1984.

[8] Tomasello, M. *First Verbs: A Case Study of Early Lexical Development*. Cambridge: CUP, 1992.

[9] Tomasello, M. Do young children have adult syntactic competence? *Cognition*, 74:209–253, 2000.

[10] Tomasello, M. The item-based nature of children's early syntactic development. *Trends in Cognitive Sciences*, 4(4):153–163, 2000.

[11] Tomasello, M. *Constructing a Language: a Usage-based Theory of Language Acquisition.* Harvard University Press: Harvard University Press, 2003.

[12] Wexler, K. Very early parameter-setting and the unique checking constraint: a new explanation of the optional infinitive stage. *Lingua*, 106:23–79, 1998.

[13] Xiao, L., Cai, X., Lee, T. The development of the verb category and verb argument structures in mandarin-speaking children before two years of age. Tokyo, Japan: the Tokyo Conference on Psycholinguistic, 2006. March 17–18.

[14] Yang, X., Xiao, D. The ba construction in early syntactic development of chinese: A case study. In *The Monterey Institute of International Studies.* California, U.S.A: NACCL17, 2005. June 24-26.

[15] Zhang, J., Fang, L., Lee, T. and. Yang, X. The acquisition of sentence final particles in beijing chinese. Leiden University: IACL13, 2005. June 9-11.

The Philosophy of Language in the Pre-Qin Era[1]

Zhou Jianshe

Capital Normal University

zhoujs@mail.cnu.edu.cn

ABSTRACT. The study of pronunciation, characters, vocabulary, syntax, and rhetoric is the task of linguistics. The traditional study of language describes the phenomena of languages, which produces descriptive linguistics. If someone wants to know the deep meaning of a language phenomenon, they attempt to find the essentiality that lies in the back of language, it's result is a linguistic theory called the philosophy of language. The philosophy of language is the study of philosophical problems of the language adopting the method of philosophy. The philosophy of language in the Pre-Qin era has been divided into many schools, including Confucianism, Mojia, Mingjia, Fajia, Daoism, Zajia, Yinyang, and so on. Most thought on the philosophy of language was recorded in books written by Mojia and Mingjia. Many basic projects or concepts in the philosophy of language have been studied. They have been classified as referring theories, classifying theories, pragmatic theories, and interpretation theories.

The study of the philosophy of language in western countries is well known, but it has just started in China. So we should first try to understand the concept of the philosophy of language. Language is a symbolic system built by pronunciation, vocabulary and syntax. Pronunciation, vocabulary and syntax are three basic elements of a language system. Language use is called speech. Characters are symbols in which speech is written. Rhetoric shapes speech to produce the desired effects on auditors. The study of pronunciation, characters, vocabulary, syntax, and rhetoric is the task of linguistics.

There are many ways that people have used the regularities of language to improve the quality of human life which have resulted in different research approaches to the study of language. The traditional study of language describes the phenomena of languages, which produces descriptive linguistics. If someone wants to know the deep meaning of a language phenomenon, they attempt to find the essentiality that lies in the *back* of language. The result is a linguistic theory called the philosophy of language. Describing or asking about the phenomena of language sometimes has little ambiguity. Some people try to formalize language and then find relationships between sentence forms that are abstracted from natural language. The result of this is a formal logic or formal linguistics. People hope information communicates at great speed. The regularities are become a technology, which changes the language laws into a series of principles that can be understood by computers. This research produces computational linguistics. In spite of appearing multi-directional, the field of language research does not

---

[1] The author is with Capital Normal University, Beijing 100037 China.

intend to express whether a language is better or worse. All of them have a common purpose, to support people to recognize the regularities of language for the purpose of making communication more efficient.

There has not been equal progress in all the multiple directions of research in the field of language study. Some areas are still weak, such as philosophical, formal and technical studies, especially philosophical research in the Chinese language is less well developed than descriptive research. In fact, strengthening research efforts in the weak areas is of great importance. If and only if the main fields of language study have been developed can the regularities of language use be apprehended.

The philosophy of language is the study of philosophical problems of the language adopting the method of philosophy. The philosophical problems are basic problems which lie at the root of the language phenomenon. For example, why is the pronunciation of "a" louder than "u"? This is not a philosophical problem. It is a simple phenomenon of language. Why does a word have meaning? This is a philosophical problem which entails a cognitive law about language. The philosophical method is a method which is used to recognize the essentials of language rationally. The philosophical method usually operates based according to logical reason. In other words, the philosophy of language is the study of rational knowledge of philosophical problems which lie in the background of the language phenomenon.

The sciences are not distinguished by nation and time. The names of nations or times are often given on behalf of the sciences just to explain that science is studied in special regions or times. According to this classification, people know where and when achievements in science come from. The philosophy of language in the Pre-Qin era is not an independent philosophy of language. The expression represents only that it belongs to the same science called the philosophy of language. Many ideas and theories have their origins in the Pre-Qin era of China's history.

Pre-Qin thought has been divided into many schools, including Confucianism, Mojia, Mingjia, Fajia, Daoism, Zajia, Yinyang, and so on. Most thought on the philosophy of language was recorded in books written by Mojia and Mingjia. Many basic projects or concepts in the philosophy of language have been studied. They have been classified as referring theories, classifying theories, pragmatic theories, and interpretation theories.

The referring thoughts in the philosophy of language from Pre-Qin times are as follows–language symbols have functions referring to realities. The ancient people say there are names that name things, and all things can be given names. People refer to things by using symbols; things get a relationship to people by symbols. That is to say, all things are referred to or referenced. There is a relationship between the thing, the name and the reference. Everything exists in the world and is an object named or referred to by language symbols. Referring is a process in which people refer to things, as well as the kind of names that are used during the referring process. The referring name and referring process do not exist. The Pre-Qin philosophers called this name and process *nothing*. The referring name is a signifier; the subject related to the signifier is the referent. The signifier may partly mean the referring process, but it does not have the features of the referent, so that the philosophers in the Pre-Qin Dynasty don't think that the signifier is the referent. The Pre-Qin philosophers believe that names do not have necessary relationships with things. The names of things are established by usage and convention.

There are certai principles in the naming of a thing. The name must correspond to the thing; the same kind of things must be named with similar names; if a simple name signifies a thing, it is not allowed to use a complex name. The contents of names are not congenital, which stem from senses about realities. People can check the relationships between the content of names and the things that are referred to by the names. There are two errors which people make which make the relationships inconsistent: First, there is a real name, which does not correspond to the content; second, there is a fallacious name which is used to refer to a real thing.

Philosophers in the Pre-Qin Dynasty found many characters for names. Collecting the same names that have similar characters together in one class, they classified various kinds of categories for names. From the perspective of governing nations, they divided the names into four parts: the names of criminal law, the titles of nobility, the names of culture, and the other names, which are called the governing names. In terms of the qualities of things, the names are divided into three kinds: the names of things, the honor and disgrace names, and the good and evil names. According to the shapes of named things, the names were divided into object-names and unobject-names. The object-names refer to concrete things that are easy to be apprehended with the senses. The unobject-names are the names that refer to abstract objects. From the relationships between names and their content, the philosophers divided names into two situations–different names refer to the same content or the same names refer to different realities. In other words, one name has multiple objects; multiple names have only one-object. From the extension of concepts, the Pre-Qin philosophers thought that there were three kinds of names: the general name, the class name, and the individual name. The general name refers to common things and is called a category in the field of philosophy. "Thing" is an example for the name of a category, which refers to all realities; there is no higher class above it. The class name refers to a group of things which have common characteristics. The individual name refers to one object that does not include other elements in it. From the perspective of the features of Chinese characters, names can be divided into the pointing thing's name, the hieroglyph name, the shape-sound name, and the understanding name. On the features of groups of things, the types of names include the people's name, the names of buildings, the name of tools, and so on.

Classifying names is important to enhance the recognition of things through language. The Philosophers in Pre-Qin times attach great importance to the roles of names. Their enthusiasm is greater than that of philosophers in other periods because they thought names had a tight relationship with the control of nations. Confucius' thought is representative in this respect. Confucius says, if a name is not used correctly, then speech is also not standard; people will not be successful in their efforts to do anything based on informal speech. If their speech is not formal and correct, they will fail to create the nation system; if the nation does not have a system, people will not know what they should do. Yin Wen, Deng Xi, Gongsun Long have similar ideas. They claim that one of the responsibilities that super governors have is to check whether a name is consistent with its content or not. They even think this is the criterion for deciding the governor's abilities.

Using the symbols of language should obey three principles. First is the principle of consistency claimed by Xun Kuang. He divided the elements of speech into three parts: the Dao, namely the natures, the heart, the content, and the speech, namely the judgments and reasons. The principle of consistency means that the content of names or speeches are consistent to the natures that were referred by the names or speeches. Meanwhile, the contents of literal names or speeches are inconsistent to what one said. The second is the clear *principle*. Xun Kung has explained it as, when one hears a name, one should understand its referent. In other words, the contents of the name should be clear. If it is, the name is useful to communication. The third regards one's reference. Gongsun Long says that if one thing is not here, then one cannot say that it is here, and if one thing is not there, then one cannot say that it is there.

The Pre-Qin philosophers found a series of tactics or prescriptions for speech acts. The speaker should speak something in terms of the situation of audience. If the audience has deep thoughts, the speaker should talk about *extensive* knowledge with him so as to spread his field of vision. If the audience is slow-witted, the speaker should help the audience to know what is right or sensible. The second tactic is that speeches should be appropriate to the audience's ability and customs. Because names often have a content which includes ethics and customs, this is why a son cannot be called *dog or thief*. The third tactic of speech acts is that the structure of sentences should be accustomed to grammar. The Pre-Qin philosophers noticed that there are sentences which have the same structure. Some of the sentences are not permitted to be spoken. For example, *there is a thief*; the sentence *the thief* is a person who is allowed to say, however, if *there are many thieves*, the sentence *there is many persons* will not be allowed to say. Because the two sentences are not consistent in the aspects of their semantic structures although the latter has same grammatical structure as the first sentence. The final tactic is that the Pre-Qin Dynasty philosophers advised that the speaker should avoid saying sentences which are not necessary because some sorts of sentences offend the audience, and even produce dangers.

The theories of interpretations in the Pre-Qin era deal with many aspects of knowledge. The interpretations of the meaning sources are based on materialism. Xun Kuang thought that hearts were able to think things; the sources of things thought by the heart were from sense. Gongsun Long affirmed that meanings are gained by sense. He says, when one sees a stone, he can grasp the white color of the stone from his visual sense; touching the stone, he feels the solidity of the stone from his tactile sensation, so the color and solidity of the thing are separated.

The interpretations of semantic structure try to grasp the meanings by analyzing internal relationships of terms. The philosophers in Pre-Qin Dynasty find that names are composed of different elements, which have particular characteristics of semantic. Yin Wen gives a term "*good ox*" to explain that the name of quality can widely restrict the name of matter, for example, the "good" restricts the "ox".

Analyzing qualities of things is an important method of interpretation. The Pre-Qin philosophers believed that the key to this method is to find the qualities of things. For example, according to the number of things, horse and cattle all have four feet, so they belong to the same kind of animal, that is to say, horse and cattle are from one class or category—"one". On the basis of other conditions, we should say that the horse and cattle are "two," namely; they are two sorts of animals. The symbol as a whole does not/should not decompose or else the meanings of words will be sink into ambiguity.

The theoretical interpretations of definitions and enumerations are not given, but there are many used examples in Pre-Qin times. Yin Wen says, the names are symbols referring to things. This is a definition of the function of the name. The purpose of definition is to reveal the quality of things. The goal of enumerations is to understand the quantities of things. Tai Gong says, a general has five abilities: bravery, wisdom, humanity, faith, and honesty. This is an enumeration, from which people can understand the field of the discussion. There is another method in which the definition and the enumeration are often used at the same time.

> Equal proofs reveal one equal relationship between one symbol and the other with mathematical precision and method. This way is used by Gongsun Long to prove the proposition "WHITE HORSES NO HORSES" (Bai ma fei ma). This thought is similar to the thought of the great philosopher of language Gottlieb Frege.

Bibliography

[1] Carnap, R. *Meaning and Necessity*. University of Chicago Press, 2nd edn., 1956.
[2] Davidson, D. *Inquiries into Truth and Interpretation*. Oxford University Press, 1982.
[3] Martinich, A. P. *The Philosophy of Language*. Oxford University Press, 2001.
[4] McCawley, J. P. *Everything that Linguists Have Always Want to Know about Logic, but ashamed to ask*. Oxford: Blackwell, 1981.
[5] Miller, A. *Philosophy of Language*. University College London Press, 1998.
[6] Palmer, H. *Words and Terms*. Cambridge University Press, 1986.
[7] Parkinson, G. H. R. (Ed.). *The Theory of Meaning*. Oxford University Press, 1968.
[8] Parkinson, G. H. R. (Ed.). *An Encyclopedia of Philosophy*. Routledge, 1988.

Context and Relevance[1]

Wang Lihui

Beijing Normal University

warnleehui@sina.com.cn

ABSTRACT. Human knowledge is contextually related. Epistemological research often evades the problems of context and relevance, but in order to deal with commonsense knowledge, artificial intelligence has to address the issue of the context of human language. Therefore, both epistemological research and cognitive science inevitably involve the study of epistemic relevance. Relevance, a central problem in philosophy, logic and linguistics, occupies a very important position in current epistemological study. With special reference to the technical treatment of contextual formalization in artificial intelligence, this paper attempts to illustrate the importance of context and relevance in modern cognitive science.

1 Two different theories of context

Context is a notion with many definitions. The famous definition we all know is Frege's "context principle" in philosophy: "An expression has a meaning only in the context of a sentence". Another well-known sociolinguist, Bronislaw Malinowski distinguished the "context of situation" and the "context of culture". His argument is that the meaning of any single word is to a very high degree dependent on its context, and an utterance has no meaning at all if it is out of its context of situation. In this paper, we concur with the viewpoints of Frege and Malinowski; however, we do not have a definition for context which is not subject to doubt. Before we look at our research in AI, we should look at two different theories: the object theory of context, where a context is a set of features of the world, and the subjective theory of context, where a context is a speaker's or a agent's cognitive background with respect to a situation.[10]

In semantics, context is a set of the features of the world — "Context is a package of whatever parameters are needed to determine the referent of the directly referential expressions."[6] Here it means that every parameter in a sentence has an interpretation as a natural feature of the world. When we try to understand a sentence we actually try to understand the meaning of the world which was expressed by words. But often we don't consider context of our feelings. In this way, the context is a set of assumptions about the world. In 1993, John McCarthy showed in Notes on Formalizing Context: "...context is a group of assertions closed (under entailment) about which something can be said." In the same year, another AI researcher Giunchiglia said: "a theory of the world

---

[1]The author is with College of Philosophy and Sociology, Beijing Normal University, Beijing 100875 China.

which encodes an individual's perspective about it"[4]. So we have two very different interpretations of context: context is a feature of the world (semantics) or a representation of features of the world (AI). Apparently, the second theory is wider than that of the former: it is concerned with any feature of the world, not just with the limited set devised by Kaplan.

In this paper, we will discuss the definition of context from the perspective of AI and point out that we may need a third definition of context. The third definition may be philosophical. It may be expressed in the form of a question–is that context reality?

2 Is Relevance Represented?

AI has two goals, the ultimate goal and the technical goal. The ultimate goal is to make machines think and act like humans. The technical goal is to make machines becoming valid tools for stimulating and modeling human thinking. "The problem of relevance" ("frame problem") in AI is closely related to the ultimate goal of AI. To make machines think and act like human means, computers have to deal with things (including unexpected things) in the same way a person does. In order to make computer systems more intelligent to users, computers have to be more sensitive to the context–in other words, the computers' actions may be dependent on such contextual factors as time, place, or the history of interaction.

The issue of relevance and its meaning seems to be intuitively clear. Broadly speaking, a piece of information is said to be relevant if it is of consequence to the matter in hand or if it interferes with some of our beliefs or actions. Thus, we can describe the "problem of relevance" as the problem of identifying and using properly all the information that should exert an influence on our beliefs, goals, or plans.

Like context, relevance is also a general question in science and philosophy. In the following subsections, we review some of the formal treatments of relevance in AI. Their goal is to explicitly represent relevance by logic in human cognition. For example, a smile could be understood either as reassuring or as a cautionary, depending on the context in which it is made. It is not the case that one first recognizes a smile and then interprets it as either reassuring or cautionary. The CYC Project and McCarthy are all working on this.

The CYC Project is a large plan to solve questions around the issues of knowledge representation. In this project, scientists look at context as an agent. In 1991, Lenet and Feigenbaum maintained consistency at the cost of context, and Haugeland developed the following expression:

1. An instance of I, in context C, would be (or count as) an R.
2. Here is an instance of I; and it is in context C.
3. So, here is an R.[5]

To Haugeland, the conclusion R is not really recognition at all, as is usually the case in human cognition.

In spite of this reasoning, the CYC Project endlessly adds entries to limit context. The sequent outcome is that the knowledge deposit becomes bigger and bigger until the machine can not load it. We all know, in the world facts often happen before we become aware of them. This is the true feather of world and contexts are always same. In other words, the facts (or phenomenon) and the context are joint, not separate entities. As such, the context determines what a smile means as much as the smile defines and reinforces the context. But in AI or computer science, there is no valid technique for satisfying this requirement.

Formalizing context. John McCarthy has done a lot to formalize the context. He uses *Ist and Value. Ist* (c, p) proposotion p is true in the context c; Value(c, e)—-designating the value of term e in the context c.

He gives a famous illustration that when we say a sentence, it may have a different indication.

ist
ist(context-of("U.S. legal history"),
"Holmes is a Supreme Court Justice"
ist (context-of("Sherlock Holmes stories")
"Holmes is a detective"
value
–value(context-of("Sherlock Holmes stories"),
"number of Holmes's wives=0
–value(context-of("U.S. legal history"),
"number of Holmes's wives)=1

This illustration is very simple and essential. It becomes more complicated when we enter an ideographic thing. Saying that the context is normal rich, McCarthy suggests that we can never exhaust the features of the context.

As limited agents we cannot reach truth per se or objectivity per se. As agents in a community we have the right and the duty to fight to achieve them. For that purpose, the most important thing we need is a study of the relations among contexts in order to more accurately represent the complex web of views. The real novelty of the logic of contextual reasoning in AI is the analysis of the different kinds of rules which regulate the relations among cognitive contexts.

AI research has begun to take seriously the role of context in recent years, but "relevance" remains a major challenge for AI research. Dewey used the notion of "situation" to also talk about relevance: "The existence of the problematic situation to be resolved exercises control over the selective discrimination of relevant and effective evidential qualities as means"[2]. Relevance, Dewey concluded, is not inherent but accrues to natural qualities by virtue of the special function they perform in inquiry. The impress of the individual inquirer on the context does determine relevance. Given that there is a close relation between context and relevance, what has been ignored in AI? We should seek the answer in a philosophical and scientific tradition of objectivism and subjectivism. As Dewey notes, there is bad and good subjectivity. Confounding them has been the source of major mistakes in science and philosophy.

3 Context as reality

Context and relevance are intertwined aspects of thought and intelligence. We cannot assume that we have reached the definitive representation of the structure of reality; there is no ultimate outer context. Logic is strictly constrained to "explicit" forms of knowledge representation. In other words, nothing could be taken for granted whether it is part of the context or pat of the subject matter. The traditional AI models including the CYC Project and McCarthy's work have a hard time dealing with context because every fact with a remote chance of relevance has to be codified and explicitly represented. But relevant facts are limitless.

In this paper, we don't carry out the explanation at the logic level, but we do take a general survey of context. We cannot avoid the phenomena that the axioms cannot always be valid in the world. We can always find a wider context where the axioms are not valid. In AI, this question confuses researchers, McCarthy brings forward a theoretical way to solve the question, but in my opinion, he cannot successfully finish his wok. In order to avoid being puzzled by context, we may use operations or rules among contexts, as McCarthy says:

-entering and exiting a context,

-discharging some sentence, true in some contexts, but false in a wider context,

-lifting some sentence true in some contexts into another context, verifying in this way different kinds of compatibility among contexts.
[8]

His saying is still a home truth. The meaning of a sentence depends on speaker, place and time. The value of these parameters (speaker, place and time) must be represented as part of the speaker's cognitive state and the observer's cognitive state.

We can summarize some features of context: 1. Context most often is not explicitly identifiable. 2. There are no sharp boundaries among contexts. 3. The logical aspects of thinking cannot be isolated from material considerations. 4. Behaviour and context are jointly recognizable. In dealing with these features, we educe that context in AI poses a major difficultly. John Dewey said, "We grasp the meaning of what is said in our language not because our appreciation of the context is unnecessary but because context is inescapably present."[2] Indeed, context is the reality in our world; we must consider context as a reality.

Bibliography

[1] Carlo, P. Objective and cognitive context. In Bonquet, P., et al (Eds.), *CONTEXT '99. LNAI 1688*, 270–283. 1999.

[2] Dewey, J. *Context and Thought.* 1960. Edited by Richard Bernstein.

[3] Ekbia, H. R., Maguitman, A. G. Context and revelance: A pragmatic approach. In Akman, V., et al (Eds.), *CONTEXT '2001. LNAI 2116*, 156–169. 2001.

[4] Giunchigia, F. Contextual reasoning. In Penco, C., Dalla, C. (Eds.), *Linguaggi e Macchine, special issue of Epistemologia*, vol. 16, 345–364. 1993.

[5] Haugeland, J. Pattern and being. In *Having Thought: Essays in the Metaphysics of Mind*, 1998. Cambridge: Harvard University Press, 1993.

[6] Kaplan, D. Afterthoughts. In Almog, J., Perry, J., Wettstein, H. (Eds.), *Themes from Kaplan.* Oxford: Oxford U., 1989.

[7] Lenet, D., Guha, R. *Building Large Knowledge-based Systems: Representation and Inference in the CYC Project.* Addison-Wesley, 1990.

[8] McCarthy, J. *Formalizing Commonsense: papers by John McCarthy.* Ablex Publishing Corporation, 1990.

[9] McCarthy, J. Notes on formalizing context. In *Proc IJCAI '93*, 555–560. 1993.
[10] Penco, C. *Three Alternatives on Context.* URL http://www.dif.unige.it/epi/hp/penco/.
[11] Vassallo, N. Contexts and philosophical problems of knowledge. In Akman, V., et al (Eds.), *CONTEXT '2001. LNAI 2116*, 353–366. 2001.

Pragmatics and the Frame Problem[1]

Wang Zhidong
Tsinghua University
wzd05@mails.tsinghua.edu.cn

ABSTRACT. Pragmatics provides an important foundation for establishing an innovative research direction and methodology for investigating the frame problem in AI. The basic theory of pragmatics can be helpful in understanding intelligence from the point of view that the nature of language consists in its usage. The AI frame problem is essentially a puzzle. Its action in real time shows no effect when an agent using formal arithmetic is embedded in a complex situation. The purpose of this paper is to share my research on how to liberate the frame problem from its bondage to logicism and to propose a new interpretation and approach to the problem based on pragmatics theory as the basis for intelligent action rather than on static arithmetic. I use pragmatics rules to construct and respond to complex cognitive situations which contain different language usages.

1 Pragmatics and Speech Act Theory

Since Wittgenstein proposed the idea that language and games bear a family resemblance, he made known the difference between a perspective based on language reference (semantic) and a perspective based on language usage (pragmatics). From a semantic perspective, the nature of language consists in its reference to the real world. The research into language reference opens out the whole meaning of language. Syntax and semantics all belong to the category of the reference perspective. In contrast, from a pragmatics perspective of usage, the nature of language inheres in its usage. The reference of language is a component only of language meaning. In the process of using language, context, speaker's mind, hearer's mind, time and place are important components of language meaning. Pragmatics then belongs to category of usage. The object of research not only includes reference but also includes the index word, context and the cognitive states of the speaker and the hearer.

Pragmatics theory provides an important foundation for research into how intelligent actions in real time adapt to the world situation. It can provide significant direction in understanding and solving the frame problem.

The frame problem in AI is essentially a puzzle. Its action in real time shows no effect when an agent is using formal arithmetic in a complex situation. This paper describes my research on how to liberate the frame problem from its bondage to logicism and proposes a new interpretation. In my new interpretation, I rely

---

[1] The author is with Institute of Science, Technology, Society, Tsinghua University, Beijing 100084 China.

on pragmatics theory as the basis for intelligent action rather than on static arithmetic. I use pragmatics rules to construct different complex cognitive situations with different language usages.

Speech acts theory was formulated by J.L.Austin and inherited and developed by J.R.Searle. It has had an important effect on linguistic philosophy and the philosophy of mind. The theory claims that 2saying something is doing something, and speech is an act2.Indeed, this not a behaviorist philosophy. In cognitive research, speech acts, as special acts, have more important foundation effects than physical actions. J.R.Searle argued that 2the minimal units of human communication are speech acts of a type called illocutionary acts 2 [13]. In general, an illocutionary act consists of an illocutionary force F and a propositional content P which has the form F (P) in which a propositional content P can be analyzed in terms of its syntax and semantics. Speech act theory pays close attention to analyzing the components and logical structure of illocutionary force.

2 INTRODUCTION OF THE FRAME PROBLEM

2.1 WHAT IS THE FRAME PROBLEM

The frame problem first arose in the context of McCarthy and Hayes's Situation Calculus.[6] A robot needs a theory of common-sense reasoning when it works in a world situation. They attempt to formalize in first-order logic the common-sense reasoning, but they encounter an inextricable difficulty because the frame axiom is unending. The frame problem originated in this paradox.The enumeration of frame axioms is evidently hopeless: there is no end to the list of properties which are not affected by an action or event. Were the listing of frame axioms tractable, there would have been no frame problem. So the frame problem can be considered to be how to deal in a tractable way with the non-effects of an event or action.

2.2 RESEARCH INTO THE FRAME PROBLEM

To solve the frame problem, many AI researchers and logicians have introduced solutions. The idea for a solution surely appears to be right when humans can cope with the non-changes by mostly ignoring them. One family of approaches, typified by the work of Marvin Minsky [8] and Roger Schank [12], derives its ignoring power from the attention-focusing power of stereotypes. The inspiring insight here is the idea that all of life's experiences, in all their variety, boil are nothing but variations on a manageable number of stereotypic themes or paradigmatic scenarios–'frames' in Minsky's terms or 'scripts' in Schank's terms. Now there is an artificial agent with a well-stocked compendium of frames or scripts appropriately linked to each other., Such systems obviously perform creditably when the world corresponds and co-operates with their stereotypes. Even with anticipated variations in them or when their worlds turn perverse, such systems typically cannot recover gracefully from the misanalyses into which they lead. When these embarrassing misadventures occur, the system designer can improve the design by adding provisions to deal with particular cases. It is important to note that the system does not redesign or improve itself or learn from its own experience. It rather must wait for an external designer to select and implement an improved design which can cope with each new situation. In fact, the script or frame approach is an attempt to pre-solve the frame problem the particular agent is likely to encounter. When people avail themselves of stereotypes, however, they are at least relying on stereotypes of their own devising. To date no one has been able to present workable ideas about how a person's frame-making or script-writing machinery might be guided by its previous experience.[1]

Some other solutions have also been proposed to the frame problem. Drew McDermott and Jon Doyle, for example, have developed a 'non-monotonic logic' [McDermott and Doyle,1980]. Ray Reiter has developed a 'logic for default reasoning' [10], and John McCarthy has developed a system of circumscription, a formalized 'rule of conjecture that can be used by a person or program for 2 jumping to conclusions 2' [5]. McDermott has offered a "temporal logic for reasoning about processes and plans " [7]. From a traditional non-deductive perspective, Kevin Korb has offered a "causal induction solution"[4]. Unfortunately, none of these proposed solutions is, or is claimed to be, a complete solution to the frame problem, but they might be components for such a solution.

3 PRAGMATICS PERSPECTIVE: THE FRAME PROBLEM SOLUTION

3.1 PRAGMATICS AND LOGICISM

The frame problem finds its radical origin in the research principles of logicism. It confronts the exponent-exploded dilemma in which the computable space emerges in a complex situation. It is one of the most important problems in AI. Research in logicism primarily exists in formal syntax and formal semantics. It maintains the primary standpoint that formal syntax is a completed and consistent system and that formal syntactic structure is irrespective of semantic concerns. Consequently, it provides an aura of metaphysics to the frame problem. This methodology plays a dominant role in AI research. In spite of the competition from neural networks [11], genetic algorithms [3] and information theorety machine learning [9, 15], these theories essentially still have the characteristics of formal arithmetic. The fantasy ability to perform infinitely many logical inferences in finite time has often been held to be the quintessential condition for achieving AI's goals. Therefore, modern AI always has expanded its research horizon under a frame of logicism.

Pragmatics on the other hand originates in the critique of syntax and semantics. It shifts the foundation of logicism away from syntax and semantics and creates a new theory which incorporates such factors as context and human cognition. It claims that formal arithmetic doesn't determine semantic things by contraries. Rather, the usage of symbols determines the syntax and meaning. Just as Wittgenstein determined, language usage is like a game in which the different game rules create different games. There are no consistent and unitive rulesto govern usage, so usage instead precedes rules. Although pragmatics is opposed essentially to logicism, it doesn't deny the principle of logicism completely.It emphasizes that the usage of symbols precedes the structure of those symbols. While the structure of the symbols exerts an influence on usage, the structure changes when different usages of symbols do. It is not a closed system. It would be a hopeless effort to solve all the problems which originate in logicism, including the frame problem, so long as we remain captive to the research frame of logicism.

3.2 SPEECH ACTS: AN IMPORTANT FOUNDATION OF COGNITIVE SITUATION

J.R.Searle considered that illocutionary acts are the minimal units of human communication and form the foundation for the construction of social reality. In fact, by analyzing the components of illocutionary force, we discover that illocutionary acts are the direct foundation for the construction of cognitive situations, but the construction of social realities and world facts are indirect, because the illocutionary point or purpose is the most important component of illocutionary acts. Fig.1 shows that the effects of illocutionary acts don't directly change world situations and directly change speaker and hearer cognitive situations. Therefore, by illocutionary acts, a speaker not only transfers and changes the cognitive situation but also constructs the cognitive situation himself.

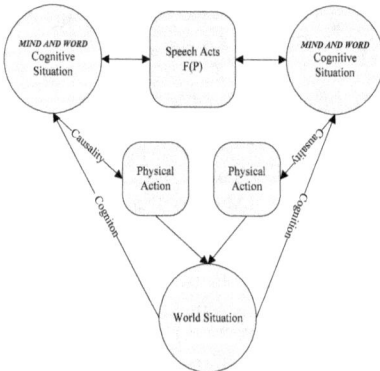

Figure 1: Relation of speech acts and cognitive situation, physical action and world situation

Cognitive situations exert influences on the world situation by their physical actions, and the construction of the world situation inclines to be identical with the cognitive situation. Certainly, the relationship between this cognitive situation and the physical action isn't a necessary one like the relationship between a robot's actions and arithmetic, but it exists as a causal relation, which is what Hume describes as a common-sense causal relationship, not a logical causal relationship. Because of the construction character of the cognitive situation by means of free will, events and speech acts in cognitive situations exert influences on the cognitive situation, but don't make certain determining effects on the final cognitive situation. Therefore, the relationship between the speech acts and the final cognitive situation or intention isn't a necessary relationship. It is just that these causal but not necessary relationships result in human intelligence.

Investigated from the whole relationship chain as a mirror of world situations, cognitive situations construct many complex possible cognitive situations as the effect of speech acts, then select the favorable final cognitive situation and exert an influence on the world situation by physical actions with causal relationships. This finally inclines the world situation to be identical with the cognitive situation, so the changes in the world situation are favorable. The whole process shows intelligence and it is valid.

4 THE FRAME PROBLEM: SOLUTIONS BASED ON PRAGMATICS THEORY

The operational control of robots uses a formal arithmetic mechanism whose working principle is illustrated in Fig.2.This mechanism can be effective in special areas such as the toyness of environment. The early version of the SRI robot SHAKEY, for example, operated in a simplified and sterile world in which it had few problems to worry about. It could make effective actions with recourse to an exhaustive consideration of frame axioms. But in a complex and changing world, arithmetic is not capable of exhaustively evaluating all the possible changes in a given situation because of the frame problem. As a result, many na?ve and invalid actions appear.

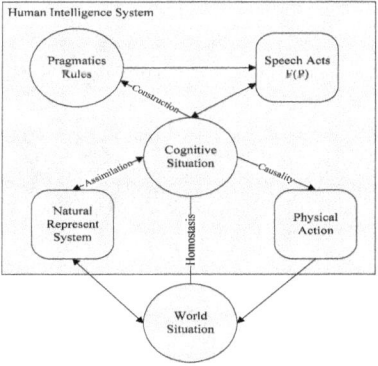

Figure 2: Principle of Robot AI System

The embodiment of intrinsic intelligence characteristics is an effective action in a complex world situation. It can make the situation inclined to favor changes. Indeed, the appearance of this type of intelligent action relies on a cognitive situation which can represent accurately the world situation. The formation of this cognitive situation is based on the use of natural language and pragmatics and not on formal arithmetic with its foundation in logicism. Therefore, we need to do research into intelligence and the frame problem from the direction of the operational principles of pragmatics and the cognitive situation. A robot controlled by formal arithmetic and driven by a limited intelligence system is cannot generate effective and intelligent actions. J.R.Searle demonstrated this point in his famous 'Chinese Room Argument'[14]

What then are the necessary conditions for the implementation of intelligent actions? First, we need to learn how humans with intelligent representation operate within the field of intelligent action. Fig.3 illustrates the operation principle of the human intelligence system. Humans are able to construct all possible cognitive situations through speech acts in the language world. In this process of construction, all the rules of pragmatics come into play. These rules include some open and uncompleted constructions in addition to unitive and completed ones. The system can adjust dynamically to changes in the cognitive situation. On the other hand, the construction of the cognitive situation and the changes which are introduced into it by the use of speech acts based on the dynamic use of pragmatics rules creates a dynamic and interactive bi-directional relationship between them. This is different from the one-directional relationship which exists between arithmetic and actions performed by a logical robot. We can conclude, therefore, that the necessary conditions required for the implementation of intelligent actions must take into account simultaneously the internal cognitive situation and the external speech acts and dynamic pragmatics rules.

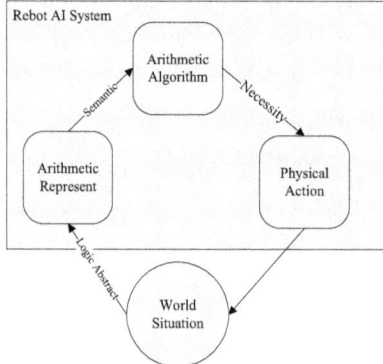

Figure 3: Principle of Human Intelligence System

What then are the necessary conditions for the implementation of intelligent actions? First, we need to learn how humans with intelligent representation operate within the field of intelligent action.FIG.3 illustrates the operation principle of the human intelligence system. Humans are able to construct all possible cognitive situations through speech acts in the language world. In this process of construction, all the rules of pragmatics come into play. These rules include some open and uncompleted constructions in addition to unitive and completed ones. The system can adjust dynamically to changes in the cognitive situation. On the other hand, the construction of the cognitive situation and the changes which are introduced into it by the use of speech acts based on the dynamic use of pragmatics rules creates a dynamic and interactive bi-directional relationship between them. This is different from the one-directional relationship which exists between arithmetic and actions performed by a logical robot. We can conclude, therefore, that the necessary conditions required for the implementation of intelligent actions must take into account simultaneously the internal cognitive situation and the external speech acts and dynamic pragmatics rules.

How do we implement intelligent action under these three conditions? In fact, the most important operation mechanism of an intelligent agent is the mechanism called the 'memory-prediction principle'. If that operation mechanism is a computed mechanism, even the simplest action requires a huge computing capacity. Many psychological investigations show that most human actions are controlled by our common experience stored in our memory. The world situation in which we live our ordinary lives is large and complex. It is impossible to preserve all the details of our common place experience in our finite neural units. One effective way of overcoming this limitation is to substitute long term memory and short term memory of direct experience for symbol memory. In this way, on the one hand, we can free up memory space for expanded memory information and on the other hand, we can use speech acts based on symbol memory effectively to construct complex cognitive situations and bring into being advanced intelligent actions. In this way, we can implement intelligent action. We need to be able to create a system in which not only speech acts based on symbol memory exist and where we can construct cognitive situations based on dynamic pragmatics rules. But we must also be sure that there exist instinctive actions in genetic schema which can directly generate basic stereotypes for adapting the world situation. So long as we can make research progress toward a pragmatics of cognition, we will be able before long to unlock the secret of intelligent action.

The frame problem and the implementation of intelligent actions problem are essentially two sides of the same problem. The implementation of intelligent action does not mean the frame problem, and the existence of the frame problem does not mean the impossibility of intelligent action. Hence, by research and analysis into the mechanism of intelligent action based on pragmatics characteristics, we can dissolve the frame problem and substitute for it an approach based on pragmatics theory.

Bibliography

[1] Dennett, D. C. Congnitive wheels:the frame problem of ai. In Hookway, C. (Ed.), *Minds,Machines,and Evolution:Philosophical Studies*, 129–151. Cambridge University Press, 1984.

[2] Ginsberg, M. L. (Ed.). *Readings in Nonmonotonic Reasoning*. Morgan Kaufmann.

[3] Goldberg, D. E. *Genetic Algorithms in Search, Optimization, and Machine Learning*. Addison-Wesley, 1989.

[4] Korb, K. B. The frame poblem: An ai fairy tale. *Minds and Machines*, 8:Kluwer Academic Publishers, 1998.

[5] McCarthy, J. Circumscription-a form of non-monotonic reasoning. *Artificiall Intelligence*, 13:27–39, 1980.

[6] McCarthy, J., Hayes, P. J. Some philosophical problems from the standpoint of artificial intelligence. In Meltzer, B., Michie, D. (Eds.), *Machine Intelligence*, vol. 4, 463–502. Edinburgh University Press, 1969. Reprinted in [2].

[7] McDemott, D. A temporal logic for reasoning about processes and plans. *Cognitive Science*, 6:101–55, 1982.

[8] Minsky, M. *A Framework for Representing Knowledge*. MIT AI Lab., 1981. Memo 3306, Quotation from excerpts repr. in [?].

[9] Quinlan, J. R. *Programs for Machine Learning*. Morgan Kaufmann, 1993.

[10] Reiter, R. A logic for default reasoning. *Artificial Intelligence*, 13:81–132, 1980.

[11] Rumelhart, D. E., McClelland, J. *Parallel Distributed Processing*, vol. 1. MIT, 1986.

[12] Schank, R. C., Abelson, R. P. *Scrips,Plans,Goals,and Understanding:An Inquiry into Human Knowledge*. Hillsdale, NJ: Erlbaum, 1977.

[13] Searle, J. R. *Speech Acts:An Essay in the Philosophy of Language*. 1969.

[14] Searle, J. R. Mind,brains, and programs. *The Behavioral and Brain Science*, 3:417–424, 1980.

[15] Wallace, C. S., Freeman, P. R. Estimation and inference by compact coding. *Journal of the Royal Statistical Society, Series B*, 49:240–252, 1987.

A Philosophical Exploration into the Cognitive Construction of Scientific Discovery[1]

Wang Xiaohong

Xi'an Jiaotong University

amandawxh@yahoo.cn

ABSTRACT. This paper firstly analyzes the epistemological origins and the basis of theories in the field of cognitive construction of scientific discovery in which accomplishments derive benefits from philosophy and history of science and cognitive psychology, and embody the integration of diverse approaches. Secondly, from the methodological perspective, this paper discusses the three approaches of existing computational models of scientific discovery: (1) simulating the real scientific discoveries; (2) engaging in discovering new scientific knowledge; (3) modeling and instantiating ideas of philosophy of science. The three approaches have distinct goals and different backgrounds but share some common ideas.

Generally speaking, there are four main practical fields involved in the cognitive construction of scientific discovery (CCSD) and creative thinking. They are computational modeling, neural network modeling, cognitive experiments, and historical case studies. Our philosophical exploration selects the computational model as the entrance to CCSD. Currently, computer scientists, philosophers of science and cognitive scientists have given us very helpful results for understanding the processes of SD and creative thinking. Early in this field, such philosophers as Bruce Buchanan, Lindley Darden and Paul Thagard simultaneously claimed that the philosophy of science could derive very useful ideas from work being done in AI. A few years ago, a number of philosophers demonstrated again that "the 'experimental' aspect of computational modeling has been seen as a particularly important addition to philosophical methodology". [5] The accomplishments of computational models of SD derived benefits from the theories in philosophy of science, case studies in history of science, experiments of cognitive psychology, and the neural network models in neuroscience. So the computational models embodied the integration of the above diverse approaches to scientific discovery.

---

[1] The author is with The School of Humanities & Social Sciences, Xi'an Jiaotong University, Xi'an 710049 China.

1 Computational Discovery and the "Philosophy of Discovery"

During the pioneering works regarding computational modeling of creative thinking that Simon's group had done, there happened an obvious change of the methodology from 70's of last century. The general approach of developing "general problem solving" programs was given up and changed to a special/individual approach, or a "divide-and-conquer strategy"[12]. According to Simon's argument [15], none of the present discovery programs could complete all of the aspects of discovery. One has to decompose discovery into different subtasks. That is, designers develop different programs separately (not integrated) focusing on performing special subtasks of SD respectively (one by one). From then on, we can see, this special approach was adopted not only by Simon's group, but also by other scientists and philosophers who engaged in computational modeling of SD. I classify the programs according to the main subtasks they performed: (1) to find the rules of data: BACON[9], FAHRENHEIT, IDS[12] etc. (2) to find the qualitative laws and concepts: GLAUBER, STAHL, DALTON[10], AM; (3) to fulfill the representation or explanation: GELL-MANN, BR-3[6], BR-4[7], PAULI[17], MECHEM[16], ECHO[14] etc.. (4) to design the experiment and make theory revision: KEKADA[8], COAST etc.. (5) to detect error and recover: STAHLp, AbE, TRANSGENE[3] etc.. (6) to make analogy: Drama, ACME, Copycat, SME, LISA etc..

This classification does not mean each program's real performing strictly limited in only one subtask, that is, in fact, impossible for any program to run. Some subtasks are more fundamental than others, and some subtasks are more intimately assistants to each other in special situations. We emphasize that this is a technical classification tending to focus on the main achievement of each program in the techniques of AI or ML.

Some interesting results come from the exploration of the epistemological origins and theoretical basis of this field. On the one hand, the occurrence of computational modeling of SD gave "friends of discovery" a great excitement. Performing the discovery tasks through AI programs strongly supported the arguments that scientific discovery has the logic and the processes of discovery can be rationally reconstructed. Meanwhile, on the other hand, the designing ideas of AI discovery programs were intimately influenced by "philosophy of science".

A widely influential analysis of the history of the "philosophy of discovery" was proposed by Larry Laudan[11]. Over a long period of history, the dominant epistemology was known as "infallibilism". The majority of philosophers believed that knowledge was infallible, "be true and known to be true"[11]; moreover, the post hoc evaluation of the consequences of a theory was inconclusive to verify the truth of a theory. The logic of discovery, in such a situation, was regarded as the sole completely adequate justification for a scientific theory. Indeed, an infallible LOD would automatically justify the validity of its productions. So in the age of infallible epistemology, the logics of post hoc theory testing were "rendered redundant and gratuitous"[11]. Conversely, when fallibilism replaced the dominance of infallibilism, all knowledge became fallible. There was no guaranteed LOD at all. The logics of post hoc testing, in this situation, became the sole effective way to build up the plausibility of a theory. As a result, the philosophy of discovery was rooted out from the ground of fallibilism. However, the decline of LOD was accompanied by an epistemic aim. There should be at least another reason for the existence of LOD. Laudan proposed the reason, but he was very doubtful of its application. He described that besides the epistemic aim of the LOD, some methodologists had been engaging in the heuristic and practical aim of LOD: to provide efficient rules of inventing new ideas so as to speed the growth of science[11].

However, Laudan doubted the current reasons for studying LOD: "[O]ne must ask what is specifically philosophical about studying the genesis of theories. Simply put, a theory is an artifact, fashioned perhaps by certain tools (e.g., implicit rules of 'search'). The investigation of the mode of manufacture of artifacts (whether clay pots, surgical scalpels, or vitamin pills) is not normally viewed as a philosophical activity ... If this essay provides a partial answer to the question 'Why was the logic of discovery abandoned? It poses afresh the challenge: Why should the logic of discovery be revived?'[11]."

We think the analogy Laudan made ignores the essential difference between the two kinds of artifacts. The theoretical artifact is completely different from the material artifact regarding the mechanism of generation or the mode of manufacture. The former generates from innovative thinking, and the mechanism of generation is far from being well-known. The exploration itself is a very challenging enterprise which is a new trans-disciplinary field of research–cognitive science. Philosophy is one of the disciplines involved according to a widely accepted argument[1]. But the latter generates purely mechanical tools or procedures which are designed by human beings in the first place. The design problems are under the control of engineers. If we take human brain as a special machine, the question is how to design the mechanism for generating creative thinking? We are at the beginning of the enterprise to design and understand the mechanism of generation of theory. As the same as any discipline' beginning, philosopher plays the necessary role among the explorers. So our claim is: the heuristic and practical aim is a reasonable and profitable reason to revive the philosophy of discovery in the background of fallibilism.

On the other hand, the computational models of SD displayed the feasibility of studying LOD with the aim. Simon was the earliest pioneer in this field, and one of the main advocators of revival of LOD in the era of fallibilism either. Simon absorbed the ideas of discovery from a few important methodologists or philosophers, Francis Bacon (1561~1626), Charles Sanders Peirce (1839~1914), Norwood Russell Hanson (1924~1967), Thomas S. Kuhn (1922~1996), and very possible, Larry Laudan either. Being the theoretical ground of designing AI models of SD, Simon's discovery theory absorbed the nourishing ingredients coming from positivism of Bacon and Hanson, historism of Hanson and Kuhn, and showed the high agreements on the stand of pragmatistic positivism with Laudan's ideas[18].

2 Methodological Classification of SD Models

From the methodological angle I classify the existing computational models of SD: (1) simulating the processes of the scientific discoveries in real history, typified by the programs of BACON, FAHRENHEIT, IDS STAHL, STAHLp, KEKADA, COAST, BR-3, BR-4 etc.; (2) engaging in discovering new scientific knowledge, exemplified by the work of GELL-MANN, PAULI, MECHEM etc.; (3) instantiating and verifying relating philosophical ideas of SD such as TRANGENE, Drama, ACME, ECHO etc.. The three approaches have distinct goals and help us to understand the discovery from different angles of view.

The features of simulation on the simulation approach embody as followings: to simulate or tend to simulate the heuristics in real discovery. These strategies were really used by scientists, or might be thought of use to scientists; to use the data in the real discovery as input; to gain the same results as that in real discovery. The goal of the approach is to conclude the reasoning mechanisms (heuristics) generally accounting for a variety of historical achievements, consequently, to help human scientists to do discoveries more efficiently.

This approach emphasizes the historical and psychological adequacy of a discovery theory. KEKADA models the heuristics Hans Krebs used in the discovery of urea cycle. Simon and Qin did related cognitive experiment[13] and concluded the behaviors of the successful problem-solvers were guided by the heuristics very similar to BACON's. Langley and his group especially simulate a characteristic of history of science, that scientific knowledge is in growth and the development of science is incremental, demonstrated by STAHL, KEKADA, BR-4/3, which all involve modeling an extended period in the history of science, rather than isolated events.

The achievements of the discovery approach are few but very significant in methodology in the field of CSSD. The designer claims the evidence that a novel finding in particle physics was enabled by the machine discovery program PAULI proves the core strategy of heuristic search used in this field is complete to account for a salient aspect of human reasoning, that is discovery in science. The three programs have generated interesting findings that differ from those found by human scientists, and their results and search methods are less plausible as historical accounts than simulations. The three programs all rely the common predominance which computers have over human beings–super strong search and calculating capacity. But this does not contradict heuristic search. They implement the exhaustive search for the simplest mechanism guided by heuristics. The programs came up with findings that human beings could not come up with, so the new findings demonstrate the approach is methodologically feasible or necessary in SD. Lindley Darden evaluated the work of the discovery approach precisely when she said, "Cognitive science research has shown that humans have a tendency to focus too rapidly on one hypothesis before doing a systematic search of the hypothesis space. Discovery programs that are more systematic and more thorough than humans are an aid to the scientist."[2]

The instantiation approach has been put into practice by Paul Thagard and Lindley Darden. Darden's TRANSGENE system models the reasoning strategies in scientific theory revision, "anomaly-driven redesign tasks". The epistemological approach of TRANSGENE is opposed Dehem-Quine holism. According to the Duhem-Quine thesis, it is impossible to test a scientific hypothesis in isolation. Instead, physicists can only test the whole hypothesis group. If the predictions fail the test, what the physicist knows is that one hypothesis at least in the group should be revised; however, the test does not indicate which one should be revised. In logic, the Dehem-Quine holism is better than related theories from previous empiricism and positivism. But in fact, scientists do localize the fault component of a theory/hypothesis and redesign and remove the anomaly.

Scientists' real practices are not limited by Dehem-Quine holism. Additionally, Darden argued TRANSGENE's different versions demonstrated "the efficacy of dependency directed backtracking versus random trial and error." This strategy other than Popper's trial and error method is useful in fact.

Thagard's ECHO system was built to instantiate Thagard's theory of explanatory coherence (TEC) and to some extent to reject Popper's falsificationism. ECHO does not give up a promising theory just because of its empirical problems. Also, ECHO opposes Kuhn's argument on the incommensurability or untranslatability between competing theories and claims the criterion of explanatory power.

Thagard thinks the cognitive mechanisms by which new conceptual systems in science are constructed include conceptual combination (just as PI implements) and analogy. He and his colleagues built a few computational models of analogical thinking[4], such as ACME and Drama. Both programs are based on the theoretical background of "Multiconstraint Theory" developed by Thagard and his colleagues. Through the mechanization, it is specified "how the constraints of similarity, structure, and purpose can be jointly optimized to yield a coherent set of correspondences between a source and target". Like other philosophers of science who have preference on cognitive construction of philosophy of science, such as Ronald Giere and Nancy Nersessian, Thagard's philosophical reconstruction of science is tightly based on the results of cognitive experiments and historical case studies. But he does deeper computational modeling than other philosophers do and incorporates research of neural models into his simulation of scientific discovery. His computational simulation of SD embodies the integration of the other three practical fields in CCSD.

Note: The research for this paper was supported by Fulbright Grant 06-07 VRS. Program

Bibliography

[1] *Stanford Encyclopedia of Philosophy*. 2004. URL plato.stanford.edu/entries/cognitive-science. Apr. 20.

[2] Darden, L. Recent work in computational scientific discovery. In Shafto, M., Langley, P. (Eds.), *Proceedings of the Nineteenth Conference of the Cognitive Science Society*, 161–166. Mahwah, New Jersey: Lawrence Erlbaum, 1997.

[3] Darden, L. Anomaly-driven theory redesign: Computational philosophy of science experiments. In Bynum, T. W., Moor, J. H. (Eds.), *The Digital Phoenix: How Computers are Changing Philosophy*, 62–78. New York: Blackwell Publishers, 1998.

[4] Eliasmith, C., Thagard, P. Integrating structure and meaning: A distributed model of analogical mapping. *Cognitive Science*, 25:245–286, 2001. From Thagard's homepage.

[5] Floridi, L. *Blackwell Guide to the Philosophy of Computing and Information*. Blackwell Publishin, 2003.

[6] Kocabas, S. Conflict resolution as discovery in particle physics. *Machine Learnin*, 6:277–309, 1991.

[7] Kocabas, S., Langley, P. An integrated framework for extended discovery in particle physics. In Jantke, K., Shinohara, A. (Eds.), *DS 2001*, 182–195. LNAI 2226, 2001.

[8] Kulkarni, D., Simon, H. The processes of scientific discovery: The strategy of experimentation. *Cognitive Science*, 12:139–175, 1988.

[9] Langley, P. Data-driven discovery of physical laws. *Cognitive Science*, 5:31–54, 1981.

[10] Langley, P., Simon, H. A., Bradshaw, G. L., Zytkow, J. M. *Scientific Discovery: Computational Explorations of the Creative Processes*. Cambridge, MA: MIT Press, 1987.

[11] Laudan, L. Why was the logic of discovery abandoned? In Nickles, T. (Ed.), *Scientific Discovery, Logic, and Rationality*, 173–184. Dordrecht, Netherland: D. Reidel, 1980.

[12] Nordhausen, B., Langley, P. An integrated framework for empirical discovery. *Machine Learning*, 12:17, 1993.

[13] Qin, Y., Simon, H. Laboratory replication of scientific discovery processes. *Cognitive Scienc*, 14:281–312, 1990.

[14] Schank, P., Ranney, M. The psychological fidelity of echo: Modeling an experimental study of explanatory coherence. In *Proceedings of the Thirteenth Annual Conference of the Cognitive Science Society:*, 892–897. 1991.

[15] Simon, H. A. Machine discovery. *Foundation of Science, 1995/1996*, 2:171–200, 1997.

[16] Valdes-Perez, R. Machine discovery in chemistry: New results. *Artificial Intelligence*, 74:191–201, 1995.

[17] Valdes-Perez, R. E. Algebraic reasoning about reactions: Discovery of conserved properties in particle physics. *Machine Learnin*, 17:47–67, 1994.

[18] Wang, X. A study of the problems in scientific methodology arising out of the BACON system, 2004. Doctorial dissertation, Peking University.

Can a Machine Lie?[1]

Wang Simin
Tsinghua University
joycewongszeman@gmail.com

ABSTRACT. Lying is a special kind of speech act. Some writers have singled out the ability to lie as one of the criteria that distinguishes human beings from other animals. Is lying unique to human beings? In this paper, we're interested in questioning whether a machine can lie. Though the topic is not unfamiliar –not only can we find the robot's lying in science fiction literature and films–but Alan Turing's 1950 classic paper considers the question of whether machines can think. We want to review this question from a new perspective by defining lying in its relation to consciousness.

1 Prelude: Lying and human language

In Sir John Eccles' *Evolution of the Brain–Creation of the Self*, the author, a distinguished scientist and Nobel Prize winner, states that the most comprehensive scope of all that can be subsumed in the category of language is that formulated by Buhler and further developed by Popper. According to the Bühler-Popper classification there are four levels of language: "expressive (or symptomatic) function" and "releasing (or signaling) function" are two lower forms of language that animal and human languages have in common; and "descriptive function" and "argumentative function" are two higher forms that may be uniquely human. It is assumed that the possibility of lying is implicit on the third level (descriptive function) of language. The unique feature of it is that the statements may be factually true or factually false. It is important to recognize that the two lower functions of language are associated with utterances that are both expressions and signals.[10, p. 72]

In the past, Descartes proposed that there is a qualitative difference between man and animals, as displayed by language. Animals are automata lacking anything equivalent to human self-consciousness. Human beings are guided by reason, animals by instinct, and to Descartes human language is an activity of the human soul.[10, p. 76] John Searle, a contemporary philosopher, supplies the essential thing about human beings when he argues that language gives them the capacity to represent. Humans can lie in representing something as being the case even when they know it is not the case[12]. That's why human language can be numbered among the unique adaptations of our species which is not shared with any of the rest of the animal kingdom. But as technology progresses, the usage of human natural language no more appears only between human and human communication. It is also a tool in human-computer interaction. The ability to

---

[1]The author is with Tsinghua University, Beijing 100084 China.

lie, we assume, is one criteria that separates human beings from other animals. What happens when we compare the ability to lie between human beings and machines?

2 Is a machine lying?

Here are two cases found from the internet.

Case 1: A computer user posted a question titled "My hard drive is lying" on one tech-support website[2]: "I just got a new 120GB external hard drive for my laptop. I got it to store all my music and DVD ISO files. It has 98.7GB free space on it, but it says there is not enough disc space to copy a 4.35GB file across. It still copies other files across. I've tried everything, defrags, cleanups and creating new ISO's. The kicker is, there are already some DVD sized ISO files on it and they move across with no trouble. Can anybody help before I go mad?"

Case 2: There was an article "Your PC Is Lying To You"[3] posted on the Wall Street Journal in Dec of 2006. The author, Jeremy Wagstaff, who is a technology columnist, mentions that "the computer is a liar and a cheat. Blame it on the progress bar." As we know, the progress bar is the little indicator that pops up when the computer user is installing software, downloading a file or opening a program. It usually looks like a fuel gauge, filling from left to right. Sometimes there's more information, such as how long the task is likely to take. E.g. 'This task is 56.12% done.' 'There are 10.37 minutes remaining.' It offered a glimpse of new vistas of productivity, where the user knew exactly how long before the computer was ready for them. But the progress bar does not always work properly. Chances are you've seen a progress bar telling you something is going to take ten minutes, go down to nine minutes,...three minutes, two minutes then either stick there or creep back up to three minutes – or thirty minutes! Jeremy Wagstaff said the progress bar has turned us all into dupes and fools.

Nevertheless, is the computer really a liar? Does it really tell lies? Before answering the above questions, let's seek first for a definition of lying.

Although "lie" is a common word and people usually think they know what it is and how to deal it, many scholars have examined it and some state that "the very great vagueness in the concept of 'lie' and the many inconsistencies in the common usage of the word 'lie' "[2]. The meaning of "lie" in the Oxford English Dictionary is: "a false statement made with the intent to deceive."[1] But the definition here seems not clear and ambiguous, for what does the falsehood refer to?

Usually, when people try to define lie, the first thing that comes to mind is probably the idea of saying something false. If a child broke the vase carelessly and denied his behavior to his parent, it's obvious a lie. The false statement, however, is not adequate, since people frequently say things that are false but which nonetheless are not called lies. E.g. a child says, "London is the capital of China." when he did not study his lesson well and believes this to be true. So, if the speaker is sincerely trying to convey what he believes to be true information, such honest mistakes and innocent misrepresentations (actually occur frequently)

---

[2] http://forums.techguy.org/hardware/567891-my-hard-drive-lying.html
[3] http://www.chuguolu.com/a/95518.html

Can a Machine Lie?

are not labeled lies. Augustine had already distinguished this in his article "On Lying " written in the 4th century, "For not every one who says a false thing lies, if he believes or opines that to be true which he says."[4] Then, must a lie be a false statement? Consider below situation, if a child has a test she hasn't studied for and so she doesn't want to go to school. She says to her mother, 'I'm sick.' Her mother takes her temperature, and it turns out to her own surprise that her temperature higher than normal. And she really is sick, later that day developing the measles[4]. The child says the true information to her mother, her sick is not a false statement, is it a lie? In Augustine's words, "a person may say a true thing, and yet lie, if he thinks it to be false and utters it for true, although in reality it is so as he utters it."[4]

It's important to notice there are two aspects falsehood in an utterance: objectively and subjectively. Below table is revised from Kang Lee's article "Lying as doing deceptive things with words", it illustrates the different situation between several verbal communications.[5]

	Word		intention	subjective belief	objective Factuality
Forms of verbal communication	Literal meaning	Deeper meaning			
honest, accurate statement	△	△	△	△	△
honest mistake	○	○	○	○	△
verbal error	○	△	△	△	△
irony sarcasm	○	△	○	△	Not ○
line	○	○	○	△	irrelevant

Table I: Five factors of speech acts

Augustine's view of the liar is, "For from the sense of his own mind, not from the verity or falsity of the things themselves, is he to be judged to lie or not to lie. Therefore he who utters a false thing for a true, which however he opines to be true, may be called erring and rash: but he is not rightly said to lie; because he has not a double heart when he utters it, neither does he wish to deceive, but is deceived." Moreover, Augustine claimed the fault of him who lies, is, the desire of deceiving in the uttering of his mind; no matter he actually achieves deception or not.[4]

In Chisholm and Feehan's well-known article "The intent to deceive"[7], they give a philosophic account of lying. They believe with St. Thomas "a false statement uttered with intent to deceive is a lie". But Chisholm and Feehan think the expression "He intends to say what is false" must be taken to imply not that there is something false that he intentionally states but rather that he intends it to be the case that he states something that is false. Their definition of a lie is: "L lies to D=df There is a proposition p such that (i) either L believes

---

[4] An example cited from [9]
[5] The table is revised from Kang Lee's "Lying as doing deceptive Things with words" which is edited in [3]

that p is not true or L believes that p is false and (ii) L asserts p to D." And their definition of a assertion is: "L asserts p to D=df L states p to D and does so under conditions which, he believes, justify D in believing that he, L, not only accepts p, but also intends to contribute causally to D's believing that he, L, accepts p."

Now, we go back to the situation of the "machine is lying" mentioned above. In case 1, the computer seems saying falsehood when it has far more free space left in the hard drive, and the user thought he is cheated. But the reality is, the user's file system indeed is FAT32 but he didn't know the FAT32 system has a 4GB limit before, so he just blamed the machine lying under his ignorance. It's obviously not a machine lying case at all. In case 2, it's actually our mistake, too. We're at fault to think the progress bar measures progress well as in how much progress was being made in the task you had set the computer. For all computers' internal chippery and circuitry aren't good at measuring time, there are just too many internal tasks going on for them to make an accurate assessment of when they will all finish. As Jeremy Wagstaff says also in his above article, "people get bored easily and then start up other programs, it will further confuse the progress bar's calculations". Maybe we sometimes feel being cheated by the progress bar, but it's better to say that the liar is the computer programmer rather than computer itself.

3 Can a machine lie?

This question is easily associated with the celebrated Turing's test or game. In remembering the "Turing's test" we could say that deception is the original, foundational challenge and proof for AI. According to David Gelernter, an AI researcher at Yale University, "Keep in mind that the whole basis of the Turing test is lying. The computer is instructed to lie and pass itself off as a human being. Turing assumes that everything it says will be a lie. He doesn't talk about the real deep meaning of lying... it's certainly not the case that the computer is in any sense telling the truth. It's telling you something about its performance, not something about facts or reality or the way it's made or what its mental life is like."[6]

It seems the standard of Turing test can be considered in two aspects: whether Turing intended the machines just have the humanlike communicative ability will be credited with intelligence or the machines have to indeed cheat the human by telling lies. If the answer is the first one, i.e., if no amount of questioning or conversation allows you to tell which it is, then we have to concede that a machine can think. Some argue that today and even more in the future we do not know and cannot so simply understand whether a given answer comes from a person, or from a data base or expert system, or from an intelligent personal assistant, etc. So, a weak form of the Turing's test is now a daily and common experience on the net, and not so sophisticated and intelligent agents or dialogue systems are needed for this.[6] But if the answer is on the ability of telling lies, we think it's the question about machine consciousness and at least the issues of

---

[6]From the debate of "Machine consciousness" between Ray Kurzweil and David Gelernter on 30th Nov2006 in a special MIT event, creativity: the mind, machines, and mathematics. http://www.kurzweilai.net/meme/frame.html?main=/articles/art0688.html

intentionality, freewill and creativity are related to it.

"Lying is essentially intentional," Charles Fried, a professor at Harvard Law School points out. "Speech, communication, and therefore perversion of these in lying are paradigmatically intentional. One can only lie intentionally–it is not possible to lie inadvertently or as the known but unwanted side effect of some other purpose".[11] Furthermore, we've already discussed what a liar is from the sense of his own mind, not from the verity or falsity of the things he says themselves. It's obvious to see that intentionality plays an important role in telling a lie.

As for "freewill", J.A. Bames' book A Pack of Lies, Towards a Sociology of Lying mentions Hannah Arendt, a German Jewish political theorist, who proclaims that our ability to lie belongs among few obvious, demonstrable data that confirm human freedom. The same sentiment can be found in Ekman, the famous psychologist, who also points out the intent of the liar is one of the criteria for distinguishing lies from other kinds of deception. Just as we mentioned above, he agrees that liars may actually tell the truth, but that is not their intent. Truthful people may provide false information, but that is not their intent. So, it's important that the liar deliberately choose to mislead the target. The liar has the choice; the liar could choose not to lie. By his definition, presumably, a pathological liar is compelled to lie and therefore is not a liar.

For the lie unlike the truth has one face, the reverse of truth has a hundred thousand faces. The creativity also often is needed in creating a lie. In the sixteenth century, the writer Bartholomaeus Ingannevole wrote that never to lie admits of no imagining which is all that God did give man to distinguish him from the beasts of the field. It is perhaps the perception of lying as evidence for freedom and imagination that explains the attraction of compilations of lies, both scholarly and demotic.[5]

Actually, there've been some articles studying the issue of machines lying. For example, a paper "Toward the lying machine"[8] was published recently (July 27, 2007) by scholars in psychology, cognitive and computer science. Their goal is to develop a lying machine that manipulates human beliefs through what they call 'sophistic lies': psychologically persuasive yet fallacious arguments that lead humans incrementally into false (erroneous) beliefs while simultaneously confirming their naive intuitions about what is true. They emphasize they are not concerned with the metaphysical or philosophical question of whether a machine can have the requisite mens rea for meta-cognitive acts of mendacity. They are not concerned with whether a machine can "believe" at all. They claim they are doing philosophical AI, not the philosophy of AI. In contrast, what we're interested is the foundational problem. And, this paper is just the beginning of our investigation.

Bibliography

[1] *Oxford English Dictionary*. Oxford: The Clarendon Press, 2nd edn., 1989.

[2] Arnold, I. Deontology and the ethics of lying. *Philosophy and Phenomenological Research*, 24(4):463–480, 1964.

[3] Astington, J. W. (Ed.). *Minds in the making. Essays in Honor of Davis R. Olson*. Blackwell Publishers, 2000.

[4] Augustine. *On Lying*, chap. last. URL http://www.newadvent.org/fathers/1312.htm.

[5] Barnes, J. A. *A pack of Lies, Towards a sociology of lying.* Australian National University: Cambridge University Press, 1994.

[6] Castelfranchi, C. Artificial liars: Why computers will (necessarily) deceive us and each other. *Ethics and Information Technology*, 2:113–119, 2000.

[7] Chisholm, R. M., Feehan, T. D. The intent to deceive. *Journal of Philosophy*, 74:143–159, 1977.

[8] Clark, M., et al. *Toward the lying machine.* 2007. URL www.rpi.edu/~clarkm5/papers/NA-CAP2007.pdf.

[9] Coleman, L., Kay, P. Prototype semantics: The english word lie. *Language*, 57:26–44, 1981.

[10] Eccles, J. C. *Evolution of the brain: creation of the self.* London, New York: Routledge, 1989.

[11] Fried, C. *Right and Wrong.* Cambridge, Mass: Harvard University Press, 1978.

[12] Searle, J. *Social Ontology: Some Basic Principles.* 2004. URL http://socrates.berkeley.edu/~jsearle/AnthropologicalTheoryFNLversion.doc.

On Engineering Rationality[1]

Li Bocong and Wang Nan
Graduate University of the Chinese Academy of Sciences
libocong@gucas.ac.cn

ABSTRACT. The problem of rationality is important in the philosophy of science and the philosophy of engineering. Engineering rationality, however, is different from scientific rationality. It is not the traditional rationality of thinking by nature but a kind of "practical rationality". This paper compares engineering rationality and scientific rationality in four aspects and draws the conclusion that engineering rationality should be regarded as a form of collective rationality, machine-users' rationality, local rationality, and situational rationality. Finally, the paper discusses the problem of the research methodology used in the study of engineering rationality.

In the realm of philosophy, the word 'reason' used to be a popular one. It was replaced by the new word 'rationality', which was first used by Max Weber and has become more popular than 'reason' over the course of the last century.

Both reason and rationality are polysemous words. In different contexts, 'reason' has different meanings, and 'rationality' is similar in its usage. Some philosophers regard reason as objective, while others think of reason as subjective. As far as rationality is concerned, scholars often talk about substantive rationality, instrumental rationality, procedural rationality, bounded rationality, and so on.

It is interesting that the word 'rational' is an adjective which corresponds to both 'reason' and 'rationality'. The usage of 'rationality', however, is connected only with human activity. One cannot say, for example, that mountains and rivers are rational, but one can say that a man is 'rational' or 'irrational'. Scholars in economics and ethics, and scientists generally, have different understandings of the meaning of 'rationality' and the relation between 'reason' and 'rationality'. They all regard rationality as a characteristic human activity. The topic of this paper is to discuss the meaning of of 'rationality'.

The concept of rationality applies to many different kinds of human activity, especially to economic activity and scientific activity. Consequently, many scholars have mainly been concerned with economic rationality and scientific rationality so far. Although engineering practice is an important human activity, few scholars regard 'engineering rationality' as a particular kind of rationality. I think differently. In my opinion, 'engineering rationality' is the essence and focus of rationality.

In order to discuss engineering rationality, I want to discuss first the concept of engineering. Engineering (in Chinese characters 工程) often denotes different

---

[1]The authors are with The Research Center for Engineering and Society, Graduate University of the Chinese Academy of Sciences, Beijing 100047 China.

objects. Sometimes engineering, for example, can mean electronic power engineering, genetic engineering, construction engineering, and so on. 'Engineering' means a practical theory or a system of technological knowledge which is taught in technical or engineering colleges. I am not going to discuss the meaning of 'engineering' that is often connected with subjects in the curriculum. At the same time scholars and laypersons often refer to 'engineering' as a particular kind of practice, especially productive activity. Most Chinese engineers and scholars regard engineering as a large-scale productive activity. Carl Mitcham, a famous philosopher of technology, also says:

> The hypothesis here is that the genus for understanding engineering is human activity–that is, that engineering is to be defined not so much in terms of special types of objects or of certain kinds of knowledge or of unique volitional commitments as in terms of a particular species of behavior. However, given the extreme diversity of human activity– from speaking, playing, politicking, and praying to human hunting, gathering, agriculture, eating, and art–this is not so much a genus as a super ordinate class within which one can distinguish at least doing and making. With Aristotle, such a distinction will be accepted on the basis of 'other discourses,' and the specific difference for engineering activity will be sought not within human activity as such but within human productive activity, that is, making.[1]

This essay focuses mainly on engineering as a productive activity. As a kind of practice, engineering often means engineering projects. People do engineering project by project. So the project is often taken as the concrete unit of engineering. Because 工程 or engineering often means a project, the Chinese characters 工程 have been translated into the English word 'engineering' or 'project'.

Some people often confuse not only science with technology, but also technology with engineering. In fact, science should not be confused with technology, and technology should not be confused with engineering. The essence of scientific activity is discovery, the essence of technological activity is invention, and the essence of engineering activity is creativity or the making of artifacts. The fact that discovery is different from invention is parallel to the fact that science is different from technology. It is obvious that the relation between technology and engineering is similar to the relation between science and technology.

As a kind of practical activity, engineering has many dimensions–a technological dimension, an economic dimension, a social dimension, a managerial dimension, an ethical dimension, a political dimension, a psychological dimension, and so on. Therefore one should regard engineering as a multi-dimensional practical activity. Even though technology is a component of engineering, engineering should not be confused with technology.

Engineering is both a multi-dimensional practical activity and a progress which goes through many phases. Generally speaking, a project begins with a plan or a design phase, a manufacturing phase, a sales phase, a use phase, and finally an ultimate disposal phase. Scholars should not neglect all the phases.

According to the notion that rationality is connected with human activity, we can connect scientific rationality with the scientists' activity and engineering rationality with the engineers' activity. Entrepreneurs and engineers often behave quite differently from scientists when they carry out a project. Therefore, in this paper, I want to argue that engineering rationality and scientific rationality are two different kinds of rationality from four aspects.

First, because engineering projects are carried out collectively, engineering rationality must be regarded as a collective form of rationality, which is quite different from individual rationality. While the scientific community consists of scientists, the engineering community consists of entrepreneurs, managers, engineers, investors, workers and stakeholders. The members of the engineering community must cooperate with each other in carrying out their engineering activity, but they often at the same time conflict with each other. They bargain and negotiate according to the collective rationality.

Second, engineering projects cannot be completed without workers, and workers must carry out engineering projects using machines. Engineering rationality should be regarded as a machine-users' rationality.

Third, entrepreneurs and engineers develop their rational ability by gaining experience from engineering projects. They have to enlarge and enrich their engineering knowledge which consists of scientific knowledge, technical knowledge, market knowledge, managerial knowledge, and so on. Scientists generally neglect local knowledge because local knowledge is useless from a scientific viewpoint. But entrepreneurs and engineers pay great attention to local knowledge because it is important from an engineering point of view of. Hence, engineering rationality can be seen as local rationality.

Finally, from the point of view of scientific rationality, the special conditions are unimportant and unnecessary. But from the point of view of engineering rationality, the special conditions are crucial. Engineering rationality should be treated as situational rationality.

The relation between the universal and the individual is important. While some philosophers focused their attention on the universal, others focus on the individual. Robert K. Merton advanced a famous thesis on an ethos of science which consists of communism, universalism, disinterestedness and organized scepticism. Both scientific laws and scientific concepts are ubiquitous. There are a lot of questions in science, and there are a lot of questions in engineering too. It is interesting that the answer to a scientific question is unique and the answer to an engineering question is not unique. Engineers and managers can answer the same engineering question differently.

Billy Vaughn Koen considers engineering method a heuristic method. He says, "Engineering design, or the engineering method, is the use of heuristics to cause the best change in a poorly understood situation within the available resources." "A heuristic does not guarantee a solution, it may conflict with other heuristics." [2] I think that the most important thesis is that "everything in engineering is heuristic". From what has been mentioned above, we know that the scientific method is different from the engineering method and that scientific rationality is different from engineering rationality. Engineering rationality is not the traditional rationality of thinking in nature. It is a kind of "practical rationality" by nature.

When we do research into engineering rationality, we should learn from the research achievements of scientific rationality, the rationality of economic activity, and the rationality of the law from our predecessors. We need to make a deep investigation and study of actual engineering activity and the actual problems encountered in engineering practice even more. Furthermore, we should use the linguistic and etymological analysis, but we need to pay attention to the method of "facing phenomena directly, and meaning grasped words forgotten (words are employed to convey ideas; but when the ideas are apprehended, men forget the words)". The ancient Chinese philosopher Chuangtse advocated the method of "meaning grasped words forgotten". Almost four hundreds years ago, Bacon warned people not to get caught up in the four idols, especially the idol of the market which is caused by the use of language and the most difficult to overcome. Modern philosopher Husserl proposed the phenomenological method of going "back to the thing itself". Their methodological opinions are noteworthy. It is not easy for many modern scholars to "face with phenomena directly" through the "lingual fog" which is caused by many factors and to find the way out of the "lingual maze" and find the way to "meaning grasped words forgotten".

Chinese scholars have published several books about rationality in recent years, such as Huihua Hu's *The Problem of Rationality*, Wei Wu's *Practical Rationality*, Lei Ma's *Progress, Rationality and Truth* and Shizhong Zhou's *Rationality of Law*. I hope more and more Chinese scholars can pay attention to the new field of engineering rationality as a guideline for attaining more and more research achievements.

Bibliography

[1] Koen, B. V. The engineering method. In Durbin, P. T. (Ed.), *Critical perspectives on nonacademic science and engineering*, 80–81. Bethlehem: Lehigh University Press; London: Associated University Presses, 1991.

[2] Mitcham, C. Engineering as productive activity: philosophical remarks. In Durbin, P. T. (Ed.), *Critical perspectives on nonacademic science and engineering*, 33–59. Bethlehem: Lehigh University Press; London: Associated University Presses, 1991.

A New Interdisciplinary Study of Economics and Logic[1]

Qu Maisheng

Tianjin University of Commerce

qms@tjcu.edu.cn

ABSTRACT. Economic logic, a new interdisciplinary study combining economics and logic, has caught the attention of theorists from China and abroad. This paper elaborates the basic concept and content of economic logic in terms of its object, nature, and features and explains its theoretical basis and characteristic features. Finally, the paper proposes a new approach to such interdisciplinary study.
This paper was sponsored by the 2006 China National Social Sciences Foundation Program entitled Study of Economic Logic (Approval No 06BZX050).

1 Economic logic, a new interdisciplinary subject of economics and logic

Theory refers to a set of simple logic systems for the interpretation of phenomena. Economic logic refers to a simple theoretical system that can be applied to the exploration of the structural features, laws and methods of thinking by economic man in the course of his economic activities and research (of economics) in certain social situations. Economic logic, simply put, is a subject that studies economic reasoning and methods. Instead of using simple logic or a popularized form of "traditional logic" (similarly unlike modern formal logic), economic logic focuses on the study of thinking about true and effective access to the dynamics and mechanism in which economic value is created and realization of further abstraction and construction of its universal form.

Economic logic mainly focuses on (1) the definition of economic logic, which covers mainly the object, nature, significance and method of economic logic; (2) the laws of economic thinking, i.e. the integral logic, which is mainly composed of certainty logic in the non-normative mode, the arrangement logic of normative mode, the argument logic in the confronting mode, and the enrichment logic of advantages and disadvantages; (3) the procedures of economic thinking, i.e. ontology, which is mainly about the logical analysis of the process of thinking about choices (as choosing links all economic activities, the core of economic study therefore is choosing. To wit, almost all phenomena in the scope of study in economics is related to choosing. This part is the key to the study of economics. It is mainly composed of the logic of economic problems, the logic of

---

[1] The author is with The Research Center for Economic Logic and Renovation, Tianjin University of Commerce, East Entrance of Jinba Road, Beichen District, Tianjin 300134 China. The paper is translated by Ge Yajun.

economic information, the logic of economic forecasts and the logic of economic decision-making.); and (4) the methodology of economic thinking, which consists mainly of economic hypotheses, economic paradoxes, the methodology of logical interpretation, the theoretical development and defense of economics, the methodology of scientific logic, the methodology of dialectical logic, the methodology of pragmatic economic logic(inclusive of rhetoric), the methodology of game logic, and the methodology of fuzzy logic.

Economic logic these days has attracted the attention of researchers in logic and economic theory from China and abroad. It is recognized as a frontier problem as well. Particularly, the in-depth study and contributions of world famous masters of economics have set off a new surge in international studies in this field. For instance, in October 2005, the Nobel Prize in economics was awarded to Thomas Schelling, an American economist, and Robert Aumann, an Israeli economist, in recognition of their contributions to the promotion of understanding of conflict and cooperation. Starting with John Nash, many economists have been awarded the Nobel Prize in game theory. Game logic also constitutes an important part of economic logic. It is said, therefore, that being awarded the Nobel Prize in economics is a great victory for economic logic. A second victory for game theory clearly demonstrates the necessity of economic logic, which pushes to new highs the study and popularization of economic logic.

There are six main concepts in the study of economic logic: the concept of dialectic logic [2], the concept of scientific logic [6, p. 103], the concept of game logic [6, p. 74], the concept of pragmatic logic [6, p. 58], the concept of symbiosis logic [6, p. 460], and the concept of integrated logic, which is namely the macro logic concept, which maintains that the foregoing five are incomplete, though correct, but need to be integrated, supplemented and mutually perfected. The macro logic concept of economic logic, broadly defined, is a mode of logic system behavior that contains a number of logic branches. As a huge modality, the economic logic system is unlike a system which simply borrows, mechanically applies or patches-up various unrelated logical systems. Economic logic is a new organic system developed with scientifically concluded features, laws, methods and modes of thinking in epistemological logic (i.e. deduction logic, induction logic and analogical logic), dialectical logic, scientific logic, game logic and pragmatics logic, which is directed at all aspects of the study of economics and economic activity. [6, p. 6]

Economic logic has the following features:

a. Disciplinary features. Economic logic is an interdisciplinary subject of economics and logic, manifesting an interdisciplinary particularity.

Economic logic is an interdisciplinary study which is wide in scope and complex in target. First, economic logic considers the tools for the study of economic activity and behavior, which in turn requires the study of theoretical succession and the renovation of logic as a subject tool. Second, in studying logic structure, method and the laws of thinking in economics, economic logic considers both the methodology of economics and the ontology of each school of economics, namely the economic phenomena which reflect its special interdisciplinary characteristics and enabling its theoretical framework to remain fresh and vigorous.

b. Features of the object of study with economic reasoning and its method as skeleton

Rather than abstract reasoning in sheer logic, economic logic is a practical reasoning or pragmatic reasoning in the study of economics and economic phenomena. Economic reasoning, in particular decision reasoning or strategic reasoning, focuses mainly on how to offset the various risks arising out of any error and ignorance with the gain of knowledge, and how to offset any gain likely to be lost with success likely available. As a whole, the purpose and intention of economic reasoning lies in the realization of the balance of minimized cost and maximized return in coping with risk.[6, p. 18]

c. **Features of the theoretical system: both static logic and dynamic logic constitute its theoretical system, with dynamic logic prevailing.**

The criterion for "validity" varies in static logic and dynamic logic. Static logic or formal logic takes the true-false relation of proposition forms as the criterion for validity. The rules or principles for valid behavior of logic are subject to limitation and "intervention" by such principles as the (human) priority of value, the carefulness of behavior, the cost-effectiveness of behavior (limited rationality), and the cooperativeness of action.[5, p. 119]

2 Basis for the study of economic logic

2.1 Social and historical sources

Economic logic has ancient sources and a profound tradition in society and history. According to Karl Marx, the form of human existence changes with economic development. Wherever there are human beings, there will be economic activities, wherever there are economic activities, there will be unique thinking; wherever there is unique thinking, there be unique forms of logic. Economic logic is this unique form of logic. The idea of economic logic has had a long and sustained development.

From the Chinese perspective, economic logic originated in China over 2500 years ago in the Spring and Autumn and Warring States periods. The then logic of Mohism was developed by Mo-tse and his disciples by summarizing scientifically the features and laws of the then world-leading economic thinking in economic activities in China. It can be said that Mohist logic is the earliest economic logic in China. Its purpose is to pursue success and appropriateness, and its value is to distinguish advantages from disadvantages and resolve suspicion. "QU", the Chinese character in DA QU and XIAO QU, the two versions that carry logic, means pursuing advantages and avoiding disadvantages,. "QU" refers to "seeking for advantage and avoiding disadvantage" in action, an action combining motive with effect, which means giving rise to advantages and avoiding disadvantages. "QU", therefore, is "behavior logic"; it is about Reasoning. The Mohists advocated "TUI" (the Chinese character), which refers not only to reasoning in the sense of formal logic (which focuses on induction and analogy), but it also contains the undeniable requirements of action, cost-effectiveness (value), rationality, justness or appropriateness. The Mohists advocated a form of Reasoning which serves the values of identification, choosing, acceptance or rejection of "JIE YONG", "FEILE", "advantages" and "disadvantages". The Mohist economic logic plays a vitally important role in controlling the social, political, and economic activities of society. It was a necessary thinking tool for distinguishing right from wrong, order from disorder, and similarity from difference. It was a major economic tool for making such distinctions as concept and entity, coping with advantages and disadvantages, and determining suspicion in political and economic activities.[7, p. 2]

Master Sun's *The Art of War* in China's Warring States period claims to depict war strategies and tactics, in which the idea of decision-making (game theory) reflects the apparent beginnings of game logic in economic logic. As an American management scholar put it, , the fundamentals of game theory were already being applied to military affairs in China over two thousand years ago.[3]

2.2 Demands of the times

Why is economic logic being proposed definitely and independently as a separate field of study? While logic is closely linked to social politics, mathematics and language, we have unfortunately separated logic and economics. They should be closely tied up. General logic obscures economic logic, and the universality of logic wipes out the individuality of economic logic. Consequently, we are hardly aware of the existence of economic logic, which eventually leads to our failure to apply logic voluntarily to achieve success by pursuing advantages and avoiding disadvantages. Another reason is that in establishing and implementing the concept of scientific development and constructing a harmonious society in globalized economic activities, we need a way to sum our experiences in the form of economic logic. Meanwhile, there are many misunderstandings in us that urgently need to be addressed through economic logic. These times call for an important role to be played by economic logic.

For this purpose, we duly established the China National Society for Economic Logic in Xian in 1983 and initiated a systematic, organized study of economic logic. Over the past two decades, economic logic researchers have authored a large number of papers, writings and textbooks. A good case in point is QU Maisheng's *Work and Logic* (Beijing University Press, 1987) with 80,000 volumes already sold. Having successfully gained domestic and international fame, this book reveals the features, laws and methods of thinking in economic logic and was awarded the JIN YUE LIN Academic Prize, the supreme award in the field of logic in China. *In Pursuit of Transcendence - A Click at Economic Logic*, a book of 480,000 Chinese characters published by Hong Kong East and West Culture Enterprise Ltd. in April 2004, crystallizes the collective wisdom of the China National Special Committee of Economic Logic and epitomizes the research results from economic logic in China. This book reflects the academic peak of the study of economic logic in China today.

2.3 Theoretical study of economics in need of the birth of economic logic

Economists with innovative insight attach considerable importance to the study and application of economic logic. Adam Smith's great contribution is one of the foundation texts of classical economics. It helped to create economics as an independent subject and provided a tradition for economic logic within the study of economics. Since Adam Smith, economic logic has been an important tradition in mainstream western economics since David Ricardo. For instance, 1970 Nobel Prize winner in Economics Paul A. Samuelson cautioned against economic man in his popular textbook Economics (17th edn.) Chapter I was entitled "The Logic of Economics". Samuelson warned that we must be on guard against various thinking fallacies common in economic reasoning, that many different variables are involved due to the complicated nature of economic relationships. He argued that economic man is likely to confuse the true reasons behind an event and the impact of government policy on the economy. In particular, he analyzed three common fallacies: the post hoc fallacy, the fallacy of failure to hold other things constant, and the fallacy of composition.[8, p. 3]

Many other famous economists and business managers deem logic as something indispensable in economics. According to John Maynard Keynes, economics is a branch of logic, a mode of thinking [1, p. 100]. According to British economist David Smith, thinking like an economist means analyzing problems by means of logic and substituting analysis for affirmation with. [9, p. 5]. Chinese economist LIN Yifu argues that in founding a theory of economics, a clear definition must be provided for targeted questions and given conditions; reasoning from premise to conclusion must comply with the strict norms of formal logic. [4, p. 105]

3 Problems and focus in the study of economic logic

In the study of economic logic, four problems stand out. The first one is the object and nature of economic logic, the second is the selection of the cut-in of economics and logic, the third is the combination of scientificity and practicability in economic logic, and the forth is the coordination of diversity and unification as a tool in the theoretical framework of economic logic. As an instrument theory, logic is various in category and rapid in development, so effort needs to be exerted on which logic to take as the study tool and how to deal properly with its formality and informality. Further efforts need to be made on how and from which aspect to select the object of economic study. This is because, for object theory, there are many schools of economics which have achieved rapid development. These are the four major problems in the study of economic where breakthroughs need to be made in the future.

Science develops on a combined path of polarization and integration. Academic division of work tends to create barriers between subjects which hinder discovery. As Karl Marx said, problems are a sign of the times. Having emerged in response to the problems of the times, economic logic is itself growing rapidly as it responds to the economic problems of human beings.

Bibliography

[1] Blaug, M. *The Methodology of Economics*. Beijing University Publishing House, 1990.

[2] Gui, Q. Deng xiaoping's "inequality of economics" and dialectic logic. In *Proceedings of the National Dialectic Logic Conference*. Guilin, 2005.

[3] Hoch, S. J., Kunreuther, H. C., Gunther, R. E. *On Making Decisions*. Shanghai Jiaotong University Publishing House, 2003. Translated by WU Hong.

[4] Lin, Y. *On Economics Methodology*. Beijing University Publishing House, 2005.

[5] Pan, T. *Introduction to Behavior Science Methodology.* Central Translating & Editing Publishing House, 1999.

[6] Qu, M. (Ed.). *In Pursuit of Transcendence - A Click at Economic Logic.* Hong Kong East and West Culture Enterprise Ltd, 2004.

[7] Qu, M. Economic logic and construction of harmonious society. *Journal of Anhui University, Issue 5,* 2006.

[8] Samuelson, P., Nordhaus, W. *Economics.* People's Post-telecom Publishing House, 2004. Chief translated by XIAO Chen, Version 17.

[9] Smith, D. *Free Lunch: Easily Digestible Economics.* Electronic Industry Publishing House, 2004.

A Model for the Construction of Institutional Facts by Speech Acts[1]

Song Chunyan

Hunan Academy of Social Sciences

scysts@gmail.com

ABSTRACT. This paper presents a model for the construction of institutional facts by speech acts based on John Searle's concept of social ontology. Social ontology has to do with how society constitutes itself in social facts, especially institutional facts, a special class of social facts–such as government, money, education. One of the most fascinating features of institutional facts is that a very large number of them can be created by explicit performative utterances, which are members of the class of declarative speech acts. In conclusion, I define a function which models the construction of institutional facts by declarative speech acts.

1 Introduction

John Searle's social ontology is described first in his book Speech Acts: *an Essay in the Philosophy of Language* and further developed in *The Construction of Social Reality*. In writing *Speech Acts: an Essay in the Philosophy of Language*, he found that speech acts not only represent facts but also construct facts. After making a distinction between brute facts and institutional facts, he observed that a large number of institutional facts can be created by explicit performative utterances[6, p. 34], which are members of the class of declarative speech acts.

Institutional facts are the fundamental fabric of society. They exist only because human beings believe them to exist - things for example like money, marriage, and government. Institutional facts have three important characteristics[3]: collective intentionality, status function, and constitutive rule. Institutional facts must be accepted collectively, so the speaker of a speech act who is constructing an institutional fact must be generally recognized to be empowered or vested with the authority to represent the collective intentionality. The status function means that the institutional facts have been invested with a new function because they have been assigned the new status, e.g. money acquires a new function in commodity exchange. The constitutive rule means that every institutional fact has the constitutive rule of "X counts as Y in C".

Searle thought that collective intentionality can be represented by language, which can stand for the collective acceptance. Moreover, we always construct direct institutional facts by declarative speech acts. Sometimes we also need

---

[1]The author is with Hunan Academy of Social Sciences.

commissive speech acts to construct indirect institutional facts, but for the purpose of this paper, we will only discuss declarative speech acts.

2 A macro-level model for the construction of institutional facts

We live in one world, not two or more. We communicate with each other every day. In Searle's eyes, the basic unit of communication is the speech act. In every speech act, each person is not just a speaker but also a hearer. By speech acts, we link each other together in a network. We can see each speaker-hearer pair as a node in the network. A series of nodes makes up a chain which can stand for a possible world of institutional facts. In each possible world, every institutional fact can be constructed by one speaker and one hearer. As a result, on the one hand, there are no institutional facts which exist independently. Every institutional fact must exist in an institutional network; on the other hand, every institutional fact can be constructed by the successful performance of corresponding speech acts. Therefore, the speaker in the construction of one institutional fact can be the hearer in the construction of another institution fact; the network of speech acts also can stand for the network of institutional facts in the real world. At last, we get a macro-level model for the construction of institutional facts as Fig.1.

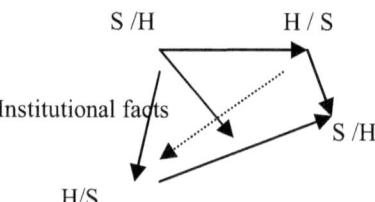

Figure 1: A macro-level model for the construction of institutional facts S: speaker; H: hearer

3 A micro-level analysis for the construction of institutional facts

Institutional facts have a status function which is imposed on them by the human beings who comprise a speech community. The status function has a deontic power which takes the form of moral responsibility, obligation, and rights. One of the most fascinating features of institutional facts is that a very large number of them can be created by explicit performative utterances, which are declarative speech acts. In the declaration, the state of the entity is represented by the propositional content of the speech acts. Institutional facts are constructed by the successful performance of speech acts. Declaration brings about some alteration in the status or condition of the referenced object or objects solely by virtue of the fact that the declaration has been successfully performed. This feature of declarations distinguishes them from the other categories[5, p. 17]. Therefore the construction of institutional facts is a process of imposing status by creating things in words. The words transfer the status from the speech acts themselves to the institutional facts which had no such status function before. Moreover, the construction process is a process which creates and conveys new power. In conclusion, the status power of the speaker is represented as the illocutionary force of the speech act, which is also transferred to the institutional facts, where they are represented as the deontic power of the status function. The status of the speaker conveys and assigns the status to the status function of the institutional facts. For example, when the National Treasury Department declares that this paper in my hand is 20 yuan, then this paper becomes legal tender money valued at 20 yuan for no other reason except that the National Treasury Department says that it is. From its creation as an institutional fact, the money itself then takes on new functions, such as a medium of commodity exchange. We can use it to buy food or exchange it for other goods or services. From Fig.2 we see a micro-model depiction of the construction process.

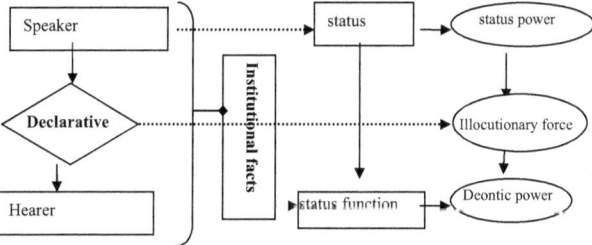

Figure 2: A Micro-model of the Construction of Institutional Facts facts

4 A logical framework for the construction of institutional facts

In Fig.2, we have made a deliberate analysis for the construction of institutional facts by declarative speech acts as a hypothesis. The hypothesis is that we take the successfully performed declaration for granted. In fact, there are several conditions which ensure its success. I would like to call attention to the feasibility condition here to stand for it (I will make further study of it later).

J. L Austin[1, p. 109] is the founder of speech act theory. He said, first we distinguish a group of things we do in saying something. In doing this we perform locutionary acts and illocutionary acts. We may also perform perlocutionary acts. Only by successfully performing the declaration, the perlucutionary acts happen. Here we have three, if not more, different senses or dimensions of the 'use of a sentence' or of 'the use of language' as a speech act (and, of course, there are others also).

For the construction of institutional facts by declaration, we need to make an analysis of the three types of speech acts: (1) a locutionary act refers to a speech utterance "I declare + clause"(or I verb NP1+NP1 be pred), with a certain sense and reference, which is roughly equivalent to 'meaning' in the traditional sense ; (2) an illocutionary act refers to the fact that the speaker believes himself to have the institutional power to make such a declaration; his intention in performing this speech act informs the hearer that it is obligatory for the hearer to act in a prescribed way; (3) a perlocutionary act refers to two levels of effect: first, the hearer understands the speaker's meaning, called effect P; second, the hearer finishes the construction of the institutional fact by carrying out the intended action, called effect Q. The successful performance of illocutionary acts is the premise and foundation of perlocutionary acts. The first perlocutionary effect P is the premise of the second perlocutionary effect Q.

Only with a successfully performed declaration, Austin argued, the perlucutionary act happens. Searle developed and refined Austin's theory. He defined four premise rules for "how to promise"[4, p. 66–67] : the propositional content rule, the preparatory rule, the sincerity rule, and the essential rule. Austin and Searle both enlightenmented us on how to analyze the preconditions of perlucutionary acts that should be true if the declaration execution is to succeed. Here I would like to provide some notes about the preconditions (S: speaker; H: hearer; A: acts to be performed; p: the content of the utterance):

1. S and H can understand each other, und (S, H);
2. S and H can believe each other, bel (S, H);
3. S believes himself to have institutional power to declare A, power (S, A);
4. S's intention in performing this action is to make H obliged to do A, obi (S, H, A);
5. S intends to declare A, intend (S, A);
6. Unless this declaration is accepted, H can do nothing, apt (H, A);
7. A didn't happen before this utterance p, before (p, A);

Then we can get a conclusion: preconditions (A): =und(S, H) $\wedge$ bel(S, H) $\wedge$ power(S, A) $\wedge$ obi(S, H, A)$\wedge$intend(S, A)$\wedge$apt (H, A)$\wedge$before (p, A) and before (p, A)$\rightarrow$ intend(S, A). The most important part of prediction is bel (S, H), standing for speaker and hearer must believe each other. Austin also agreed with this solution. If there is no historical default about someone, we can trust others. We must believe in each other, and then we can construct institutional facts.

After we grasp the preconditions, we can go on with our study. The institutional facts have such constitutive rule as "X count as Y in context C". X imposed a new status function to Y by virtue of speech acts. Then Y gets new power "S does A"(S: A single individual or a group, A: the name of an act, action, or activity, including a negative or positive event). As institutional facts must be accepted collectively, so the speaker of the speech act in the construction of the institutional fact must represent or stand for the collective intentionality. Searle once gave the structure of the collective intentionality as: we accept (S is enabled (S does A)). If the activity is a positive event, e.g. a marriage, we can represent the form precisely: we accept (S is enabled (S does A)); for a negative event, such as divorce, we can represent it as: we accept (S is required (S does A)). On the basis, we can draw out the general form of the construction of the institutional facts: We accept (the agency creates S is enabled (S does A)). The negative of this formula is: We no longer accept (the agency creates S is enabled (S does A)). In one word, we can use the function accept (S, A) to describe this process.

In addition to what has been outlined above, the construction also requires another important condition, which is called "background ability" by Searle. Background ability refers to the set of non-intentional or pre-intentional capacities that enable intentional states to function. We can describe background ability as functional background (S, H).Finally, the last condition is that every institutional fact must exist as the premise for others which have something to do with it, so we can make function ins (A) to describe this relationship. As a result, we can make a function to describe the construction of institutional facts by speech acts: <S, H, A, preconditions (A), accept (S, A), background (S, H), ins(A)>, and P→Q.

Bibliography

[1] Austin, J. L. *How to do Things with Words*. Oxford: Oxford University Press, 1962.
[2] Demolombe, R., Louis, V. Speech acts with institutional effects in agent societies. In Goble, L., John-Jules, C. M. (Eds.), *Deontic logic and artificial normative systems*. Springer Press, 2006.
[3] Searle, J. *Social ontology and political power*. URL http://www.california.com/~rathbone/searles.htm.
[4] Searle, J. *Speech Acts: An Essay in the Philosophy of Language*. London: Cambridge University Press, 1969.
[5] Searle, J. *Expression and Meaning: Studies in the Theory of Speech Acts*. Cambridge University Press, 1979.
[6] Searle, J. *The Construction of Social Reality*. New York: Free Pres, 1995.

Laws, Causality and the Intentional Explanation of Action[1]

Xu Zhu

Tsinghua University

xuzhu05@mails.thu.edu.cn

In studies of the philosophy of the social sciences, there are two fundamental problems about the intentional explanation of action which have caused many controversies: (1) should intentional explanation involve universal laws? (2) should intentional explanation be considered as a type of causal explanation? In this paper, I will argue that a valid intentional explanation doesn't have to involve a law, but even if this is true, the explanation can still be a causal one.

Karl Popper and C. G. Hempel both argue that explanations of action must involve universal laws. Popper believes that the explanations in history are based on those laws, provided by the general sciences, which are too trivial to attract enough attention. Hempel [5, p. 235] insists that, if the explanations of action want to be counted as scientific ones, they should involve general laws: "Historical explanation ...... is not prophecy or divination, but rational scientific anticipation which rests on the assumption of general laws." W. Dray, however, has done a crucial critique of the Popper-Hempel thesis. He thinks that thesis is totally wrong because, there is a need for an infinite number of elements to predict by means of laws in the socio-historical field. Thus, that sort of prediction really cannot be made; on the other hand, the so-called "laws" which seem to be able to provide predictions usually don't meet Hempel's requirement for a "law like" sentence. In terms of the first reason, MacIntyre [7, p. 91] claims that generalization in the social sciences does not have the precision conditions of validity as laws do in the natural sciences. As Bohman [2, p. 24] put it, it is a "trade-off between truth and generality" in Hempel's model. Dray [4, p. 132] insists that the aim of intentional explanation is reconstruction not of arguments, like in explanations of natural phenomena, but of practical reasoning.

Hempel's [5, p. 470] response, from his Schema R, that Dray's model can only achieve "partial explanation", since it "affords grounds for believing that it would have been rational for A to do x, but no grounds for believing that A did in fact do x". However, only when he presupposes intentional explanation has the same aim as those in natural science, Hempel's point can be effective. But that premise is exactly what Dray has refuted. Hempel hopes that empirical studies will tell us how to make a good prediction in the social sciences. Nevertheless, we have indeed many effective explanations but no precise general laws in the social sciences. I think Davidson [3, p. 17] is right to say, "it is an error to think no explanation has been given until a law has been produced". Therefore, a valid intentional explanation doesn't have to involve general laws.

---

[1] The author is with Center for STS, Tsinghua University, Beijing 100084 China.

Laws, Causality and the Intentional Explanation of Action 325

No matter whether intentional explanation needs general laws or not, the regularitist theory of causality, first founded by David Hume, has been accepted factually by both exponents and opponents of the Popper-Hempel thesis. The exponents believe that intentional explanation is causal since it involves laws, while the opponents hold the opposed claim by arguing it doesn't need laws. However, this premise has introduced many problems into the discussion. Anscombe [1, p. 16] claims the mental cause of intentional action is in the non-Humean sense. One main distinction is that the repeated experience, or constant conjunction in Hume's terms, is not necessary to show mental causation [8]. Though holding the principle of the nomological character of causality, Davidson [3, p. 215] emphasizes that this principle doesn't mean "every true singular statement of causality instantiates a law".

Therefore, we actually have two alternatives. Following the regularitist theory, intentional explanation will be identified non – causal, and the only acceptable causality is on Humean sense. However, I insist that we should reject the regularitist theory in order to argue for the causality of intentional explanation. The reason is that causality can conveniently meet the need of justification of the explanation.

W. Salmon [11, p. 323] defends the intuitive relation between causality and explanation. Intuition tells us that it is justifiable to use causes to explain effects, which is also applicable in the explanation of action. Davidson [3, p. 9] makes a clear distinction between reasons had by a person for an action and those why s/he did it. Only the latter can provide a justified explanation of that action. I suggest that should be interpreted as using cause to explain effect, since, obviously, causality is a convenient conceptual framework to provide justification for the intuition.

Certainly this kind of causality will be in the non-Humean sense. The realist, or ontic in Salmon's terms, conception of causality can be a possible alternative. For Salmon [11, p. 325], the causality of explanation cannot, as claimed by Hempel, be reduced to universal or highly probable laws, but it is based on some ontic mechanism. Different mechanisms can be specified especially in a particular domain of objects. Then, "what constitutes an adequate explanation may differ from one domain to another in the actual world" [11, p. 326].

If this is right, it is rational to believe that explaining the action A with a reason R can be justified if causation between *R and A can be shown*, which is a special mechanism in the social domain and may not be reduced to regularity. As Davidson [3, p. 215] put it, "causality and identity are relations between individual events no matter how they are described. But laws can be linguistic; and so events can instantiate laws and hence be explained or predicted in the light of laws, only as those events are described in one or another way". Even though we don't have the means to employ laws, therefore, the intentional explanation can still be causal in a realistic conception which provides justification for that explanation.

From this point of view, many seemingly plausible arguments can be defeated. First, G. Ryle [10, p. 113] thinks reason cannot be a cause because it belongs to disposition, while every cause should logically be an event. Following from Wittgenstein, Ryle believes that, if motive and reason are treated as events, then we'll permit some private mental objects–definitely a bad philosophy.

Davidson [3, p. 12] replies to two aspects of this view. One is that disposition can actually serve as a cause in explanation. The other is that even if reason itself cannot be a cause, the "onslaught" of reason, which means coming to be in the mental state or acquiring the reason [9], can be considered as a causal event. That kind of event doesn't need to presuppose any private objects to be acquired but is only a description, with mental words, of observable action.

Second, as P. Winch [12, p. 125] implies, reason usually has an "intrinsic" relation to action, while the natural causal relation is an extrinsic one. The intentional explanation can seemingly be a re-description of an action with mental words. Intentional explanation cannot be causal, therefore, since there will only be an analytic relation between the two descriptions of the same thing while the causal relation should be synthetic.

However, Davidson [3, p. 10] claims we can redescribe events with their causes. For instance, "someone was injured" can be redescribed with "he was burned". A more powerful criticism, proved by Quine, is that there is no absolute distinction between analytic and synthetic. As Davidson [3, p. 14] points out, "The truth of a causal statement depends on what events are described; its status as analytic or synthetic depends on how the events are described". If we accept a realistic conception of causality, there will be no longer any necessary requirement for a causal statement to be synthetic, since those are on a different level. It is not a big deal even if mental causation appears to be analytic.

At last, Winch [12, p. 92] suggests justifying the intentional explanation with social rules rather than the causal model. "One can act 'from considerations' only where there are accepted standards of what is appropriate to appeal to." Interpreted by MacIntyre [6, p. 214], Winch actually provides a "two-stage model": "An action is first made intelligible as the outcome of motives, reasons, and decisions; and is then made further intelligible by those motives, reasons, and decisions being set in the context of the rules of a given form of social life." The second stage of Winch's model is supposed to provide the justification for the explanation. Since the primary reason for defending causality in the intentional explanation is to meet the needs of justification, we may have to accept that intentional explanation is non-causal. If Winch can prove that with social rules only, then can we justify the explanation?

However, Winch is not successful. As Davidson puts it [3, p. 10], "it is an error to think that, because placing the action in a larger pattern explains it, therefore we now understand the sort of explanation involved". Corresponding to that comment, MacIntyre [6, p. 217] claims that an important distinction should be clarified between rules "which agents in a given society sincerely profess to follow and to which their actions may in fact conform, but which do not in fact direct their actions", and other rules "which, whether they profess to follow them or not, do in fact guide their acts by providing them with reasons and motives for acting in one way rather than another". Unfortunately, Winch totally ignores that difference. Only the second kind of rules, which actually show a causal relation between explanatory reasons and actions, can justify the explanation,. Therefore, it seems necessary to employ the causal model to justify the explanation.

In conclusion, intentional explanation can be causal, though it may not involve any universal laws. The causality relevant to reason and action doesn't mean any reduction in regularity from observation but only reveals a certain ontic mechanism in the non-Humean sense. From the arguments above, three points can be specified as necessary characters in that kind of causality: (1) the event that "agents have got those reasons" can be a cause of their actions; (2) the causality in intentional explanation can usually be shown or described as analytic relation; (3) if explanation involves some social rules, they should also show some causal relation between reasons and actions.

Bibliography

[1] Anscombe, G. E. M. *Intention*. Cambridge: Harvard University Press, 2000.

[2] Bohman, J. *New Philosophy of Social Science*. Cambridge: Polity Press, 1991.

[3] Davidson, D. *Essays on Actions and Events*. New York: Oxford University Press, 1980.

[4] Dray, W. *Laws and Explanation in History*. London: Oxford University Press., 1957.

[5] Hempel, C. *Aspects of Scientific Explanation*. New York: Macmillan, 1965.

[6] MacIntyre, A. *Against the Self-Images of the Age*. Notre Dame: University of Notre Dame Press, 1978.

[7] MacIntyre, A. *After Virtue*. Notre Dame: University of Notre Dame Press, 1984.

[8] Macklin, R. Action, causality, and teleology. *The British Journal for the Philosophy of Science*, 19:301–316, 1969.

[9] Macklin, R. Explanation and action: Recent issues and controversies. *Synthese*, 20:388–415, 1969.

[10] Ryle, G. *The Concept of Mind*. London: Hutchinson House, 1949.

[11] Salmon, W. C. *Causality and Explanation*. New York: Oxford University Press, 1998.

[12] Winch, P. *The Idea of a Social Science*. London: Routledge, 1990.

A Probe into the Question of Ethical Restrictions on Military Technologies[1]

Yan Wei

Dalian University of Technology

yanweidl@sohu.com

Liu Zeyuan

Dalian University of Technology

ABSTRACT. In recent years, rapid developments in hi-tech and information technology have driven even more rapid innovations in military technology. Faced with many new ethical challenges and the problems posed by new military technologies spread by globalization, we have to define and understand their ethical implications and then modify or change their affects. Information Warfare, for example, has emerged as a decisive factor in 20th and 21st Century modern warfare. It requires that we carefully develop a new morality regarding military technology. The paper explains the need for ethical improvements with respect to military technology and proposes some techniques which can be used to enforce ethical restrictions. The paper then analyzes some methods for realizing a new military ethics.

Given the dual possibility of doing good things or evil things with the modern technology at our disposal, it is necessary that traditional ethics must be extended and applied to new concrete areas in the use of the technological means at our disposal. As once described by the American technique philosopher Mitcham, the concepts and principles of traditional ethics need to be extended and applied in such new areas as nuclear research and technology, environmental science, and the ethics of calculation.[5] This paper discusses ethical concerns for the restriction of military technology development based on war ethics.

1 The development of war and ethical improvements in military technology

Though the statistics are admittedly sketchy, there were 4500 wars from 3200 B.C. to 1999 which greatly influenced human culture and society. Due to wars, 7,000,000,000 people died. The theory, the meaning and the methods of war were especially evident in the trend of the development of war. Driven by rapid innovations in science and technology, the destructive power of weapons increased each day during the 20th Century. When B-17s dropped atomic bombs on two Japanese cities, it marked the beginning of the era of nuclear warfare. Nuclear weapons became the strategic foundation for the army, navy and air force. Today, guided missiles can carry 10-15 independently targeted nuclear warheads and launch a nuclear attack on any target in the world. In the nuclear age, because

---
[1] The author is with Box 808 No.1 XiaoLong Street, Zhongshan District.

of the total destructive capability of nuclear weapons, space and time factors which could prevent destruction in the past have lost completely their defensive function. Because of "nuclear winter" theories, defense experts and strategists now recognize that it makes common sense to curb the offensive and defensive use of nuclear weapons.

Any act of war can result in injury to a certain extent to both hostile parties. Any worthy argument in support of war has to obtain effective support from considerations of ethics and politics ethics. The act of war has close ties with ethics from its creation. War is "politics by another path."[2] War is armed conflict of nation and nation, race and race, rank and rank or political group and political group. War is the ultimate means for resolving mankind's conflicts. Ethical restrictions on war are the result of what happens as mankind's civilization evolves.

The military techniques for conducting war have two kinds of meanings: the narrow sense means mankind has the ability to develop weapons and can realize ethical and political means and the methods through weapons. It is mainly mankind's technique for applying the weapon technology. The broad sense refers to the means and the method of mankind's thinking, the bloodshed way of solving benefits conflict. The essence of military technique is the means which solves an interpersonal extreme antinomy embodied in interpersonal relationships. Military technical developments are closely tied to ethics by all means. Nuclear weapon technology brings the art of war to a peak. The weapon development which adheres to military technologies causes a series of ethical problems: the problem of value direction of weapons development, the problem of limiting weapons during the period of application, and the problem of ethical stipulations for weapons development. The ethics of military technology development should adapt to the requirements of mankind's civilization continually to enrich its contents.

2 Techniques for ethical restrictions on military technologies

Presently, some people think that ethics in military technology is not reasonable ("justice war theory"), but that the target objective should determine the rationality of its existence. The ethics of military technology studies how the methods of pursuing war should be restricted by morals or why rules of morality will impose reasonable limits on the use of military technology in the end. As a kind of moral power, ethics develops with social improvement. Military morals and studies have new meaning in this era of nuclear weapons–the atomic bomb, hydrogen bomb, the electronic jam, bio-chemical weapons, and laser weapons. Military technology has exerted a great influence on the development of human society and interpersonal relationships. Information technology has emerged as the technical condition which decides and limits the methods for pursuing war.

In the mid-20th century, especially in the 70's and 80's, taking the microelectronics and the computer technology as the standard, new technology began to sweep the world. Information technology has to do with information gathering, examination, transformation, display, saving, processing, transportation, identification, withdrawal, control and the application of technique. Actually it, as a kind of technology cluster, includes micro-electronics, photo-electronics, laser technology, fiber-optics, superconductive technology, etc. With constant progress in modern science and technology, drawing on information technology as a core, extensive application of biology, aerospace, and material technology in military research directly drove the rapid development of reconnaissance, automatic direction, military aerospace, electronic information and nuclear bio-chemical technology, etc..

Each age makes its own unique demands on military technology. There are always conditions for the occurrence of military technique transformation and its completion. It doesn't only need higher scientific technology to guide it. It also requires higher economic development and a strong industrial foundation. The emergence of missile nuclear weapon is inseparable with the development and breakthrough of the atomic physics, chemistry, aerodynamics, mechanical engineering, systematic science and rocket technology, electronic technology, and automatic control technology. War has become an information-based enterprise. Taking information technology as its core, military research has applied many new technologies and provoked a series of technological revolutions.

3 Methods for the realization of a military ethics

When high technology greatly enhances the destructive power of the war making capacity, it poses a series of new ethical questions. It mandates that people in power increasingly need to restrict the development of military technology on the basis of ethics. In order to accomplish this, we need to work through international societies and non-governmental organizations. And we need to make full use of information technology. We need to regulate military technology and weapons, lower the destructive capacity of war to human life, property, and the environment, and gradually reduce the frequency of war, and ultimately even exterminate war.

3.1 Intervention of international societies

In recent years the development of international disarmament exercises symbolizes the rise of international humanitarianism. This spirit of humanitarianism has already regulates to some extent the means by which war is waged. In 1899 and 1907 The Hague Convention Article 22 stipulates: Concerning the methods which are used in order to hurt the enemy, there are limits for battling nation's power. In 1977, the Geneva Agreement lists "war means and methods" as its first principle and subject of concern. With the development of international society, the United Nations started to take moral responsibility for solving destructive disputes. The International Arms Control Treaty forbids or restricts some weapons of mass destruction such as chemical weapons and restricts research and development of these kinds of weapons.

4 Intervention of non-governmental organizations (NGOs)

Information technology and globalization enable people in the world to see clearly the terrible images of war. More and more people request that we should take action to stop war and regulate the methods used to wage war. The politicians are bound to have a reaction. If they succeed, it will promote their reputations and probably can transfer focus from them onto the failure of nations to take such positive steps. The reason for the need for non-government organization intervention is that to some degree, war is subject to international convention. When these conventions are not obeyed, non-government organizations can assert international pressure and force conflicting nations to obey these conventions. In order to intervene effectively, the non-governmental organizations have to maintain an on-going dialogue with all parties to conflicts.

4.1 Gentler techniques soften war

Recently, many high-tech regulating weapons have been developed which have the destructive force of small nuclear weapons. With significant developments in scientific technology, war has increasingly presented a "softening" trend. The revolution in scientific technology has come about by harnessing the power of machine technology, electronic technology, and nuclear technology. It has led to a "gentle technique revolution" (taking information technique as an example). It makes it possible to realize ethical restrictions on warfare. In the past, annihilating the enemy was the goal of warfare. It was the core of Napoleon's military strategy: "I will do my best to annihilate the enemy". Along with progress in military technology, the purpose of annihilating the enemy now has to be balanced against the goal of saving ourselves. We need to learn to control the enemy and protect ourselves at the same time. It is called "win without war". The appearance of benign weapons is evidence for the argument that gentler more humanitarian techniques can soften war's destructive power.

Bibliography

[1] Cai, Q., Shen, P. *Background to Social Problems under Globalization.* Beijing: Beijing University Press, 1997.

[2] Clausewiz. *War Theory.* Beijing: Commercial Press, 1978.

[3] Huang, H. *The Report of the World's New RMA.* Beijing: People's Press, 2004.

[4] Liu, J., Zeng, H. Ethics of war: A perspective of the nation about the world. *Studies in Ethics*, 2006. April.

[5] Qiao, R. *Technique Philosophy: General Outline.* Tianjin: Science Technique Publishing, 1999.

[6] Shi, H. The influence of the progress of battle on military techniques. *Defense Science and Technology*, 2005. July.

[7] Wu, R. *Information-based War Theory.* Beijing: Military Science Press, 2004.

[8] Xu, L. *Technique Philosophy.* Shanghai: Fudan University Press, 2005.

The Ethical Connotations of I. Bernard Cohen's Thoughts on Scientific Revolution[1]

Niu Junmei

Southeast University

niujunmei1981@126.com

The main purpose of this essay is to explore the ethical connotations of I. Bernard Cohen's thoughts on scientific revolution. In his book *Revolution in Science*(1985) [2], Professor Cohen, the eminent historian of science, amply documents the chronological history and successive transformations of the word 'revolution' from the scientific to the social and back to the scientific spheres through the last four centuries and analyzes the stages by which revolutions in science progress from inception of a revolutionary idea to the acceptance and use of a new science by other scientists, individuals, and their communities. Based on his historical and analytical study, Cohen also develops a set of criteria for the occurrence of scientific revolutions and uses them to examine critically major revolutions which have occurred in science such as the Copernican Revolution and the Newtonian Revolution.

Absorbing insights from STS into his history of science research, Cohen's book transcends the "internalist" - "externalist" dichotomy and begins from the assumption that a complex societal system is composed of science, both natural and social, and many other equally important social structural and cultural elements, including economic and political factors on the social side, and religion, values, and ideology on the cultural side. In other words, as a social institution, science is only one important aspect of a complex system and, therefore, has reciprocal relations and interacts with all the other components of society.

According to this assumption, in Cohen's view, the occurrence of scientific revolutions not only involves the mutual relations of the natural and social sciences but also the close interactions of both of them with many other single or sets of social and cultural factors. And it is the multi-factor, non- determinant, non-monistic nature of scientific revolutions that makes the combination of the study of revolutions in science with the study of ethics possible. By contrast, if either the natural sciences or the social sciences is the simple determinant of the other, or is simplistically determined by any other social structural or cultural factors, that is, if science as a whole is either wholly independent of or dependent on other factors, the ethics of science and technology has no objects and thereby its existence is inconceivable. So his discussion helps to explore some of the most crucial issues relating to the role of scientific revolutions in the history of science

---

[1]The author is with Department of Philosophy & Science, China University of Mining & Technology.

and in daily life. Generally speaking, the ethics of science and technology focuses on the partly interdependent relations among science, society and nature and always needs to answer such basic questions as: how does science impact society and how does society affect the development of science and the knowledge it produces? How are objects constructed by subjects during the unfolding of a scientific revolution? How are the ethical relationships among subjects of scientific activity formed? Why are the responsibilities of subjects indispensable and what are the social and moral responsibilities of scientists, enterprises, citizens and the country?

Professor Cohen's research is useful to ethics scholars who explore the above questions because it provokes a unique theoretical approach to the study of how to unveil the ethical relations and ethical essence of scientific revolutions so as to make science useful for humanity. In this paper, I will mainly discuss three aspects of the ethical connotations of I. Bernard Cohen's thoughts on scientific revolution:

First, Cohen's thoughts on scientific revolution indicate the ethical essence of the foundations (the aims and the methods) and the process of scientific revolutions. In his view, there are at least two novel and characteristic internal criteria for the occurrence of a revolution in science: methods and aims. On the one hand, not only the newly formed method, which combines experiment and observation with mathematics, is of special value for scientific epistemology and the democratic nature of knowledge, but it also shows the scientists' state of moral cultivation. Cohen claims that "anyone who understood the true methods of scientific enquiry and had acquired the necessary skill to make experiments and observations could have made the discovery in the first instance-provided"[1, p. 6]. On the other hand, the widely shared conviction that the true goal of the pursuit of the scientific understanding of nature must be to better everyday life through applied science constitutes an important feature of revolution. Hence, the transition of aims, another sign of revolution in the internal aspects of science, reveals the historical distinctions among attitudes towards the ethical relationships among scientific revolution, humans, society and nature, of which the proper ethical relationship between science and nature is very critical nowadays as the environmental crisis has become more and more acute and global in scope.

As for the process of a successful scientific revolution, Professor Cohen claims that there are four major and clearly distinguishable and successive stages in a scientific revolution. First there is the 'intellectual revolution', the 'revolution-in-itself'.[2, p. 28] According to Cohen, the first stage of revolution comes about when "one or more scientists are always found to accomplish at the beginning of all revolutions in science. It consists of an individual or a group creative act that is usually independent of interactions with the community of other scientists. It is complete in itself." The second stage is 'a commitment to the new method, concept or theory'.[2, p. 29] This phase still remains private. The third stage is 'revolution on paper', "in which an idea or set of ideas has been entered into general circulation among members of the scientific community."[2, p. 29] The last stage is a successful revolution in science—-one that influences other scientists and affects the future course of science. On this basis, Cohen believes that the scientific tradition and scientific revolution could keep a reasonable tension through a series of gradual transformations in the course of revolutions.

Additionally, he considers the citation by contemporaries and successors, both positive and negative, as an important criterion of success in a scientific revolution. From the angle of ethics, each phase of Cohen's scientific revolution is filled with moral dilemmas which are the concrete manifestations of value conflicts among scientists, scientific communities, society, nature and humanity.

Second, Cohen's brilliant study of the effects of scientific revolutions in comparison with political revolutions implicates the value-dependence of scientific revolutions and testifies to the impossibility of the perfect neutrality of science. In Cohen's opinion, revolutions in science have two main effects, one social and the other scientific. On the social side, Cohen agrees with Kuhn that many scientific revolutions are not only revolutions in science but also revolutions in value systems and social ideology. Hence, as the most important impetus of technology and economy, scientific revolutions have a significant influence on society and its ideology and even political transformation. At the same time, a revolution in science encounters numerous social obstacles and can fail at any one of these four stages for a variety of reasons. On the scientific side, there are various moral conflicts between scientists and their communities during the course of a scientific revolution, such as scientists' conversion from conservative ideas to innovative ideas, disputes over the priority of discovery, and the demurral to the scientific transformations and revolutions from the perspective of religion and so on. Among them, a primary challenge to the social sciences and humanities from Scientific Revolution which produced modern science was to accommodate a new science to the "exact sciences" such as mathematics and physics. In his research into interactions between the natural sciences and the social sciences since the Scientific Revolution, which was Cohen's other academic interest, he holds that the great accomplishments of the natural sciences has led to the emulation of the natural sciences by the social sciences and humanities, which consequently bring promotion and demotion to science and society simultaneously, thus creating crises of legitimacy in the social sciences and humanities as well as the value crisis of its own validity. So he advocates that scientists should respect the others' achievements and scientific revolutions should be conducted in a proper way so that others easily could understand and accept them.

Finally, Cohen's investigations into specific criteria for determining whether or not there has been a revolution and the creative factors in producing a revolutionary new idea are very enlightening for us. They define the research rights of the subjects and their responsibilities for the consequences. The climate, inside and outside of science, today influences on a large scale the development of the four phases in the unfolding of a scientific revolution, and individual scientists or groups of scientists are always part of a wider social environment or context with which they are in constant communication and which has strongly shaped their knowledge, skills, and resources, and so on. In consequence, not only scientists and their communities but also all the related subjects and their organizations should take corresponding responsibility for scientific revolutions. And since ethical responsibilities flow from all human relationships in scientific revolutions, from the interpersonal to the social and professional, it is believed that the motivation and consequences and the harmonious development among humans, nature and society must all be taken into consideration. Only the individuals and groups that voluntarily shoulder the burden and accept personal or collective responsibility for the entire society could be considered responsible organizations and communities.

Cohen's explorations uncover nothing less than the nature of all scientific revolutions. His analysis is one of the most impressive surveys of the history of science ever undertaken and provides a unique perspective for defining and understanding the ethical-moral essence of revolutions in science and solving the ethical and moral dilemmas posed by contemporary science. The purpose of the ethics of science and technology is primarily to offer a set of values, principles, and standards to guide decision making and conduct when ethical issues arise in the process of scientific activity, but obviously it does not provide a set of rules that prescribe how the concerned subjects of scientific revolutions should act in every situation, so specific applications of ethical codes must take into account the context in which they are being considered and the possibility of conflicts among science, society and nature.

Bibliography

[1] Cohen, I. B. *The Newtonian Revolution: with Illustrations of the transformation of scientific ideas*. Cambridge: Cambridge University Press, 1980.

[2] Cohen, I. B. *Revolution in Science*. Cambridge: The Belknap Press of Harvard University Press, 1985.

[3] Cohen, I. B. (Ed.). *Puritanism and the rise of modern science: the Merton thesis*. New Brunswick (N.J.): Rutgers University Press, 1990.

[4] Cohen, I. B. *Interactions: some contacts between the natural sciences and the social sciences*. Lodon: MIT Press, 1994.

[5] Cohen, I. B. *The Natural sciences and the social sciences: some critical and historical perspectives*. Dordrecht: Kluwer Academic Company, 1994.

[6] Cohen, I. B. *Science and the founding fathers: science in the political thought of Jefferson, Franklin, Adams and Madison*. New York: W.W. Norton Company, 1995.

[7] Mendelsohn, E. *Transformation and tradition in the sciences: essays in honor of I. Bernard Cohen*. New York: Cambridge University Press, 1984.

Ethical Issues Posed by Nanotechnology in the Workplace[1]

Zhu Fengqing

Harbin Institute of Technology

zhufq54@163.com

ABSTRACT. There is no scientific clarity regarding the potential health effects of occupational exposure to nano-particles. Many situations that have ethical implications for affected workers have been identified. These situations include the (1) identification and communication of hazards and risks by scientists, authorities, and employers; (2) workers' acceptance of risk; (3) selection and implementation of controls; (4) establishment of medical screening programs; (5) investment in toxicological and control research; and (6) promoting respect for people. As the ethical issues are identified and explored, options for decision-makers can be developed. Additionally, societal deliberations about the workplace risks posed by nanotechnologies may be enhanced by giving special emphasis to small businesses and adopting a global perspective.

Ethical issues

1 Identifying and communicating hazards and risks.

The "hazard identification" stage of risk analysis is the basis for risk management decision-making. The output of this stage is often highly debated, since the reasoning process is primarily qualitative and the results trigger other stages of analysis and decisions about preventive action [1]. Interpreting scientific information about the hazards of nano-materials is basic to communicating about the hazards and risks posed to workers. Interpreting and communicating hazard and risk information is an integral part of risk management by employers. The employers' decision-making focuses on deciding which preventive controls should be used to assure a safe and healthful workplace.

The principlist ethical approach focuses on principles such as nonmaleficence and autonomy but fails to assess the social and organizational context of occupational safety and health and the role of practitioners in relation to the corporate structure. With regard to nanotechnology, the contextual pressures on practitioners and authorities arise from a company's or society's needs and desires for nanotechnology to grow and develop. Mention of potential health concerns may be seen as alarmist, unfounded, and detrimental to the growth of the field. Nonetheless, the counter position is that conflicting demands on practitioners

---

[1]The author is with College of Humanities and Social Science, Harbin Institute of Technology, Harbin 150001 China.

for acting both as agents of a company and as autonomous professionals constitute a social and structural problem rather than a problem of individual ethics. One solution is that health pronouncements should be made independently of promotional concerns for nanotechnology.

2 Workers' acceptance of risk

Acceptance of risk is a relative concept that includes judgment about the certainty and severity of risk, the extent of the health effects, the voluntary nature of the risk, the risks and advantages of any alternatives, and compensation for undergoing the risk. It is a false premise to assert that workers have free choice in terms of which work and working conditions to accept. Although some component of self-determination is present, economic and social conditions exert the greatest influences on the workers' selection of work, level of risk tolerated, and ability to participate in risk management. Worker participation in risk management is not a static concept and has increased over the past 35 years with the implementation of team approaches, management systems, corporate responsibility, and the right to know and act movements [6]. Nonetheless, workers generally cannot universally refuse work they consider hazardous and still keep their jobs.

Conformance with the principle of autonomy depends on the extent to which workers have input into risk management at their worksites and the degree to which they are at risk after controls have been implemented. Justice is also related to worker decision-making. At issue is the extent to which workers are exposed to greater risks than the general public–or, stated another way, whether it is appropriate to exchange incentives such as wages or hazardous duty pay for additional risk from exposure to nano-particles [9]. This issue may be less significant if nano-particle controls reduce the workers' risk level to that of the general public, if conceivably both are known. Clearly, society accepts that some jobs are inherently riskier than others. However, in many countries the societal goal is to provide a safe and healthful workplace for all workers.

3 Selecting and implementing controls

The critical ethical question related to control of nano-particles is whether sufficient controls are being implemented to prevent injury and illness. If not, worker exposures may result in an increased risk of harm or actual harm. The central scientific fact is that the risk posed by nano-materials is not well established. However, preliminary information suggests that at least the same level of concern afforded to industrial fine and ultrafine particles should be extended to engineered nano-materials and that a commensurate level of protection should be instituted for them (Royal Society and Royal Academy of Engineering [8]). Any risk posed by exposure to ultrafine particles is a function of their potential toxicity and the extent of exposure. Based on limited toxicological evidence of risk and a heightened level of concern, the best approach might be to treat engineered nano-particles as if they were potential occupational hazards and to use a prudent health-protective, risk-based approach to develop interim precautionary measures consistent with good professional occupational safety and health practice (Royal Society and Royal Academy of Engineering [8]).

Such interim precautionary measures could include guidelines for conducting workplace exposure assessments, implementing engineering controls, designating work practices, and developing process or industry interim exposure limits as core elements. If the focus of exposure control is airborne particles of respirable dimensions, such approaches may be useful and reflect the professional judgment of experienced practitioners. If skin absorption is also a likely route of exposure, guidelines should be developed for preventing skin exposure. Unfortunately, the data are insufficient to make a strong risk-based assessment to inform these decisions.

4 Establishing medical screening programs

Medical screening is the application of tests to asymptomatic persons to detect those in the early stages of disease or at risk of disease. Medical screening in the workplace differs from medical screening in the general population because of the specific nature of the occupational conditions and the responsibilities of employers [3]. A wide range of ethical questions has been identified regarding the medical screening of workers and the use and implications of the findings. These questions address the rationale for screening, the voluntary nature of the screening, the action that will be taken for workers with positive tests and individuals who will have access to test information.

The ethical questions that apply to the medical screening of workers pertain to whether the screening is voluntary, who will have access to the results, and what the purpose of such access would be. Screening generally requires diagnostic confirmation; and for positive cases, screening requires timely treatment. Who is responsible to pay for these procedures? Ethical issues can also arise in the use of screening results to label or stigmatize workers or to remove them from a job. Screening results may also create psychological burdens. Resolving such ethical issues will depend partly on the degree to which the worker has been informed about how the results will be used.

5 Assuring adequate investment in toxicological and control research

Ethical issues cannot be adequately addressed for nanotechnology without sufficient knowledge of the hazards involved. Because limited information is available on the safety of an ever- growing number of nano-materials, an ongoing research effort is needed to comport with the principles of autonomy, beneficence, and nonmaleficence. In addition, research is needed on the extent of exposure and the effectiveness of controls. Internationally, such research is underway. However, the question of the level of funding of this research has ethical implications, since much of the current control guidance is precautionary and is not based on strong quantitative risk assessments. Further research is the only way to address this lack of appropriate information.

6 Promoting respect for persons

Underlying the debates about nanotechnology has been the issue of tolerating the potential for harm to some in the context of anticipated benefits to society as a whole. Such thinking embodies the utilitarian point of view that harm to one person may be justified by the larger benefit gained for others [4]. This point of view contrasts with the ethical principle of respect for persons, which emphasizes the rights of the individual and is associated with the golden rule ("Do unto others as you would have them do unto you") [2]. In the workplace, this principle translates into acknowledging for each worker the right to a safe and healthful work environment. This right imposes correlative duties on the employers and governments who must secure the workers' rights to a safe and healthful workplace [2]. The objection to this interpretation is that the rights of employers, and hence the rights of society, to property and benefit resulting from nanotechnology may be (or may appear to be) in conflict with the workers' rights. When two rights conflict with each other, some rational way must be found to determine their relative priority.

Strategies for supporting ethical decision-making.

1. **Placing special emphasis on small business**

    The occupational safety and health problems of small businesses have been a major focus of concern particularly in the last decade, since most workplaces are classified as small .The frequency of occupational injury and illness in small businesses may exceed the average for general industry across all businesses in a sector, but the frequency may not be evident in an individual company NIOSH [7]. Small businesses are generally perceived to have little time and few resources for occupational safety and health.

2. **Adopting a global perspective**

    The growth of nanotechnology is a global phenomenon that requires a global approach to hazards and risks, particularly in the workplace. The world needs internationally valid standards for nanotechnology materials as well as a uniform nomenclature [5]. Without a uniform nomenclature, investigators, insurers, regulators, governments, companies, and workers could have difficulty communicating and taking concerted action.

Bibliography

[1] Crawford-Brown, D. J., Brown, K. G. *A framework for assessing the rationality of judgments in carcinogenicity hazard identification.* 1997. URL http://www.piercelaw.edu/risk/vol8/fall/Cr-Br+.htm. Risk. Available, [accessed 11 September 2006].

[2] Gewirth, A. Human rights in the workplace. *Am J Ind Med*, 9:31–40, 1986.

[3] Harber, P., Conlon, C., McCunney, R. J. Occupational medical surveillance. In J, M. R. (Ed.), *A Practical Approach to Occupational and Environmental Medicine*, 582–599. Philadelphia: Lippincott Williams and Wilkins, 2003.

[4] Harris, C. E. Methodologies for case studies in engineering ethics. In *Emerging Technologies and Ethical Issues in Engineering*, 79–93. Washington: Conational Academy of Sciences, 2003.

[5] Hett, A. *Nanotechnology: Small Matter, Many Unknowns*. Zurich: Swiss Reinsurance Company, 2004. URL http://www.swissre.com/INTERNET/pwsfilpr.nsf/vwFilebyIDKEYLu/ULUR-68AKZF/$FILE/Publ04_Nano_en.pdf. [accessed 21 September 2006].

[6] Jensen, P. L. Assessing assessment: The danish experience of worker participation in risk assessment. *Econ Ind Democracy*, 23:201–207, 2002.

[7] NIOSH. Identifying high-risk small business industries:the basis for preventing occupational injury,illness,and fatality. In *U.S.Drtment of Health and Human Servies,Centers for Disease Control and Prevention,National Institute for Occupational Safety and Health*, 99–107. Cincinnati,OH: NIOSH(DHHS)Publishion, 1999.

[8] Royal Society and Royal Academy of Engineering. *Nanoscience and nanotechnologies: opportunities and uncertainties*. London, UK: Royal Society and Royal Academy of Engineering, 2004.

[9] Schrader-Frechette, K. *Environmental Justice: Creating Equality, Reclaiming Democracy*. New York: Oxford University Press, 2002.

On Ethical Problems in Realistic Engineering Activity[1]

Qi Yanxia
Shenyang Aerospace University
qiyanxia0423@163.com

Liu Zeyuan and Li Zuoxue
Dalian University of Technology

ABSTRACT. Engineering activity, like science and technology, is a double-edged sword. Engineering activity has not only brought man social wealth and promoted productivity but has also brought with it a negative impact that should not be overlooked. Weak awareness of engineering ethics is one of the most important reasons. The paper analyzes ethical problems that should be solved realistically in engineering activity. These problems in engineering activities create value conflicts and conflicts of interest for engineers. There is also the problem of deficiencies in engineering regulations. In conclusion, four countermeasures are proposed to improve the ethical awareness of engineers, perfect the educational system, establish realistic ethical regulations, and use the power of government to bring about and enforce reforms in engineering activity.

Engineering activities from the drawing board to the final stage are all inevitably linked to ethical problems. However, in reality, the engineering profession fails to acknowledge, fully analyze and understand these problems. Engineers have only a dim ethical awareness. As a result, some carelessly implemented engineering activities threaten and even harm people's lives or welfare. Therefore, it is necessary to improve the ethical awareness of engineers and the ethical quality of engineering projects.

As ethical responsibility puts more emphasis on the consequences of actions and it is closely connected with engineering practice, it is an important component of practical ethics. Hans Jonas was the first person to introduce the idea of responsibility into the field of ethics. Since then responsibility has become the object of concern of ethicists. It has triggered a heated discussion in philosophy and in science and technology. According to this view, the project has become the core concern of engineering ethics.[2] Compared with science and technology, engineering is more practical and is linked with responsibility more closely. Thus responsibility is not only an important component of engineering ethics, but it is the core.

The implied meaning of moral responsibility is two-fold: forward-looking responsibility and backward-looking responsibility. Forward-looking responsibility can be said to be a preventive responsibility, a duty of care or initiative. It is not only necessary, but important. It is a future-oriented action. Because there are

---

[1] The author are with Humanities and Social Science Department, Shenyang Aerospace University, and Department of Management, Shenyang Aerospace University, Shenyang 110136 China, and Center of S&T Ethic and Management, Dalian University of Technology, Dalian 116024 China respectively.

many uncertainties in engineering, engineers need to anticipate the possibilities of potential disaster before the commencement of engineering design and implementation of project activities. A responsible engineer will implement a project only after he has already fully anticipated the risks and dangers which the project may pose.[9]

In backward-looking responsibility, the individual or group is evaluated ethically for a particular action and its consequences which has occurred in the past. They could be praised for good results and bear responsibility for bad results. However, the process is very complicated. Many consequences can be prevented in advance.

An engineering activity is usually very complicated, and it has its relative independence, while still closely linked to one activity or a number of other activities. So, it is also of particular complexity to investigate liability. Whether for organizational reasons or personal reasons, individual responsibility or group liability, the investigation and distribution of responsibility is all very difficult. However, personal conduct is the basis of overall activities. Therefore, it is crucial to strengthen the awareness of ethical responsibility in order to avoid adverse consequences.

Of course, it is undeniable that engineers are in an important position on the strength of their expertise in the whole engineering process. Responsibility is the basic quality of an engineer, as well as the first requirement for professional engineers. In realistic engineering activities, however, the ethical awareness of some engineering personnel is becoming more and more indifferent and the lack of responsibility is inevitable where private interests, fear, ignorance, accepting authority without criticism, self-centeredness and so on prevail. Then many engineers cannot fulfill their ethical responsibilities which can have serious consequences. Therefore, the responsibility of engineers should not be ignored.

In modern society, it is not only engineers but the engineering community that is involved in many project activities. The engineering community is composed of mainly four-parts – workers, engineers, investors, and managers.[5] In the engineering process, each of the four groups has its own unique, irreplaceable and important role. A responsible project needs the concerted efforts of all four constituents. The project is a series of activities involving not only engineers. Each project activity faces complex conflicts of values. For example, the engineering regulations require engineers to put public safety, happiness and well-being first, while employers require engineers to be faithful agents. In reality, engineers are often caught in a value conflict dilemma where professional loyalties and responsibilities conflict.

Engineering activities need a business organization to provide financial and organizational support. Engineering and business must work together to bring engineering projects to fruition. However, many ethical problems are connected to the contradictions and conflicts between managers and engineers.

At the enterprise regulation level, the project manager is in a superior position while the engineer generally is in a subordinate position to accept and implement the orders of the manager. Sometimes the orders of the manager may breach the professional ethics of engineers, which could cause many ethical problems. It is necessary to examine the different functions and views of engineers and managers. In the organization, engineers mainly use their technical expertise to create valuable products and processes for customers. However, engineers are also

On Ethical Problems in Realistic Engineering Activity 343

professionals, and they must uphold the standards identified by the ethical code. Thus, engineers have a dual loyalty: one is loyalty to the institution; the other is their professional loyalty. Most engineers are cosmopolitan. [2] Engineers observe the norms of their professional loyalty before their organizational loyalty. They give quality particular concern. From the point of view of safety, engineers tend to be cautious and err on the side of being conservative. Compared to engineers, managers tend to have a local orientation. The function of the manager is to guide the organization, including the engineers' behavior. Managers are mainly concerned with the short-term and long-term interests of an organization and feel strong pressure to keep costs low. They consider engineers sometimes excessively cautious and too concerned with security. By comparison, engineers tend to rank every factor into grades and before other factors minimum safety and quality standards should be ensured. Engineers think that it is their obligation to be the safeguard of safety and quality standards in the process of dealing with managers.

Furthermore, no matter what the decision is, they all involve a utilitarian principle. Utilitarian thinking only makes the decisions serve the best interests of the department or company and often do not consider other factors. But if the action maximizes the interests of most people, it certainly will sacrifice the rights and interests of vulnerable groups, even cause harm to their lives. That conflicts with the value of respect for human rights. In the conflict of two values, the utilitarian value has occupied a dominant position for a long time and caused serious negative effects. The situation must be controlled from an ethical view and common ethical principles need to be advocated.

There is still another problem to which attention should be paid. It is the problem of engineering regulation. Regulation is a criterion of conduct, a functional existence. Engineering ethical regulations are often haphazard, and in some special conditions they are out of control. Engineering ethical regulations cannot provide operational rules for engineers to solve all the ethical issues. However, in some fields of engineering, ethical regulations need to be made but as of now they have not yet even been formulated. Moreover, many professional ethical regulations which are in place are imperfect or too old. Thereby new problems that emerge in some areas of engineering are not covered by the ethical regulations which exist. As a result, there is an ethics void in some engineering activities due to the rapid pace of technical innovation and progress. The ethics have not yet caught up with these advances.

Four countermeasures can be tried to solve the problems mentioned above. First and most important of all, it is a major priority to improve the ethical awareness of engineering subjects. This is one way to guarantee the avoidance of negative consequences in engineering projects. I want to emphasize that it is not only engineers but all participants should heighten ethical awareness. Second, it is urgent to set up an educational system of engineering ethics in university. The university is the cradle of engineers. It is also a very important period during which to form an axiology and an outlook on life. In many universities the educational system includes courses in engineering ethics. This is an effective method for improving ethical quality and awareness. But in China, an effective system of engineering ethical instruction has not yet been founded. The third method is to make and perfect professional ethical regulations. Regulation has the function of restriction and constriction. Ethical regulation belongs to the

---

[2]http://www.ethics.ubc.ca/mcdonald/conflict.html

domain of social supervision, and it mainly refers to conduct criteria in everyday life or at basic social levels. Professional ethical regulations restrict and control the behavior of engineering personnel. They give the engineer guidelines to help him to know what to do and what not to do. The last method is administrative regulation. This is the most powerful and effective method. When problems cannot be solved by other methods, government can legislate and enforce effective regulations and oblige all parties to comply.

Bibliography

[1] Charles, E., Harris, J., et. al. *Engineering Ethics: Concepts and Cases.* Scarborough, Ontario: Wadsworth/Thomson Learning, 2nd edn., 2000.

[2] Jonas, H. *The Imperative of Responsibility: In Search of an Ethics for the Technological Age.* Chicago: University of Chicago Press, 1984.

[3] Ladd, J. Collective and individual moral responsibility in engineering: Some questions. In Johnson, D. G. (Ed.), *Ethical Issues in Engineering.* Englewood Cliffs, New Jersey: Prentice Hall, 1991.

[4] Lenk, H., Maring, M. (Eds.). *Advances and Problems in the Philosophy of Technology.* Munster: LIT, 2001.

[5] Li, B. Works in the engineering community. *Journal of the Dialectics of Nature*, 27(2):64–68, 111, 2005.

[6] Li, S. Engineering ethics and the main problems of engineering ethics, 2003.

[7] Li, W. Technical ethics and metaphysics on jonas's theory of responsibility. *Studies in the Dialectics of Nature*, 19(2):41–47, 2003.

[8] Mitcham, C., et. al (Eds.). *Philosophy and Technology.* New York: The Free Press, 1983.

[9] Zhu, B. The ethical responsibility for engineering activities. *Studies in Ethics*, 26(6):36–41, 2006.

On Ethical-Value Tension between Science / Technology and Politics[1]

Xu Zhili

School of Humanities Beihang University

xuzhiliw@163.com

ABSTRACT. The essence of ethical-value tension in the space between science-technology and politics lies in the separating-uniting of the two value-directions pulling at them. The common value-direction of science and politics is the existing base of its value tension. This kind of common value-direction regards human radical interests as its dividing line and value-unity must also answer to the question of whether it is good or not. As the same, there are good and vicious differences of value-deviation between science-technology and politics. Some conflicts are beneficial and others are harmful. There is a special compromising mechanism in the value-direction mode which operates between science-technology and politics to bring about beneficial resolution of the tensions and conflicts.

Science-technology is an activity by which humans recognize and remake nature. This kind of activity spreads out under a certain social and political background with the economical basis. Science-technology activities influence social lives profoundly, and at the same time, they cannot go beyond limitations of social elements. As a concentrated manifestation of economy and other social elements, politics restrains the interests and rights of all kinds of social phenomena including science-technology by the great power. Politics has profound influence on science-technology's developing direction, scale and even acting mode, and dominants the allocation of its resources and results. Politics and science-technology appear an inter-effective acting procession, namely they two are in interaction. However, the inter-affection between politics and science-technology has good and vicious nature. Interaction between science-technology and politics appears evidently two extremities of goodness and viciousness in the contemporary era. We should try our utmost to probe the mechanism and law of the inter-affection between science-technology and politics to diminish vicious conflicts and advance good interaction.

---

[1]The author is with School of Humanities, Beihang University, Beijing 100191 China.

1 Introduction: the background to the reality and theory, meaning and assumption in this study.

Contemporary science-technology and politics become the greatest influential two systems of human life, and they are exerting inter-affection or inter-action. The interaction between science-technology and politics can be positive or negative. Both results have great import to contemporary human society. Positive interaction can advance science-technology's sound development and result in spillover benefits in politics and improvements in society This influence has been especially notable in the modern era. It even has had an impact on the fate of states and nations. However, Negative interactions or conflicts also exert important pressures on human society. Its results can negatively impact science-technology and political development and endanger human existence and development. Obviously, it is very important to probe into the origins of conflict between science-technology and politics and to realize the laws of interaction which govern them.

There has been a significant amount of research into the relationship between science and politics in theory in foreign countries. From M. Weber, F. Hayek, J.D. Bernal, B. Barber, the Frankfurt School, and the Social Constructionists to J. Ellul, L. Winner, D. Bell, E. Morin, and political scientist D.H. Guston – they have all made significant contributions to the study of this topic. In China, however, the study of the relationship between science-technology and politics has been weak for some reason, but some scholars now have begun to probe these issues to some degree. Using the study methods and angles of foreign and domestic scholars, some researchers have probed some aspects of the macro- and micro-relationships between science-technology and politics. In general, though, a systematic and comprehensive study on the interaction mechanism between science-technology and politics has yet to be attempted.

In the interaction zone between science-technology and politics, there are three basic sources of tension: value tension, power tension and contractual tension. The purpose of this paper is to examine the laws by which these tensions come into being and operate. On this foundation, the paper goes a step further to investigate ways to harmonize tensions between science-technology and politics and promote positive interactions between them.

2 Nature, structure and tension of the interaction space between science-technology and politics.

This paper examines the space between science-technology and politics to determine if there is a connecting inter-space and the nature of this space. There are three typical views on this problem. The first, science-technology and politics in essence don't mutually connect and their fundamental differences are basic. This is the neutral view. The second, science-technology and politics cannot be divided; they are inextricably connected and weave a kind of "seamless web" in which there is in effect no distinguishing inter-space. This is the typical social-constructionist view. The third school of thought argues that the relative space may be aptly defined, and that there are obvious differences and close connections in the relative space between science-technology and politics. To the condition of the space between science-technology and politics, the absolute-space argument maintains that there is no interaction between science-technology and politics. The "seamless web" argument does not recognize a distinction between science-technology and politics. They are the same body. The relative space argument generally admits there is an interaction between science-technology and politics which breeds a kind of organic ecological space, and there exists an organic-action and "ecological" feed back mechanism develops there between science-technology and politics.

While systematically observing and studying the space between science- technology and politics, we find three basic structural elements, namely value, power and contract. These are the three elements that bring together systems of science-technology and political systems and make them inter-restrict and inter-repel. Furthermore, when we probe the origins, motive forces and mechanisms among these three elements, we can identify three basic tensions – value tension, power tension and contractual tension. What is called 'tension' simply speaking here means active forces which inter-connect, inter-restrict and inter-repel between science-technology and politics. Value tension means deviation and coincidence of value direction between science-technology and politics; power tension means the relationship of interference and independence between science-technology and politics; contractual tension means the relationship of research activities of delegation between subjects of science-technology and politics. The mechanism of the combination of these three kinds of tension constantly repels at a deep level the interacting relationship of science-technology and politics.

3 The law of separating-uniting and organic unitary effect of ethic-value tension in the space between science-technology and politics.

The essence of value tension in the space between science-technology and politics lies in separating-uniting of two value-directions of them. The common value-direction of science and politics is the existing base of its value tension. This kind of common value-direction regards human radical interests as its dividing line and value-unity also has a question of if it is good. As the same, there are good and vicious differences of value-deviation between science-technology and politics. Some conflicts are benefit and the others are harmful. There is a special compromising mechanism of science-technology and politics value-direction when it comes true.

The first, the common value-direction of science-technology and politics, is based on human interest. They respectively act as creator and agent of human interests. The second, in contemporary society science-technology reason and politics legitimacy have inner profound connection. The radical value that science-technology subject pursues is science-technology reason, which properly becomes a foot for politics subject to establish legitimacy of its own, and reflect an important politics value. The third, in actual operation, two systems of science-technology and politics appear inter-dependent nature and embody a kind value complement of structure in real acting procession.

The unity of the value-direction of science-technology and politics is a basic condition for the existence of value tension, but forming this kind of value tension must have deviation and conflict of value directions between science-technology and politics. This kind of value deviation and conflict concern many aspects and it embodies: (1) science-technology and politics respectively deviate common value-direction of human; (2) politics warps science-technology value; (3) politics interferes and controls science-technology negative value; (4) science-technology demolishes and usurps politics value.

Since politics and science-technology exerted in human society, there have been basic unity and coincidence of politics and science-technology value-direction; at the same time, the difference and deviation of science-technology and politics in value direction accompany each other all the time. This kind of condition probably keeps for a long time and only its degree and mode changes with time. Actually, political value and science-technology value do not replace each other at all. They sometimes stress its unified nature and sometimes emphasize inter-deviation and conflict, but they always achieve a kind of compromise under certain social and historical conditions.

Tension of freedom and interference and its acting law in the space between science-technology and politics.

Power tension in the space between science-technology and politics mainly exists in two extremities between freedom and interference. Science-technology activities need freedom and freedom is a kind of right of science-technology subject. However, this kind of freedom is limited, and modern science themselves need interference of politics also. Interference of politics to science-technology is inevitable and necessary. But this kind of interference must be limited, and there should still be the space of science-technology activities beyond interference.

The argument of freedom and interference between science-technology and politics has developed out of the evolution of the relationship between science and government. This argument appeared early in the 19th Century, was experienced three times in the 20th Century, and is embodied even today in the clone-human controversy at the present time. Power tension of freedom and interference exists in the space between science-technology and politics over a long period of time, and freedom research is an important extremity of it. Curiosity is what drives the development of science-technology, and academic freedom is an important meaning and value. At the same time, academic freedom in the power tension inevitably exists as an inherent limitation.

In the power tension in the space between science-technology and politics, freedom of research corresponds to political interference. The inner limitation of research freedom reflects the necessity of political interference in science-technology activities. This interference is expressed in organization, planning and standardization. Politics subjects science-technology activities to limitations. Just as research freedom has its own limitations beyond political interference, science-technology activities also have their own space, which is the connecting belt of power tension between science-technology and politics.

4 The contractual tension in the space between science-technology and politics namely the delegation relationship between politics subject and science-technology subject.

"Contractual tension of the space between science-technology and politics appears a contradictory relationship of delegation between politics subject and science-technology subject. This contractual tension is of great real meaning it embodies politics subject's admissions to science-technology's utility value, reflects the material-economy relationship between science-technology and politics by the form of the social contract for science", and also embodies market-mechanism of science-technology. Recently, D.H. Guston has presented a mode of principal-agent theory between science-technology and politics, which provides us with a good theory tool to analyze this kind contractual relationship. There is a link of integrity and productivity between principal and agent, which needs another kind of boundary organizations to adjust.

The following is an analysis of Guston's principal-agent theory mode. Guston considers: delegation relationship is the basic relationship between politics and science; research activities are complemented by depending on integrity and productivity; after world war II, a "social contract for science " assumed that integrity and productivity were the automatic products of unfettered scientific inquiry which has undergone the requirements of royalty, financial-monetary duty and scientist's technology aims; until eighties of 20th century, all kinds of science misconducts and the falling economy achievements and results break the trust between politics and science. In order to reconstruct this trust and salute the problems of delegation, scientists and non-scientists must collaborate in new boundary organizations; flexible system design can create steady partnerships between politics and science. Guston's principal-agent theory has important meaning for clarifying the inter-acting relationship between science-technology and politics, but there are still certain defects.

5 How to realize the coordination and good interaction of tension in the space between science- technology and politics according to present affairs in China.

Obviously, the tension overall must be coordinated in the space between science-technology and politics, which is the promise for realizing good interaction. This concerns at least three meanings. First, there some necessary tension in the space between science-technology should be maintained, which is a radical way to guarantee the vitality of interaction between science-technology and politics. Second, among value tension, power tension and contractual tension in the space between science-technology and politics, a moderate degree should be kept respectively. If any one of them is too big or too small, it would not be good for the interaction between science-technology and politics. Third, common coordination should be kept among the three kinds of tension. Value tension, power tension and contractual tension act in different phases and also control the inner adjustments among them.

There are some problems in coordinating tension in the space between science-technology and politics, namely, science-technology as a subject value can not realize a high level value tension; political action is improper in the power tension relationship; and especially in contractual tension, the roles of principal and agent are indefinite, and boundary organizations lack vigor. There are no powerful market-ruling and law-guarantees for keeping the contractual relationship.

In order to realize the coordination of tension in the space between science-technology and politics in China and in order to improve the good interaction between science and politics, we must define a proper way to balance political and scientific priorities. There are many things which need to be considered. The public needs to participate in enacting science-technology policy, acting as a buffer between science-technology interests and political interests. We also need to establish effective "boundary organizations", the elastic action of science-technology funds, and powerful market mechanism and legal guarantees, etc. All these mechanisms can work together to ensure that a certain tension is adjustable among value tension, power tension and contractual tension.

On Guarantee-Systems of Science and Technology Ethical Practice in the Present Age[1]

Wan Qian and Wang Fei
Dalian University of Technology
wangqiandut@gmail.com

Wang Qian
Dalian University of Technology
wangqiandut@gmail.com

Wang Fei
Dalian University of Technology
hwangfei@163.com

ABSTRACT. How to build a guarantee-system to govern Science and Technology ethical practice is an important problem which is often faced in modern times. To build an effective S&T guarantee-system requires a relevant organization to investigate and exactly assess the ethical practices, affirm the social contributions, and ensure the stabilization of the living source and space for career development. Today, when S&T activities are highly specialized, behavior against moral norms may damage S&T activities. Therefore, we need to pay special attention to developing S&T ethical practical guarantee-systems!

Scientists and technologists in their ethical practices frequently encounter the problem of conflicts between public interests and personal interests. Only the specialists know ethical problems because of the specialization of S&T practical activity. If considering the moral and ethical responsibility, the scientists and technologists bravely disclose the behaviors that breaching the morality in scientific research institutions or enterprises. They will suffer the exclusion and the attack from the boss, colleagues or craft brothers, even lose the position; and when they chose another job, they will face to the suspicion and cold reception. Therefore, we must study the questions how to establish a guarantee -system of S&T ethical practice. Here involves following questions.

1 Is a Guarantee System of S&T Ethical Practice necessary?

It is often thought that the scientists and technologists ethical practice naturally can receive people's praise and support at present, since human civilization is at a high level today. So the problem of establishing a guarantee -system of S&T

---

[1] The author are with School of Humanities and Social Science, Dalian University of Technology, Dalian 116024 China.

ethical practice needn't be discussed. But that's not the case. Whether in the developed or the developing countries, they all have the matters that scientists and technologists suffer the exclusion and attack because of ethical practice, and some cases are quite serious. Therefore, establishing the guarantee -system of S&T ethical practice technical is extremely necessary.

The scientists and technologists expose behaviors which violate morality in scientific research institutions or the enterprises to protect the interests of the public and society from the unnecessary injury .The government and the public have the duty to protect the scientists and technologists that sacrifice their self-interest for the public good.

This ethical practice of scientists and technologists has to take great risk. In the US, Jeffery Wigand paid a great price to expose the excessive nicotine content of his company's cigarettes; Karen Silkwood was murdered when she drove to meet a New York Times reporter because of the incriminating information she had uncovered against a division of Kerr-McGee Corporation which had tampered with related material on their use of plutonium fuel.[2] A lawyer once made defense for many scientific and technical worker's ethical practices: "The exposure is lonely, there is no compensation, and it is full of danger". It includes a solid retaliation risk; these challenges are very difficult and very expensive. In other words, so-called "success" may possibly only mean to retire - - the route of retreat have already cut the money compensation - - but it may not be possible to eliminate the harm to one's reputation, occupation or interpersonal relationships.[4, pp. 253] The scientific and technical workers who speak out on behalf of the public's interest report misconduct by managers and officials that violates ethical standards in enterprise management. Such actions are often regarded as a "betrayal" of the enterprise. The employer can dismiss workers for not being loyal and disobeying orders. Frequently, employees who act in this way are labeled as "informers". Systems analyst Virginia Edgerton in New York was dismissed for violating her employer's instructions. Edward Turner in Falls City, Idaho was dismissed without being told why he was terminated. [3]

In opposite to the leading positions or the powerful enterpriser, the whistleblower is a weak crowd relatively. They are frequently faced with being dismissed and ridiculed, and left out in the cold. The whistleblower needs to consider a series of ethics question before the exposure: When to expose is in the morals permits? Is it the moral duty? When to expose is not the morals or not prudent? Is the exposure to the organization disloyal behavior? Has to follow what procedure in the exposure process? Supposes the motive which the whistleblower moves is for two interests i.e., the public interests and company interests, a higher paraffin - -public interests, the opponent can point out immediately the fact is not true. The viewpoint of James M. Roche, ex-president of US General Motors Car Company is representative extremely. He believed that, enterprise's some enemies encourage the employee to be disloyal now to the enterprise. They want to make suspicion and disharmony, covets the property and the interest of the enterprise. But regardless it is pasted on any label - - industrial spy, the exposure or specialized responsibility - - it is the dispersion not harmonious and the manufacture conflict another kind of strategy oneself. [1, pp. 11–14]

In China, there is little internal exposing behavior from in the group of scien-

---

[2]The Chinese Translation of [6, pp. 206–208].
[3]The Chinese Translation of [2, p. 208].

tists and technologists. Some scientists and technologists, who from the external expose behaviors of scientific research institutions or enterprises– the enterprise makes the counterfeit or the product quality inferior–also has met the quite big resistance.

In the 1980s, "the King of killing mice" Qiu Mandun in Hebei Province once was a well-known national character. It is said that, his "invention" – the "mysterious Qiu's lure" –can kill male or female mice according to the person's desire. In 1992, five scientists jointly pointed out to the press that the Qiu's mouse toxicant used a hyper toxic substance which the national laws forbid in writing– "the fluorine acetamide" and "the poison killing mice effectively" – these toxicants once entering the soil would cause serious pollution, and elimination is difficult. The scientists also laid bare the pseudo-science propaganda, for example the claim about "luring the male or the female mice" and so on. There are many similar instances in China. As far as we know, similar cases also happen in other countries.

In 1993, Shanxi ordinary technologist Han Chenggang published related "the mineral spring pot" a series of the article that disclosed the actual facts, criticized four mineral spring pot enterprises' illegal advertisement and the false propaganda. August 20, 1994 Han Chenggang is reported to the court by the artificial mineral water appliance specialized committee. First trial Han Chenggang lost a lawsuit. Till June 28, 1996, the Shanxi Province Higher People's Court only then in the final judgment Han Chenggang to won the court case.

December 28, 1993, the Changsha intermediate people's court on was long reaches the technical limited company to sue "Electronic Newspaper" newspaper office "insurance king" to its product the right of reputation lawsuit to make the first trial decision, "Electronic Newspaper" newspaper office lost the lawsuit. Changsha "insurance king" really false until January 24, 1997 the Chinese patent office board of review official decision announced 9122064 (namely "insure king" patent number) the practical new patent is invalid when only then sees the result, however this time "Electronic Newspaper" society the flower has gone to 570,000Yuan indemnities and more than 400,000 Yuan other expenses.

In 1999, the medical department investigated 64 kinds of unqualified healthy products, these healthy products it is said include ginseng, velvet, honey, Chinese caterpillar fungus, ant and chicken highly finished products, they have many functions such as: invigorating brain and benefiting body, regaining vigor, invigorating blood and benefiting kidney, nourishing yin and strengthening yang, modulating physiology, reinstating diseases and prolonging life, improving looks. Some healthy products which propagandized overly, although scientists and technologists raised queries and questions, the judicature investigated quite is actually difficult. The products have the serious problem in the quality, seriously endangers the populace life and health "three to take orally the fluid" to be available throughout the country reaches for 3-4 years long, to the people detected it propagandizes the curative effect is illusory, appears when the trust crisis, the factory has made the foot money. Han Chenggang also was first publicly questioned three took orally the fluid the curative effect but on the defendant the court person.[5, p. 176] He files the lawsuit to suffer very many hardships. Although finally won, although he for guarded consumers' interests to make the contribution, but he have paid quite great price.[3]

Therefore, how to establish a guarantee -system of S&T ethical practice has become a significant social problem, which whether the development of modern science and technology can de healthy or not. Without such a social guarantee, the negative effects probably engendered by science and technology developments will be more and more serious and even affect the sustainable development of the whole society.

2 Is a S&T Ethical Practice guarantee-system possible?

In order to promote science and technology ethical practice, we need to encourage more scientists and technologists to disclose the immoral behaviors in the scientific research institutions or the enterprises, and protect the public interest. We need to study the feasibility of system guarantees.

In the USA there have been some successful examples. They have constituted some law to protect the ethical behaviors of scientists and technologists. Besides the public-policy exception and whistle-blowers law, they also have the relative guarantees in organizations and institutions. Government has set up awards for encouraging scientists and technologists to pay attention to the public interests. The professional communities which include the American Society of Civil Engineers and each special engineers institute have set up relative ethical committees to deal with the ethical disputant affairs. As to the public, it is chiefly to expose the main body of the wrong behavior through public media to urge it to correct its mistakes.

"The community policy exception" the law is one kind of protection of the scientific and technical workers as the employee in practice. The law constituted by American court recent years permits the employee to enter a lawsuit against the improper or the retaliatory dismiss, this is one kind of revision to the traditional common law "to hire according to the free will". Certainly it itself also during the development, itself also has many limitations, this limitation mainly comes from the court regarding the community policy exception explanation.

1. In violates the community policy the dismissal with only to affect the employee "personally" between the interests dismiss, certainly does not have a clear boundary. For example, the court usually believes that, the community policy exceptional principle should not use in to protect these employees which conscience works. The court advocated that, the community policy exceptional principle application must involve to public's interest.
2. In the situations which there are differences between employers and employees, the courts usually refuse to give the employee protection.
3. Many courts differently treat two kinds of standards, one is a regulation which promulgated by "unofficial" organization, such as the professional mass organization's regulation, the other is a regulation which promulgated by the administration and the judicial entity, such as state registration committee's regulation. In the overwhelming majority case, the court refuses to acknowledge "unofficial" the regulation takes the community policy the foundation.
4. The majority courts seek to get a balance between the public interests and the employer interests. This kind "the balance" usually is extremely sub-

jective, moreover employer's interest as if always carries a lot of weight, only if the court warrant community policy "the foundation" has encountered the clear explicit destruction.[4]

The "whistle-blowers" law is one kind of legislation protection for scientific and technical workers. It protects the rights of the employee. In April, 1981, Michigan became the first American state to pass a law for "the exposure protection method". The main stipulation of this law is: "In industry any employee, because of reporting the violation of the federation, the state and the local law behavior to the public authority encountered the dismissal or the punishment, now they may make the appeal to the state court for unfair retaliation. If the employer cannot indicate processing is standard or the effective commercial reason to the employee take the suitable human affairs as the foundation, then the court may (determine employer) to compensate (employee) the wages, restores its work, (by employer)bears the lawsuit and attorney the expense. Also possibly punishes the employer 500 US dollars fines."[2, p. 143] But, the protection of scientific and technical workers as an employee is not widespread even in USA. It mostly limits to the medical domain.

In the US, besides to maintains the public interest in the law and the policy aspect the behavior to protect, in the organization aspect, the government mainly through the establishment of award item encouraging scientific and technical worker in the technical work (invention, design and manufacture) beneficial to the environment and the public health technology and the engineering project. Compared with the Western separation of powers administrative operation mechanism, our country's government manages the pattern to be more advantageous to the government protects scientific and technical worker's philanthropic undertaking. Perhaps this means to establish one kind "the government leadership" the social security system. It's a problem awaiting discussion. But in this social security system preparation, advancement, establishment, government if can realize and positively undertake the leading responsibility on own initiative, the advancement science and technology ethics practice safeguard system development, it is a good to our science and technology work development.

Since our country's practice of science and technology ethics is a little late, we need to absorb other country's experience. Because of the different situation of the country, we can not use the experience directly, but to choose rationally. The protection in legal aspect is essential. Considering our country's social and cultural background as well as the practical interests and ethical problem after muckraking, it is necessary to give the revelator more chance to express their suggestion, even to provide political favor and protective measures.

As to the organization construction of guarantee system, government, professional co-operation and public organization should all be developed and improved. Because the interest and obligation should be uniformed, these three aspects should be the main part of science and technology ethical practice guarantee system.

Similarly, in our country, the organization construction, group's constitution and the team co-operation should be improved. Except for the CAS and Chinese engineering colleges, which have independent agencies, other vocational groups have no such functional departments. Group constitution as well as the team

---

[4]The Chinese Translation of [2, pp. 142–143].

co-operation is only an ideal format which has not melted into scientists and technologists' ethical concerns. Because professional co-operation is in such a situation, it is impossible to construct the special ethical committee which has a higher order responsibility for professional co-operation.

3 How to Maintain S&T Ethical Practice System Guarantees?

How to maintain the guarantee system of science and technology is also a problem. A guarantee system cannot solve all the problems that the scientist and technologist encounters, and some retaliation and social pressure may exert itself against them in abnormal ways, so one period and one aspect's guarantee system can not go on, even wasting all the previous efforts. So the guarantee system's lasting effort should be combined with other social factors.

With the development of science and technology, new problems in science and technology ethical practice may arise, so the guarantee system needs to be amended over time. Especially the relevant law's modification, establishment and execution should consider the interests of all parties and be adjusted in the practical process. This is a long and hard process.

In America, although some large enterprises have set up their ethics judges or survey systems for complaints, this is mostly for maintaining its image and long-run interests, which is true on most occasions. But they are forced to have renovation or reform after public scandals or mistakes are exposed, such as happened at New York Telephone Company and Texas Devices. But the efforts are not carried our thoroughly, except in few big companies, such as the above two, Pacific Bell, GE, etc. For enterprises in a market economy, we cannot have the same requirement, but we must promote the kindness principle and encourage related science agencies or enterprises to maintain an upright image, employ staff to monitor ethical codes and maintain an ethnical environment.

Though American professional co-operation has made efforts in guarantee-system for science and technology ethical practice, the limitations on implementation and acceleration cannot be ignored. The first problem is economy. Because professional co-operation's fee almost comes from big enterprise, it is difficult to decide when a scientist or technologist has conflicted with the economic fee supplier. Second, when a professional co-operation actualizes the punishment, the biggest possibility is to cancel the member qualification, but this has no material affect on the member (especially the big enterprise) who disobeys ethical constitution. The registration committee can avoid the above two hard problems, but the man and material resources are limited, so it can't research every matter carefully. Many scientists and technologists haven't registered, which also limits the scientist and technologist ethical practice.

Public's power is enormous but also potential, so the government as well as the scientist and technologist need together to maneuver public's passion and huge energy. It needs the whole society especially the government, educator, the scientist and technologist to joint efforts to awake the public's subjective consciousness, participation consciousness and hardship consciousness, and to improve their science qualification, and moral level. The scientist and technologist can make the public as their appeal object only when public colligate qualification reaches a certain level.

Above all, the above parties' joint development efforts and cooperation are needed. Ed Turner of Idaho Falls is a good example of someone who protected his own interest through existing social system guarantees. Through this example, we should see professional co-operation's approval to Ed Turner, and we also can not ignore the law's effort. After all he protected his interests through the law.[2, pp. 208–209] Guarantee systems of science and technology ethical practice need to be improved, and it requires interdisciplinary and intercultural perspectives and ways and the collective efforts of scientist and technologists, ethical researchers and government.

The study of modern guarantee-systems of science and technology ethical practice is an important part of the study of science, technology and society. Just as there are differences between societies and cultures, different countries and regions develop guarantee systems of science and technology ethical practice which are also different, but the basic principles and the systems design have many similarities, so it requires wide and deep communication and cooperation.

Bibliography

[1] De George, R. T. Ethical responsibilities of engineers in large organizations. *The Pinto Case, Business and Professional Ethics Journal*, 1(1):11–14, 1981.

[2] Harris, C. E., et. al. *Ethics: Concepts and Cases*. Beijing Institute of Technology Press, 3rd edn.

[3] Hero-loser. Ends the era of individual anti-counterfeit. *Shanxi Youth Daily*, (3):15, 2006.

[4] Martin, M. W., Schinzinger, R. *Ethics in Engineering*. New York: The McGraw–Hill Companies, Inc, 3rd edn., 1996.

[5] Tu, J. *History of Chinese Pseudoscience*. Guizhou Education Press, 2003.

[6] Weiss, J. W. (Ed.). *Business Ethics: A Stakeholder and Issues Management Approach*. China Renmin University Press, 3rd edn.

Critical Thinking and Education[1]

Wang Hailong
Yan Shan University
gkzlhw9@sohu.com

ABSTRACT. Critical thinking is the ability to engage in reflective and independent thinking, and the ability to think clearly and rationally. Critical thinking is essential for effective functioning in the modern world and essential for the mastery of knowledge. A student's critical thinking ability can be measured using standard tests.

I have been a teacher for nearly 11 years. In my school, I taught the traditional way, where the teacher is the center of the class. According to the traditional method, a good student should do whatever the teacher says and not argue or entertain a different opinion from the teacher. These traditional students usually win the teacher's favor. This is not good a good situation in China's schools. It does not encourage independent thinking or creativity.

The purpose of education is to educate students that can think and judge for themselves. Independent thinking requires the ability to think analytically and reason logically. Critical thinking is one of education's most central goals and one of its most valued outcomes.

1 What is Critical Thinking?

Critical thinking is the ability to engage in reflective and independent thinking and to think clearly and rationally. Critical thinking does not mean being argumentative or being critical of others. Critical thinking is a general thinking skill that is useful for all sorts of careers and professions. When making a decision one must ask a series of questions. These questions relate to critical thinking. There are many benefits to critical thinking in problem solving and decision making but only if people ask the right questions. Critical think involves many skills, including the ability to listen, to read and to write, to evaluate and judge the validity of arguments, and make good claims or decisions.

It is sometimes suggested that critical thinking is incompatible with creativity. This is a misconception. A creative person is someone who can generate new ideas that are useful and relevant to the task at hand. Critical thinking plays a crucial role in evaluating the usefulness of new ideas, selecting the best ones and modifying them if necessary. Critical thinking is also necessary for self-reflection. In order to live a meaningful life and to structure one's life accordingly, we need to justify and reflect on our values and decisions. Critical thinking provides the tools for successfully conducting this self-evaluation process.

---
[1]

Critical thinking includes at least two factors: (1) cognitive ability and (2) expressive ability. The first is an important part of the ability to do scientific exploration. The second has a very important impact on our everyday life.

2 Why and How should We Teach Critical Thinking?

Critical thinking is essential for effective functioning in the modern world. It is a liberating force in education and a powerful resource in one's personal and civic life. Critical thinking can help people in many fields to distinguish between facts and opinions or personal feelings, judgments and inferences, and objective and subjective perceptions.

Critical thinking does not always come naturally, but we can get better at it if we accept some guidance. [2, p. 22] Children are not born with the power to think critically. Critical thinking is a learned ability that must be taught. Critical thinking is essential for the mastery of knowledge. Knowledge should be mastered smartly, not mechanically. So you must take a critical attitude toward the progress of study. The method of mastering knowledge and the good habits of thinking are more important than simply "knowing" the knowledge. Training in critical think can improve this capacity for everyone. Critical thinking is the art of taking charge of your own mind. Its value is simple: if we can take charge of our own minds, we can take charge of our lives. Critical thinking is an important and vital topic in modern education. So we must put training in critical thinking high on the list of priorities in our education system.

Critical thinking is a skill. There are two main factors involved in learning critical thinking: theory and practice. First, we need to learn the principles of critical thinking–such as, for example, basic logic. Critical thinking is a process and way of thinking which enables us to come to logical conclusions. It is a tool we can use to come to logical conclusions and make good decisions. We believe the best way to teach critical thinking is to integrate formal and informal logic. We also need to learn how to identify and avoid the typical fallacies people make in their thinking. Merely knowing the principles, however, is not enough. We need to improve the application of critical thinking skills in our daily lives. Students need to recognize the importance of critical thinking and have the right attitude and motivation to use them.

3 Critical Thinking in Education

In education, there are two ways to train students in critical thinking: (1) we need to offer a course in critical thinking. In this course, we can explain critical thinking in a systematic way. And (2) we should employ critical thinking in the education process. That includes providing a chance for the students to discover critical thinking principles and techniques for themselves. We can pose different opinions and questions and transform the classroom into a critical thinking culture.

The purpose of specifically teaching critical thinking is to improve the thinking skills of students and thus better prepare them to succeed in the world. We should teach students how to think, not what to think. A person who thinks critically can come to reliable and trustworthy conclusions about the world. [3, p. 3] Critical thinking can give a student the tools to become a responsible citizen who can make valuable contributions to society.

Today, education in China generally fails to teach students the essential critical thinking skills. At most, students have been stuffed in school with knowledge, but few of them know how to think. In our education, we are familiar with this model: only the teacher speaks and the students only listen. But this model does not work and must be changed. The 21st Century is called the "information century" and what the teacher can teach students is too limited. This type of "limited knowledge" can quickly become obsolete and out of date. The meaning of education is to teach the students a more workable model for thinking. As a result students become more broad-minded. They will look at the world differently and see others and themselves more cautiously. They will be able to maintain a curious and probing attitude, questioning always the value of propositions proposed to them. If there is more training of critical thinking in our education, the students will be empowered to perform better in many fields.

4 The Role of the Teacher in Teaching Critical Thinking

Good teachers are doing more in their teaching roles. They help students to recognize and act upon their capabilities, and to establish a good classroom climate. For example, when students answer a question, the teacher should follow the students' answers, whenever appropriate, with a further question which asks the students why they believe this version or think their answers are reasonable, plausible or accurate. Perhaps the most important aspect is to increase the students' will and desire to think. We should encourage students to build their faculty for critical thinking.

A student's ability for critical thinking can be measured using standard tests. There are many ways to examine and measure critical thinking ability. In China, there are many questions that examine the ability for critical thinking on tests. The tests for the MPA and MBA are examples. And we can judge the ability for critical thinking in other ways. China was a feudal society for two thousand years. There was a lack of critical thinking in traditional Chinese culture.[1, p. 20] And today the education evaluation system also restricts training for critical thinking in our education. So it is an urgent task to strengthen the training for critical thinking in education. There is a sentence in the book The Revolution of Study: "critical thinking is a very important skill in every side of the life. Liberty and opening the mind are essential to solving problems. So each evolution method should encourage it. And we should not put all the people under the same model."

The art of critical thinking requires examining not only others with suspect eyes but also one's self. Each person should have his own mind. And as a citizen, he must be rational. For critical thinking, awareness of democracy and science is the essential condition for social development. So we must pay attention in our education efforts to cultivate the students' ability for independent critical thinking.

Bibliography

[1] Gu, Z. Y., Liu, Z. H. *A Course in Critical Thinking*. Peking University Press, 2006.
[2] Moore, B. N., Parker, R. *Critical Thinking*. Mayfield Publishing Company Press, 5th edn., 1999.
[3] van Eemeren, F. H., Francisca, A., Henkeman, S. *Argumentation —A Guide to Critical Thinking*. New World Press, 2006.

Rediscovery of the Connotation of Interaction Between Science, Technology and Society[1]

Wen Jianying

Jiangsu University

Jianyingwen03@163.com

ABSTRACT. The interaction between science, technology and society (ST&S) is not only a fundamental conclusion of Sociology of Society and Science of Science, but a theoretical premise of these two and other subjects as well. In socio-economic context, the influencing relationship of the particular carriers of this interaction is enbued with a new connotation, which denotes innovation directly. The "Triple Helix Model", which is taken not only as a metaphor but a depiction of University-Industry-Government (U-I-G) discloses the environmental and dynamic system of innovation. By focusing on the interaction of ST&S, rather than emphasizing basic research only, maybe, can we promote technological innovation.

1 Historical Outline of the Interaction

The real interaction between science, technology and society (ST&S), i.e., the interplay of them, emerges only after science has become a social institution. The history of science's becoming a social institution, however, is no more than 200 years since its beginning. Before this time, science (natural philosophy) was in the hands of necromancers, monks or philosophers and was the preserve of the upper leisured class – the aristocracy. They had the money and the leisure to pursue science in an amateur way to satisfy their own curiosity and interests. And the knowledge they produced by doing science was used for self-indulgence or entertainment rather than to meet the practical needs of society. Briefly, they did science only for science's sake. This state has continued, passing down from ancient Greek and Roman times to medieval times and the Renaissance. It was not until the 17th Century that the great turning-point was reached, the point at which science began to be an organized activities. Little by little, science was recognized for its capability to solve practical problems in such areas as navigation, transportation, and dynamics. As a result science began to be patronized by newly emerging capitalists.[7, p. 203]

In fact, the scientific research of this time was done almost exclusively by scientists individually. The national subsidy for science was either lip service or not dependable. However, this eventually marked a prelude to the interaction

---

[1] The author is with Jiangsu University, Zhenjiang, Jiangsu, P. R. China 212013.

between science and society. According to R. K. Merton, of all the research i done by the Royal Society, about thirty to sixty per cent of it was done in response to society's needs.[7, p. 203] Consequently, the scientific solutions to those practical issues, in turn, stimulated the development of science itself. A positive correlation between science and society was identified as the way to make progress in both areas at the same time.

This interaction was reinforced as time passed. In nineteenth century, science as a profession separated itself from natural philosophy for the first time. Scientists became socially acceptable and prominent and their role respected by the society.[8, pp. 38–74] Scientific research turned into a necessary part of the affairs of universities. Since then, scientific research has been patronized systematically by nations and industry. The institutionalization of science was eventually completed. Science, like such other social institutions as religion, economics and politics, became an integral and indispensable part of civilized society.

Now, both science and society have realized that each side's rapid development is the condition of the other's development and support. Merton maintains that the two have a reciprocal relationship. In this sense, J. D. Bernal hits the nail on the head when he argues that the prosperity of science and society rely on the correct relationship between them.[1, p. 28] Yet, the problem is how they interact.

2 The Concrete Carriers of the Interaction

In the socio-economic context, we can analyze this interaction further from the point of the concrete embodiments of each social institution, i.e. from particular carriers of each social institution.

The institutionalization of every organizational structure follows its functional differentiation. The modern forms of governments, industry and universities are all conceived in and born from the society organism. For science, generally speaking, its carriers are universities, public research agencies and national laboratories, among which the university is the typical representative. As for the economy, its function is assumed by industry. And governments at each level represent and perform the function of administration. As a matter of fact, these three parts have a long history interaction. In Merton's dissertation, he definitively denotes interaction between them. After the Cold War, the military was markedly degraded. By contrast, economic development becomes more and more important. Meanwhile, there are increased government interventions in the development of science. A new complex of Science-Industry-Government gradually replaced the old Science-Industry-Military complex. And especially in wartime, this kind of cooperation seems extremely urgent and necessary.

After researching and by using other theories of innovation, such as National Innovation System (NIS), Triangle Model, and Mode 2 of scientific knowledge production, L. Leydesdorff and H. Etzkowitz use Triple Helix (TH) Model to depict the relationship between U-I-G, and to expound the complex micro-system of innovation.

In new classic economics, the market takes the leader's position. From the viewpoint of evolutionary economics, however, industry is the subject of innovation. Leydesdorff and Etzkowitz name their theory 'New Evolutionary Economics'. For them, the subject of innovation is neither market, industry, nor the government in the Triangle Model. They believe that innovation does not take place in a fixed unit. On the contrary, they argue that innovation often percolates up in the overlapping network of the interface of U-I-G as a result of the constant changes in the relationships among university, industry and government themselves. Together it all represents a dynamic process of knowledge flow. The TH Model has three prominent characteristics. First, every structure of the helix of the TH changes ceaselessly. Second, the function of every sub-system of the TH changes ceaselessly. Thirdly, the structure of this innovation system changes ceaselessly. All in all, in TH, everything is always in flux.[5]

3 Interaction Rather than Basic Research Only

Clearly, in the socio-economic context, the meaning of the terminology of interaction which it embraces in the Sociology of Science changes vividly. It now means cooperation, communication, and collaboration. And in essence, it connotes innovation. Meanwhile, in a knowledge-based economic society, interaction possesses a new content: the flow of knowledge.[11, p. 3] Knowledge has the potential power to call all the participants together into a 'transaction space'. In this space, by conversation and equilibrium, knowledge will flow. And this flow will result in innovation. Accordingly, knowledge in this process becomes an essential condition of innovation.[10]

From the point of view of evolutionary economics, a nation is a kind of NIS. The NIS is composed of a series of knowledge producers, transmitters, users, industry networks, and support systems. And it is the formal and informal knowledge flow and the interaction between these knowledge agencies that determines the innovative ability of industry. Just as R. R. Nelson says, the interaction between a series of institutions determines the innovative performance of a nation's industry. This kind of interaction, essentially, is what we mean by synergy.[9]

In *Science: The Endless Frontier*, V. Bush highly praises the social function of science in national security, more job opportunities, a healthier body, higher level of life standards and cultural progress. But for Bush, science is not only urgently referred to natural science, but also to basic science first and foremost. On this point, he argues that basic scientific research is *scientific capital*.[2] If we permit scientists to do their research without external interference and allow them to pursue their own pure interests, he argues, we can achieve all these goals. Today, we regard this optimistic idea as "Bush's Canon".

Apparently, there is at least one rationale behind Bush's canon, i.e. the development of science is linear. We can also take this as the Linear Model of science's development. According to this model, if we continually increase the funding of basic scientific research and develop more and more scientists and knowledge, we can eventually harvest technological innovations and endless benefits from science.

Nevertheless, just as Scott et. al. say, "However, this is an area beset by high levels of complexity. The connection of science funding with the promised corresponding outcomes are not based on rigorous understandings: even if unlimited time were available, definitive analysis of many issues is simply not possible given the uncertainties involved."[4] Meanwhile, in *Basic Research and Technology In-*

novation: Pasteur's Quadrant, D. E. Stokes, using conclusions drawn from the History of Science, also proves that there is no linear cause-effect chain between basic research and technological innovation.

As someone points out, the way that we view the world determines the way that we draw our maps; the way we draw our map will determine the paths we follow in the pursuit of progress. We cannot confirm definitely that these models uncover the black box of innovation, but what we can do is nothing less than accept a kind of expectation as an answer.[6] The interaction between U-I-G proposed by TH Model cannot be regarded only as a metaphor and a heuristic way to interpret innovation. It is in fact an empirical model as well.[3] So it is important to find out the handicaps in this interaction according to a nation's unique characteristics. Broadly speaking, how we promote interaction between ST&S is the thing to be considered first and foremost by decision makers who formulate science and technology policy.

(This paper was written under the auspices of School Fund: 06JDG051)

Bibliography

[1] Bernal, J. D. *The Social Function of Science*. Peking: The Commercial Press, 1982. Trans. by Chen, Tifang.

[2] Bush, V. *Science: The Endless Frontier*. 1945. URL http://www.nsf.gov/about/history/vbush1945.htm.

[3] Etzkowitz, H., Leydesdorff, L. The dynamics of innovation: from national system and "mode 2" to a triple helix of university-industry-government relations. *Research Policy*, (29):109–123, 2000.

[4] Jr Pielke, R. A. *Science policy without science policy research*. 2006.

[5] Leydesdorff, L., Etzkowitz, H. Emergence of a triple helix of university-industry-government relations. *Science and Public Policy*, (23):279–286, 1996.

[6] Leydesdorff, L., Etzkowitz, H. The triple helix as a model for innovation studies. *Science and Pubic Policy*, 25(3):195–203, 1998.

[7] Merton, R. K. *Science, Technology & Society in Seventeenth Century England*, vol. xxii. New York: Howard Fertig, 1970.

[8] Multhauf, R. P. The scientist and the "improver" of technology. *Technology and Culture*, 1(1):38–74, 1959.

[9] Nelson, R. R. *National Innovation Systems: A Comparative Analysis*. Oxford: Oxford University Press, 1993.

[10] Nowotny, H., et al. *Rethinking Science*. Cambridge: Polity Press, 2001.

[11] OECD. *The Knowledge-based Economy*. Paris: OECD, 1996.

The 'Atomic Bomb Project': A Study Comparing China with the USA from a Nuclear Physics Perspective[1]

Sun Hongxia

Chinese Academy of Science

hongxiasun@yahoo.com.cn

Wu Zhongqun

North China Electric Power University

zhongqunw@yahoo.com.cn

ABSTRACT. Against a 'big science' background, the Atomic Bomb Project (ABP) has been discussed in many studies which describe its political, economic, cultural, and ideological influences. There have been few studies, however, which compare the ABP in China and the USA from the perspective of how these projects affected the development of nuclear physics in their respective countries. This paper analyzes the relationship between the development of nuclear physics, big science and engineering especially with respect to knowledge-spillover. Furthermore, this paper examines the mutual relationship in China and the USA respectively between the development of nuclear physics as an academic discipline and the ABP. The paper concludes by noting specific differences in the relationships among scientific development, big science and engineering in China and the USA against the background of the social, economic, and cultural differences in these two countries.

1 Introduction

In 1905, Albert Einstein published his classic paper on the relationship between energy and mass. He predicted the probability that humans would one day discover a way to release energy from the atomic nucleus. In 1938, Otto Hahn and Friz Strassmann, based on work which Joliot-Curie had completed, discovered the secret of nuclear fission. In 1939, Joliot-Curie and Enrico Fermi demonstrated the possibility of chain fission. These scientific advances over more than 100 years laid a firm foundation for the development of nuclear physics as an academic discipline and contributed to the making of the atomic bomb. But up to that time, the development of nuclear physics alone was not enough ultimately to bring about the making of the atomic bomb. There were other factors which need to be considered.

Hahn and Strassmann's finding was the fruit of pure scientific research which was conducted in a vague and practical way. For example, "By 1940 nuclear

---

[1]The authors are with China Research Institute for Science Popularization, Beijing 100081 China, and Business School, North China Electric Power University, Beijing 102206 China, respectively.

reactions had been intensively studied for over ten years. Several books and review articles on nuclear physics had been published. New techniques had been developed for producing and controlling nuclear projectiles, for studying artificial radioactivity ...... Isotope masses had been measured accurately. Neutron-capture cross sections had been measured. Methods of slowing down neutrons had been developed."[9, p. 29] As scientists tried to harness these discoveries to make an atomic bomb, they faced repeated failures: "Although the fundamental principles were clear, the theory was full of unverified assumptions and calculations were hard to make. Predictions made in 1940 by different physicists of equally high ability were often at variance."[9, p. 29] In fact, discoveries in nuclear physics before World War II had only provided a probability for making an atomic bomb. Undoubtedly, scientific studies pushed forward the development of engineering. Stanley Goldberg [1] proposed that science played the essential role as a precursor in the Manhattan Project's accomplishments. As the ABP was being implemented, all kinds of theories and methods in nuclear physics were being found and applied. As a result, the nuclear physics discipline developed rapidly. Scientists learned, for example, how to make and purify uranium, initiate and control nuclear chain reactions, build nuclear reactors, and how to understand the function of nuclear fission. This knowledge demonstrated the direct mutual relationship between the ABP and the consequent development of nuclear physics as a discipline.

This relationship between the ABP and the development of nuclear physics, however, has not been mentioned in previous studies. Henry D. Smyth discussed the manufacture of the atomic bomb and atomic energy in detail. Michael S. Minor [7] illustrated the research and production process which led to the atomic bomb, including illustrations showing the original materials and facilities which were needed. Richard Rhodes [8] focused on the process of making the atomic bomb and its effects. In his study, Johnson W. Lewis [5] provided an historical narrative of how the atomic bomb came to be made in China. Jeff Hughes [4] described and explained the origins, workings, and future of big science and demonstrated how people learn in geometric ways from the connections between scientific organization and scientific knowledge. Jifeng Liu [6] generalized the process of making the atomic bomb and the orientation of big science. None of the above studies, however, draw the important connection between the ABP and the development of nuclear physics. This relationship can in fact now be shown as a classic example of the Knowledge Spill-over Effect.

2 The Knowledge Spill-over Effect

The industrialization of nuclear science and technology plays an important role in economical development. For example, 10 billion is spent per year in the U.S for nuclear diagnosis medical science research and development and more than 10 billion for radioactive processing all over the world. Nuclear technologies produce radioactive effectiveness, and its products are the others' significant materials and components. The impact which nuclear physics makes has already spilled over and diffused to other disciplines and industries. It has become a fundamental technology for further advances in many other disciplines. The development and application of radioactive isotopes in the medical sciences and life sciences is one example. Nuclear physics is regarded as a new kind or form of knowledge spill-over since the ABP to some extent. In general, from an economic perspective, Griliches [2] clearly demonstrated that knowledge spill-over means that imitative innovation benefits from the study of the innovations imitated. P. Stoneman [10] points out that knowledge spill-over is in itself a kind of 'learning' activity, i.e., it was an application of knowledge in the initial study and exploits and generates new knowledge by combining previously learned knowledge with available knowledge and new innovations.

In China and the USA, there are some similarities and differences in the expression of knowledge spill-over. In both cases, the development of nuclear physics came from the knowledge spill-over derived from the ABP. The difference is that the social background of the knowledge spill-over is different. In the USA the knowledge spill-over happened against a capitalist market mechanism background, but it happened against government central planning mechanism in China. The former was pushed by the practical motivation of scientists, and the latter by was pushed by the needs of government. Therefore, there was a strong initiative for the development of the discipline in the USA. There exists a dialectic relationship–as science develops rapidly, we should depend on the inner laws of science itself and less government intervention to some extent. On the other hand, we need much more government intervention if science is to be swiftly transformed into applied technology.

3 The Relationship between the ABP and the Development of Nuclear Physics in China

Before the Liberation War, nuclear physics courses were taught by many excellent physicists like Zhongyao Zhao. After the war, some Chinese physicists returned to China from abroad. They did the foundation work in nuclear physics and acted as leaders of China's atomic bomb project in the 1960s. The strategic decisions concerning China's nuclear industry were made in an enlarged meeting of the Chinese Communist Party's Central Secretariat on January 15, 1955. The Political Bureau of the Chinese Communist Party quickly adopted the 02 nuclear weapons program. After the 02 Plan was implemented, a program in nuclear education was founded and developed. Qinghua University and Peking University adopted the Soviet Union's training model after inspecting and studying the Soviet Union's nuclear physics infrastructure and curriculum. The science disciplines which were set up at Peking University were adapted from the teaching programs and curricula at Moscow University. The science and engineering disciplines at Qinghua University adapted the teaching program and syllabus at Leningrad's multidisciplinary School of Engineering. A number of teachers and scientists participated in the construction and development of the education plan at Peking and Qinghua. Outstanding science and engineering students were then sent to Peking University and Qinghua University to study nuclear physics.

The nuclear physics discipline had a solid foundation at Peking University and Qinghua University. As a result, nuclear physics studies and research made rapid progress. Some aspects of this progress disclose the relationship between the development of the nuclear physics discipline and the ABP in China. Among them, for setting up the disciplines and cultivating intellectuals, China's Communist

Party issued "8 rules for developing nuclear weapons". The rules showed how the government valued the importance of developing intellectuals. The eighth rule placed equal priority on cultivating scientists and technicians and making the nuclear weapons. For example, in December 1959, *Glass's Atomic Nuclei Reactor Engineering Principles*, the first theoretical atomic reactor textbook, was published by the Science Press. It was used to train many nuclear physics scholars and experts. After six years of struggle, nuclear physics as a solid discipline was built.

There were many positive results that were quickly achieved. The quality of teaching at Qinghua University dramatically had been improved. All 21 laboratories were established and 51 courses were taught. During the process of constructing the reactor, more than 600 graduate students finished their theses, and more than 2,000 people acquired practical experience working on the reactor. Construction of nuclear reactors was an integrated technology, and it needed a variety of disciplines to cooperate with one another. In 1966, the Cultural Revolution began and introduced an atmosphere of confusion. In spite of these distractions, Premier Zhou Enlai in August made clear instructions on which experiments needed to be completed. Under the leadership of the Ministry of Machine Design, teachers, technicians and workers cooperated closely to perform 13 thermal experiments. The verification process and research tasks were completed in less than three months. Meanwhile, the Communication University of Shanghai and the Chinese Academy of Sciences developed nuclear physics disciplines with government support. From 1955 to 1983, "there were 22124 students graduated from the above universities or institutes to meet the need of carrying out the nuclear plans". In addition, General Nie not only trained but also selected the distinguished graduates from all the universities: "He thought that it was possible to in advance distribute the top students to work in the area of the nuclear industry." In China, the development of science including the development big science was also managed, coordinated and controlled by the government to meet very specific objectives.

4 The Relationship between the Manhattan Project and the Development of Nuclear Physics in the USA

Theoretical physicist J. Robert Oppenheimer and experimentalist Ernest O. Lawrence built a great American school of physics at the University of California at Berkeley in the 1930s. Prior to 1942, research groups in nuclear physics worked separately from one another. After nuclear physics emerged as an academic discipline, nuclear physics was combined with the Manhattan Project in the USA in order to meet the urgent priorities posed by World War II. Working independently, the scientists were unable to make good use of scarce resources and facilities so that the project was not effective. After the Manhattan Project was launched, the research and manufacture of the development of the atomic bomb made great progress. "The new laboratory improved the theoretical treatment of design and performance problems, refined and extended the measurements of the nuclear constants involved, developed methods of purifying the materials to be used, and, finally, designed and constructed operable atomic bombs."[9, p. 222] In the USA, the big science model was organized and controlled by government.

5 Differences in the Relationship between the ABP and the Development of Nuclear Physics against Different Social Backgrounds

In China, the ABP succeeded because of government intervention and control from the top down; in the USA, however, the Manhattan Project was built as a bottom up model, i.e. from lower to upper. The Chinese government developed its ABP in a self-sufficient and self-reliant way: "collaboration of politics, military, science and technology did not appear all of a sudden, but through the hard work of the people," General Nie observed. "Therefore, new form, organization and practice will be needed." In the USA, by contrast, scientists from all over the world collaborated with one another to create the atomic bomb under the threat and pressure of Adolph Hitler and the Nazis.

Under Chinese government control, big science and engineering were forced to develop the tools of nuclear physics quickly; in contrast, the nuclear physics discipline in the USA developed first. As Jeff Hughes has argued, science and government did not work so closely together in the early stages of nuclear development, but major breakthroughs in science were made when government and science finally combined to focus their interests through the Manhattan Project. "Some historians have claimed that the Manhattan Project was pivotal to the establishment of what we now call 'big science' ", Simone Turchetti [11] has said, "and it defined a new power relationship between scientists and the military." Compared to nuclear physics in China, the nuclear physics discipline in the USA exhibited more independence from government control and embodied its own autonomy and initiative.

6 Conclusion

After analyzing the relationship between the ABP and the development of nuclear physics in China and the USA, we can conclude that government leadership, coordination, and intervention are crucial under certain urgent circumstances to the achievements of big science and engineering. This implies that we should plan and organize to carry out large scale scientific projects in order to drive the development of scientific disciplines. The emergence of Big Science since World War II is an effective model for speeding up the development of science in other areas as well. Before big science, science developed in a haphazard, independent and autonomous way.

Bibliography

[1] Goldberg, S. *Big Science: Atomic Bomb Research and the Beginnings of High Energy Physics*. New York: The History of Science Society, 1995.

[2] Griliches, Z. The search for r&d spillovers. *Scandinavian Journal of Economics*, 94:29–47, 1992.

- [3] Halperin, Morton, H. Chinese nuclear strategy. *China Quarterly*, 21:74–86, 1965.
- [4] Hughes, J. *The Manhattan Project: Big Science and the Atom Bomb*. Cambridge: Icon Books, 2002.
- [5] Johnson, W. L. (Ed.). *China Builds the Bomb*. Beijing: Atomic Energy Press, 1990.
- [6] Liu, J. *Two Bombs and One Satellite: a Model of Big Science*. Jinan: Shandong Education Publishing House, 2004.
- [7] Minor, M. S. China's nuclear development program. *Asian Survey*, 6(1976):571–579, 1976.
- [8] Rhodes, R. *The Making of the Atomic Bomb*. New York: Simon & Schuster, 1986.
- [9] Smyth, H. D. *Atomic Energy for Military Purposes*. Princeton: Princeton University Press, 1946.
- [10] Stoneman, P. *Handbook of the Economics of Innovation and Technological Change*. Oxford: Blackwell, 1995.
- [11] Turchetti, S. For slow neutrons, slow pay–enrico fermi's patent and the U.S. Atomic Energy Program, 1938-1953. *Isis*, 97:1–27, 2006.

www.ingramcontent.com/pod-product-compliance
Lightning Source LLC
Chambersburg PA
CBHW050834230426
43667CB00012B/2001